The Digital Transformation of Logistics

The Digital Transformation of Logistics

Demystifying Impacts of the Fourth Industrial Revolution

Edited by

Mac Sullivan
Johannes Kern

IEEE Press Series on Technology Management, Innovation, and Leadership

Published by John Wiley & Sons, Inc., Hoboken, New Jersey.
Published simultaneously in Canada.

For general information on our other products and services or for technical support, please contact our Customer Care Department within the United States at (800) 762-2974, outside the United States at (317) 572-3993 or fax (317) 572-4002.

Wiley also publishes its books in a variety of electronic formats. Some content that appears in print may not be available in electronic formats. For more information about Wiley products, visit our web site at www.wiley.com.

Library of Congress Cataloging-in-Publication Data:

Names: Sullivan, Mac, editor. | Kern, Johannes (Writer on logistics),
    editor.
Title: The digital transformation of logistics : demystifying impacts of
    the fourth industrial revolution / edited by Mac Sullivan, Johannes
    Kern.
Description: Hoboken, New Jersey : Wiley-IEEE Press, [2021] | Series: IEEE
    Press series on technology management, innovation, and leadership |
    Includes bibliographical references and index.
Identifiers: LCCN 2020042818 (print) | LCCN 2020042819 (ebook) | ISBN
    9781119646396 (paperback) | ISBN 9781119646457 (adobe pdf) | ISBN
    9781119646402 (epub)
Subjects: LCSH: Business logistics–Data processing.
Classification: LCC HD38.5 .D545 2021 (print) | LCC HD38.5 (ebook) | DDC
    658.500285–dc23
LC record available at https://lccn.loc.gov/2020042818
LC ebook record available at https://lccn.loc.gov/2020042819

Cover Design: Wiley
Cover Image: © Yuichiro Chino/Getty Images

Contents

List of Contributors

Matías Aránguiz
Law, Science and Technology, Catholic
University of Chile, Santiago, Chile

John Berry
JUSDA Supply Chain Management
Corporation, Diamond Bar, CA, USA

Ira Breskin
SUNY Maritime College, Bronx,
NY, USA

Colin Cobb
Loloi, Inc.
Dallas, TX, USA

Simon de Raadt
HyperSKu, Shenzhen, China

Sam Heuck
Atlas Sourcing Partners, Hong Kong, HK

Ruben Huber
OceanX, Laufenburg, Switzerland

Cameron Johnson
Tidal Wave Solutions, Shanghai, China

Johannes Kern
Tongji University, Shanghai, China

Nicholas Krapels
SKEMA Business School, East China
Normal University, Shanghai, China

Wesley Li
Konica Minolta, Tokyo, Japan

Cory Margand
SimpliShip, Boston, MA, USA

Andrea Margheri
Cyber Security Researcher and
Consultant Specialised in Digital
Identities and Blockchain Technology
Milan, Italy

Robert Mostert
Toll, Singapore

Axel Neher
Bosch Rexroth, Shanghai, China

Ayush Pandey
Amazon.com Inc., Seattle, WA, USA

Dyci Sfregola
New Gen Architects,
Atlanta, GA, USA

Walter Simpson
NNR Global Logistics, Dallas, TX, USA

Mac Sullivan
NNR Global Logistics, Dallas, TX, USA

Jiayu Sun
Bosch, Shanghai, China

Zheyuan Tang
NNR Global Logistics, Dallas, TX, USA

Michael Teubenbacher
CPC Consulting, Beijing, China, subsidiary
of CPC Unternehmensmanagement AG

Bill Tran
Supply Chain Management and
Information Systems at Texas Christian
University, Fort Worth, TX, USA

Alex von Stempel
Freshwater Logistics Ltd,
Middlesex, England

Scott Wang
UPS, Louisville, Shanghai, China

Andre Wheeler
Asia Pacific Connex,
Perth, Australia

Lionel Willems
Odoo, Hong Kong, China

Dennis Wong
Flexport, San Francisco, CA, USA

Duoqi Xu
School of Law, Fudan University,
Shanghai, China

Jiao Xu
Flexport, Amsterdam, NL

Author Biographies

Editors

Mac Sullivan

Mac Sullivan is currently the Head of Technology and Digital Promotion at NNR Global Logistics. In this role, he oversees a portfolio of programs that leverage technology to create operational and commercial efficiency. Mac has an MBA from Hult International Business School and is pursuing his PhD from East China Normal University in the field of Political Theory. In addition to working for the past eight years in international logistics, he teaches Supply Chain Management at Texas Christian University and Cross-cultural Communication at Tongji University.

Johannes Kern

Johannes is an Affiliated Professor of Supply Chain Management at Tongji University, China and General Manager of Xiezhi Consulting. His research focuses on logistics, purchasing, and supply chain management, particularly on buyer–supplier relationships and the digital transformation of logistics. He supports international companies in China to optimize the whole supply chain, including sourcing, transportation, warehousing, and production. Prior to that he worked in various functions in purchasing and logistics at the Bosch Group in Asia-Pacific where he led teams and managed strategic projects in China, Korea, Thailand, Malaysia, Japan, and India. Johannes holds master's degrees from the University of Munich, Germany; Emlyon Business School, France; Aston University, UK and a PhD from Technical University Darmstadt in Germany.

Authors

Matías Aránguiz

Matías is a Chilean lawyer but has moved into the area of financial technology and artificial intelligence. Currently he is deputy director of the Program of Law, Science and Technology of the Catholic University of Chile and partner in the law firm Quarkz. In addition, he is currently pursuing his PhD in the Regulation of AI in Financial Markets from Shanghai Jiaotong University. He received his Master's in Finance from Shanghai University of Finance and Economics.

John Berry

John is currently the IT Director for JUSDA Supply Chain Management, a member of the Foxconn Technology Group. In this role he leads a team that develops and implements technology solutions for the global manufacturing supply chain. Prior to JUSDA, John was the IT Director for NNR Global Logistics, a non-asset-based logistics provider. At NNR, John directed the USA IT operation as well as several global technology initiatives. This included leading various software development projects that powered NNR's air freight, ocean freight, customs, and warehousing operations. With 28 years of experience in IT, John enjoys helping a new generation of technologists establish careers in the supply chain and logistics field. John blogs at http://www.johnberry.us.

Ira Breskin

Ira Breskin is a senior lecturer at State University of New York Maritime College. He teaches applied economics, expository writing, and maritime history courses.

Breskin also is the author of The Business of Shipping (ninth edition), a seminal book that addresses industry economics, operations, and regulation.

He joined the SUNY Maritime faculty in September 2003 after completing a 25-year career as a reporter/editor for several business publications. He continues to do freelance writing.

Breskin has a BA from Columbia University, an MBA from Dowling College, and a MS in International Trade and Transportation from SUNY Maritime. He has won several academic fellowships including a Knight Bagehot at Columbia University and a National Endowment for the Humanities award to study maritime history at Mystic Seaport.

Colin Cobb

Colin recently served as the Vice President of B2B, leading sales operations for nuLOOM/RUGSUSA.com, one of the largest global rugs and decor brands. He also has experience in leading the e-commerce divisions for multiple consumer goods manufacturers in the home space. In addition to leading omni-channel and e-commerce teams, Colin has over 10 years of experience in supply chain and logistics management. His contributions with various organizations have expanded into digital marketing, product development, and optimizing retail dropship networks. He has a Master's in Logistics and Supply Chain Management from Georgia College State University, as well as his six-sigma black belt.

Simon de Raadt

Simon is Vice President Europe at HyperSKU, Managing Partner at MAiNS International, and co-founder of DigiDutch. He has been based in China since 2011 and active in cross-border trade and e-commerce logistics between China and Europe. His specialty lies in e-commerce parcel deliveries from and to China. Prior to MAiNS, he has been working at TNT Post, Capgemini, and various other positions. Simon has a Master's degree in Business Economics at the University of Amsterdam.

Samuel Heuck

Sam is the founder and director of Atlas Sourcing Partners, a Hong Kong-based trading company specializing in building materials. For six years, he was based in Shanghai where he led the turnarounds of two manufacturing businesses. Sam has transactional and operating experience in over 15 countries including India, Vietnam, and Germany with extensive experience and focus on China and Asia-Pacific. At press time, he was an MBA candidate at MIT Sloan School of Management.

Ruben Huber

Ruben grew up in the trucking and warehousing business in Germany and held various management and executive roles in leading shipping lines, NVOCCs and intermodal operators across Europe, China and the Middle East. Today he is consulting on strategy, digitalization, internationalization, and logistics. He is the founder and director of OceanX, a network and think tank of leading seafreight specialist firms. He has a freight forwarding education, holds a German degree in business economics and an MBA from Aston University.

Cameron Johnson

Cameron has 20 years of experience in management across various industries in China. He drives the strategic consulting and advising initiatives at Tidal Wave Solutions. His expertise focuses on developing and executing strategy in difficult business environments, turning a business around to profitability, and operational excellence across a global company.

Cameron was a long-serving Asia General Manager at a leading carbon fiber company, where he worked in close collaboration with European and US stakeholders, suppliers, downstream customers, and regulators. He participates extensively in the American Chamber of Commerce where he served as the Chairman of the Aerospace Sub-committee in 2015–2017, and co-chaired the Future Leaders Committee from 2016–2018. He is currently Vice Chair of the Manufacturers' Business Council, and an advisor to the National IDC Industrial Technological Innovation Strategic Alliance, a China-based technology association.

Nicholas Krapels

Nic is the Managing Director for DarcMatter China. He is a leading blockchain expert in the Asia-Pacific region and an experienced financial consultant. He has an MBA from the University of South Carolina and is pursuing his PhD in Political Theory from East China Normal University.

Wesley Li

Wesley Li is an IT and service operations director with more than 17 years of combined experience in the logistics, freight forwarding, manufacturing, business consulting, and technology services industries.

Over the past few years, Wesley has been focusing on bringing business and IT together and promoting the concept of "IT Business Partner," shaping the traditional IT culture to fit the new IT business world. As a senior manager, he has been closely working with CEO and the senior management team to promote, drive, and resolve business issues by applying new technology. This great experience has given him many new business insights and turn them into innovations. He has demonstrated a capability to achieve business goals through innovative technology. He has a broad range of technical and information security expertise coupled with highly effective communications and inter-personal skills.

Cory Margand

Cory is the co-founder and CEO of SimpliShip, which is an international air and ocean freight SaaS-powered marketplace for freight procurement including a full suite of APIs. Cory has 14 years' experience in supply chain management working with companies like Adidas and Rockport. Cory has a Master's in Operations Design and Leadership from Worcester Polytechnic Institute.

Andrea Margheri

Andrea Margheri is a Cyber Security researcher and consultant specialized in digital identities and blockchain technology. Andrea holds a Ph.D. in Computer Science from the University of Pisa. Specialized in the design and deployment of access management systems, Andrea was with the University of Southampton where he coordinated research initiatives on blockchain in the context of EU H2020 projects and UK-founded industrial projects.

Robert Mostert

Robert Mostert is Head of Human Resources of the Global Forwarding Division of the Toll Group. He joined Toll in March 2018 after having worked for nine years for Standard Chartered Bank (SCB), where he was Global Head of HR of its Risk function, followed by a number of years as Head of HR of its Global Information Technology area and SCB's Global Shared Services Centers.

Prior to his time at SCB, Robert worked as Global Director Human Resources of the Biochemicals Division of CSM, as Director Organizational Development of Innodata Corporation (a BPO provider that operates predominantly in the Philippines and India), and as Senior Manager of KPMG Consulting in Hong Kong.

Robert holds a Master's degree in Law from the University of Utrecht, The Netherlands, an MBA from Bradford University Management Centre in the United Kingdom, and an Executive Master's Degree in Consulting and Coaching for Change from INSEAD. He has recently also completed the Senior Executive Leadership Program at Harvard Business School.

Axel Neher

Axel has 15 years of working experience for Bosch and Bosch Rexroth globally in various management and consulting positions in logistics and lean production. He has published several articles on supply chain management, international logistics management, and accounting. He has an MBA and a PhD from the University of Marburg.

Ayush Pandey

As a kid, the game of "Farmville" piqued Ayush Pandey's interest in the supply and demand. A supply chain enthusiast with a keen interest in global logistics, Ayush has leveraged exceptional cross-functional collaborations at the Indian Institute of Technology, Bombay, Tata Power Ltd., and Philips Lighting to drive operational efficiencies in multimillion-dollar global logistics and warehousing projects. Driving a positive change in all his professional engagements inspires

Ayush to get to work every day. Currently pursuing his Master's in Supply Chain Management (with a focus in logistics and transportation) from Northeastern University, Boston, Ayush is excited about two things in the shipping industry in the next decade: digitalization and de-carbonization. Having witnessed the birth of mega-ships, IMO 2020, and COVID-19, there is not a single day when he does not think, "What an exciting time to be a Logistics professional!" Apart from his professional engagements, Ayush has spent a good part of the last decade in 100+ community building projects, and is passionate about women empowerment, ending hunger, and cross-cultural communication.

Jens Puttfarcken

Dr. Ing. Jens Puttfarcken, 54 years old, has worked in the automotive sector for 29 years and moved from Fiat to Porsche in 1997. He has held multiple positions at Porsche AG, including Manager Sales of Porsche subsidiaries in Europe and subsequently in a worldwide capacity. In 2004 and 2010, he was appointed Vice President Customer Relations, and Vice President After Sales respectively at Porsche AG, contributing considerably to maintain Porsche customer brand loyalty and to enhance their enthusiasm for the brand. In 2015, Dr. Ing. Jens Puttfarcken was appointed Chief Executive Officer of Porsche Deutschland GmbH, where he was responsible for maintaining Porsche's stable development in the domestic market.

Dyci Sfregola

Dyci Manns Sfregola holds a Master's in Engineering Management from Kennesaw State University's Southern Polytechnic College of Engineering. She spent the early part of her career in digital and experiential marketing before transitioning to supply chain. She is currently a consultant focused on Connected Planning process improvement and leveraging cloud-based technology to achieve supply chain excellence.

Walter Simpson

Walter graduated from Clemson University in 2019 with a degree in supply chain management. After spending a summer in Hong Kong working for local 3PL Wilson Logistics Ltd., he began his career in February of 2020 with NNR Global Logistics in Dallas, TX. In his role as Process Coordinator for the Technology and Digital Promotion Team, he uses process mining and mapping techniques to analyze the current state of workflows and leads projects to implement improvements using digital solutions.

Jiayu Sun

Jiayu is a Senior Training Specialist at Bosch. She has experiences in instructional design, digital learning, internal marketing and communication and is a certified articulate storyline expert. Prior to that she worked in various functions at Bosch where she was responsible for competence management and learning innovation. Jiayu holds a bachelor's degree from Shanghai International Studies University and Shanghai Dianji University.

Zheyuan Tang

Zheyuan Tang is an Economics graduate from the University of Illinois at Chicago. He was the General Manager at Perception Jewelers where he engaged with thousands of international clients and handled logistics operations. A first-generation immigrant from China,

Zheyuan has certifications from both Peking and Fudan University. He specializes in researches on international trade and relations.

Michael Babilon-Teubenbacher

Michael Babilon-Teubenbacher is Partner with CPC AG, the German hidden Champion in Change Management providing services in the fields of change, project and HR management as well as organizational development.

Having graduated in International Business Administration from the International Management Center in Krems, Austria and spent most of the past two decades outside of his home country, Michael is truly a global citizen. Apart from project assignments in Japan, Spain, the Netherlands, and Australia he has so far spent more than six years in China ramping up and leading CPC Consulting (Beijing) Co., Ltd., the local subsidiary of CPC Group, as CEO. In this function he has also spearheaded CPC's activities in Asia.

In more than 16 years of consulting, he has gained in-depth experience in transformation, strategy definition as well as organizational development projects in highly dynamic environments. While the automotive field clearly is his home-turf, he feels just as well at ease in telecommunications, banking, and logistics where during many assignments at the interface between business and IT his ability to build bridges and thus drive successful transformation has proven a key success factor.

Bill Tran

Bill is a sophomore, studying Supply Chain Management and Information Systems at Texas Christian University. Bill is also pursuing a MicroMasters degree in Supply Chain Management at MIT. He is currently working on a startup which provides business intelligence tools for logistics companies to leverage their competitive advantages and exceed their customer expectations.

Alex von Stempel

Alex von Stempel has 30 years' experience as a commentator and trade journalist, most of which dedicated to freight transport and logistics of perishables.

Thirteen years ago, he set out on his own, devoting a significant proportion of his time to developing global and regional perishable logistics events, including Cool Logistics Global, Logistics Hub and Cool Logistics Asia, which is held alongside the Asia Fruit Exhibition in Hong Kong.

Alex von Stempel also works as an independent logistics adviser and frequently acts as master of ceremonies for third-party events.

Scott Wang

Scott is Air Gateway Manager at UPS where he is responsible for managing the daily gateway air, space procurement and hub operations in Shanghai, ensuring superior service and developing innovative logistics solutions. Prior to that, he was a program manager for two of the most important accounts at UPS globally. Here, he led a team and managed complex projects and designed programs to ensure that projects were timely on board. Scott is member of the UPS China Millennial Business Resource Group, which supports innovation in the company by bringing a diverse set of ideas and insights to the job. Scott holds an MBA from Fudan University and a BSc from Shanghai International Studies University.

Andre Wheeler

Based in Perth, Western Australia, Andre runs the Asia-Pacific Connex consultancy. With more than 20 years' experience in international business, he maintains a diverse network of personal contacts throughout the United States, Asia, SE Asia, Africa, and the United Kingdom.

Holding a B. Science (Hons) degree and an MBA, he is currently working toward his PhD on the Impact of the China One Belt One Road initiative on infrastructure and logistics in the ASEAN Region.

Author of the book: "China's Belt Road Initiative: The Challenge for The Middle Kingdom Through A New Logistics Paradigm," he has also published and presented a number of papers concerning China's Belt Road Initiative and the changing Eurasian trade paradigm.

He is recognized for his contribution and participation in the development of the Western Australian government's Asian Engagement Strategy 2019–2030.

Lionel Willems

Lionel is team leader of project managers and business analysts in the open-core ERP company Odoo. After graduating with a master's in supply chain management and international business from the Louvain School of Management (Belgium), he helped in digitalizing SMEs across Asia by working in Indonesia, Vietnam, and Hong Kong for three years. Strong in functional, technical, project management and cross-industry knowledge, Lionel's contribution to this book reflects his passion for logistics, supply chains, and IT.

Dennis Wong

Dennis has 15 years' experience in various functions at leading third-party logistics providers in the United States and China. Most recently he was the Director of Sales at Flexport Asia-Pacific Region, and previously the Director of Sales and Marketing at Kuehne + Nagel Hong Kong, where he acted as Head of Sales for Hong Kong, South China, and Macau. Dennis studied at California State Polytechnic University in Pomona.

Duoqi Xu

Duoqi is a Professor of Law at Fudan University School of Law, Shanghai Shuguang Scholar and one of The Top Ten Excellent Young Jurists in Shanghai (fifth session). She was also a Fulbright visiting scholar at Harvard law school from 2016 to 2017, was a Hauser Global research fellow at NYU from 2008 to 2009, and a visiting professor at Taiwan University in 2013. Professor Xu is the chief-editor of Internet Finance Law Review, the Director of Center for Internet Finance Law Innovation of Shanghai Jiao Tong University.

Jiao Xu

Jiao Xu is a Global Customs Analyst at Flexport in Amsterdam office. Compliance of international trade and logistics is a constant focus in both her academic and professional career. She did her master's of Global Supply Chain management and Change at Maastricht University in the Netherlands after finishing the bachelor program of Logistics Management in Shanghai Customs College. She did research in supply chain compliance and visibility in B2C E-commerce between China and Europe, discussing from the perspective of multiple stakeholders throughout the supply chain.

Foreword

The Fourth Industrial Revolution is in full swing, and we can see every day how technology is fundamentally changing businesses and our lives. Artificial intelligence in medical imaging is automating workflows and increasing diagnosis accuracy. Digital twins for aircraft engines are used to predict their physical behavior under extreme conditions. In virtual classrooms, interactive whiteboards and online content delivery are improving student outcomes while reducing teaching burden for instructors. Mobile payment platforms are helping consumers and companies to better manage their financial operations. Businesses around the world are striving to adopt new digitalization strategies to capitalize on these exciting new trends.

The automotive industry is heavily dependent on the use of technology, and it has been an exciting and challenging experience to align on digital strategies with our many supply chain partners. The future is upon us as we see autonomous vehicles become a reality that will drive the biggest revolution since the invention of the car. I can feel the pulse of change especially here in China where our customers are on average 36 years old, much younger compared with the United States or Germany where the average age is over 50. These customers demand very strong orientation toward future technologies, connectivity, and digitalization. Therefore, we are set to transform into a software-enabled car company, combining the traditional Porsche spirit with the power of new technology. A concentrated effort throughout our departments here in China together with Porsche AG is needed to bring this digital transformation to our customers.

While aesthetically pleasing, professional-grade products have been the backbone of our success. We are now having to evolve and reexamine our people, processes, and products through the lens of technology. Developing and manufacturing inspirational vehicles requires Porsche to leverage our supply chain partners in every way as we desire to be world leaders in sustainable cutting-edge supply chain excellence. As an example, together with Porsche AG, we not only adopted SAP's enterprise resource planning applications for just-in-sequence delivery of parts but also went a step further to partner and co-innovate with SAP on creating digital transformation solutions. On a global level, with our venture capital unit Porsche Ventures, we are also seeking strategic investments into future technologies such as artificial intelligence, blockchain, and virtual and augmented reality. Our strategic investment portfolio already includes over 50 innovative startups, such as "Gapless", which builds the world's first blockchain platform for car management. Executives throughout the business community need to be out there learning about these new technologies to be able to understand the complexity and potentially disruptive impacts that they may have on their industry.

Throughout my career, I have seen technology quickly upgrades our production facilities, marketing capabilities, and products. However, there has been little change when it comes to the logistics of moving our products. As vehicles are produced in a limited series and made up of more than 20 000 individual parts, distribution and logistics is an integral component to the success of Porsche globally as we strive to get the right product to the right person at the right price whenever they so desire it. We consider logistics as an opportunity to differentiate ourselves from our competitors. For example, when we distribute our sports cars, we are transporting a number of factory-new models from Leipzig to Chongqing via rail instead of ocean. This strategic logistics decision allowed us to shorten our delivery time by three weeks and allowed us to fulfill customer demand in Southwest China much more quickly. Sensors on board the train allow us to increase supply chain visibility by giving us real-time location throughout the journey, and we are also able to monitor the humidity, shock, and temperature levels. The data generated by these movements allow Porsche to further improve our logistics process and is a good example how the Fourth Industrial Revolution will change logistics to improve performance and customer satisfaction. With IoT sensors, a new world is opening up where we have abundant possibilities for new analysis and potential ways to automate to create more efficient and sustainable processes. It is exciting to think about all the areas where Logistics 4.0 will enable new business models and advanced value streams. In order to bring these new opportunities of leveraging technology, we must avoid speculation and stop using overused and distorted terms that jeopardize all the great progress being made.

From CEOs to operations clerks, there is a need to understand these cutting-edge technologies in the context of today's market, the maturity within the logistics and supply chain community, and technological interdependencies. This requirement is what makes this book so important. It demystifies an area that is clouded by hype and misinformation and provides a remarkable overview of the industry and its transformation. The book provides the reader a working vocabulary and knowledge base to give them a deeper understanding about current industry activities and support their upskilling efforts. After reading this book, I am confident that it will inspire them to think deeply about how technology will influence their job in the near future.

Mac Sullivan and Johannes Kern have managed to assemble a stellar lineup of practitioners and academics that share an impressive multifaceted perspective on this often-overlooked topic. This book contains learning points for everyone active in logistics and supply chain management, from students to seasoned experts. By offering different sections focusing on people, technology, and platforms, the authors have enabled the reader to analyze the impacts through different domains as well. Whether you want to learn more about a specific technology or want to future-proof your team by getting some tips on simplifying complexity, the book offers a wide range of content. Fundamentally, the book does a great job in creating readable content that will educate the reader in an area that is ripe for opportunistic companies and individuals. Help yourself by digging into this book as I challenge you to start your own evolution to be at the forefront of digitalization in the logistics industry.

Dr.-Ing Jens Puttfarcken, President and CEO of Porsche China
Shanghai

Acknowledgments

Johannes and I were both finishing up our PhDs in Shanghai when we first met. Ironically, I was writing my dissertation on how technology will disrupt the logistics industry, and Johannes was writing his on how relationships still play a key role in the supply chain industry. Both of us came from the business world where Johannes is a supply chain consultant and I work in freight forwarding. We were both interested in academia as well though and coincidentally both taught classes at Tongji University. Recognizing that the supply chain and logistics industry needed a hybrid book that bridged the academic and business worlds to show the opportunities arising, we set about brainstorming what the book would look like. We were conscious of the fact that a lot of books that we used in our teaching were far too academic, while many of the nonfiction digital transformation books rarely touch on the intricacies of the supply chain world.

It was at this point in 2018, I met Tariq Samad in Shanghai as I hosted a group of his MS students who were specializing in the management of technology. After I gave a quick presentation on the future of logistics, Tariq shot me an email where he mentioned this series for IEEE and Wiley. Johannes and I met him on his follow-up trip and pitched our proposed collected edition. From there, we continually worked on this project with our contributors over the past 18 months and are excited to be part of the series. We sincerely appreciate the opportunity and support that Tariq has given us throughout the process.

We want to give a big thanks to the staff of Wiley, Mary Hatcher, Teresa Netzler, Victoria Bradshaw, and Louis Vasanth Manoharan for being of tremendous help guiding us along this multi-year process.

Walter Simpson has been with us for the past year on this project, and there is no way that we could have gotten as far as we have without his help. He has put in so much effort in coordinating, editing, and writing. We are proud to have him on the team and are excited to see his career progress.

Our more than 30 contributing thought leaders on this project who are based all over the world and come from very diverse backgrounds all share the passion for knowledge and a desire to build a stronger logistics industry by leveraging technology. We want to thank the contributors for their tireless efforts as they took time away from their work, families, and spare time to make valuable contributions to this project.

Lastly, we also want to thank the Council of Supply Chain Management Professionals (CSCMP) and Descartes Systems Group for allowing us to use their glossary of supply chain and logistics definitions.

Mac Sullivan and Johannes Kern

I want to thank my beautiful wife, Rebecca, for holding down the fort while I spent hours away from the family to work on this project. I want to thank my parents, Page and Billy, for their investment in my education as they had to give up some luxuries in life to make that happen and for cheering me on throughout this writing process. To my father-in-law, Binfu, who watched the kids and helped us out during the past 2 years as we worked on this project, I say a big thank you. I would like to thank my boss and mentor, Jeff McDonald, for taking a chance on me years ago and bringing me into this crazy industry. Johannes, you have been a guiding light throughout this project, and I was lucky to have met you as you provided the structure and diligence needed to see this through.

Mac Sullivan

In addition to all the great people mentioned above, I would like to extend my thanks to Mac Sullivan who drove this project full of passion and dedication. Mac, your tenacity and unstoppable can-do attitude made the book what it is today. Finally, I must thank my future wife Li Chunyou. Without her love, care and patience, this edited volume could not have been written.

Johannes Kern

A Note from the Series Editor

Welcome to the brand-new Wiley-IEEE Press Series on Technology Management, Innovation, and Leadership!

The IEEE Press imprint of John Wiley & Sons is well known for its books on technical and engineering topics. This new series extends the reach of the imprint, from engineering and scientific developments to innovation and business models, policy and regulation, and ultimately to societal impact. For those who are seeking to make a positive difference for themselves, their organization, and the world, technology management, innovation, and leadership are essential skills to hone.

The world today is increasingly technological in many ways. Yet, while scientific and technical breakthroughs remain important, it's connecting the dots from invention to innovation to the betterment of humanity and our ecosphere that has become increasingly critical. Whether it's climate change or water management or space exploration or global healthcare, a technological breakthrough is just the first step. Further requirements can include prototyping and validation, system or ecosystem integration, intellectual property protection, supply/value chain set-up, manufacturing capacity, regulatory and certification compliance, market studies, distribution channels, cost estimation and revenue projection, environmental sustainability assessment, and more. The time, effort, and funding required for realizing real-world impact dwarfs what was expended on the invention. There are no generic answers to the big-picture questions either; the considerations vary by industry sector, technology area, geography, and other factors.

Volumes in the series will address related topics both in general—e.g., frameworks that can be applied across many industry sectors—and in the context of one or more application domains. Examples of the latter include logistics and transportation, smart cities and infrastructure, energy and environment, and biomedicine and healthcare. The series scope also covers the role of government and policy, particularly in an international technological context.

With 30 years of corporate experience behind me and about five years now in the role of leading a Management of Technology program at a university, I see a broad-based need for this series that extends across industry, academia, government, and nongovernmental organizations. We expect to produce titles that are relevant for researchers, practitioners, educators, and others.

I am honored to be leading this important and timely publication venture.

Tariq Samad
Senior Fellow and Honeywell/W.R. Sweatt Chair in Technology Management
Director of Graduate Studies, M.S. Management of Technology
Technological Leadership Institute | University of Minnesota samad@ieee.org

Section I

Introduction

1

Demystifying the Impacts of the Fourth Industrial Revolution on Logistics

An Introduction

Mac Sullivan

NNR Global Logistics, Dallas, TX, USA

Introduction

Future of Work in the Fourth Industrial Revolution

Throughout history, during periods of agricultural and industrial reform, society was worried that most of its population will be out of a job as a new technology is developed (Manyika et al. 2017). These fears are reinforced by documented events in modern history where technology has led to mass layoffs. A transition from a post-industrial era into a knowledge era has brought forth the same old argument that automation and artificial intelligence (AI) are going to cause widespread disruption. Until this point, white-collar job has been dominated by the technologies created in the past four decades with the rise of the personal computer, Internet, and widespread business software applications. There has not been a large shift in the business-to-business (B2B) world in terms of reallocation of white-collar labor resources and skillsets, which is surprising given the pervasiveness of mobile, e-commerce, and sharing economy trends that are driving consumer behavior.

Some signals show that the world is at the brink of a new technological revolution, now referred to as the Fourth Industrial Revolution (4IR), where the convergence of new and old technologies promises to redefine and transform the future of work and more. The intricacy and extent to which this transformation will happen will be unlike anything that has previously occurred (Schwab 2015). Implementing and realizing the results of the technologies laid out in the 4IR remain theoretical outside a few select companies. However, breakthroughs in AI and machine learning are poised to have a direct impact on our daily lives (Lee 2018). In some ways, the 4IR is a continuation of digitalization brought about by the Third Industrial Revolution as seen in Figure 1.1. However, it is unique in how it is blending the physical and cyber worlds through the prevalence of several technologies: Internet of Things (IoT) devices, cheaper cloud computing, AI, and automation (Marr 2018).

The computerization of rote tasks once done by clerks, the displacement of traditional business models through platform software application and websites, and the automation

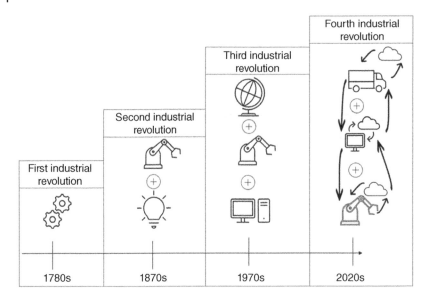

Figure 1.1 The Fourth Industrial Revolution.

of manufacturing and physical movement of goods were all witnessed in most parts of the developed world in the early twenty-first century. Looking toward the future, by 2022, companies have the potential to boost revenues by nearly 40 % by investing in AI and human partnership that would translate to $800 billion in new profitability (Simkin 2016). What the steam engine was able to do in decades for muscle power, computers, and digital automation was done for brain power by fundamentally reshaping the way that we think and the environment we operate in (Brynjolfsson and McAfee 2014). Much like its predecessors, the 4IR is changing the way that humans interact with the world and opening new opportunities for businesses to optimize their workflows.

In his book, *Hit Refresh*, Microsoft's CEO Satya Nadella states that the origin of the 4IR can be identified by "A confluence of three breakthroughs—Big Data, massive computing power, and sophisticated algorithms," and that it is "accelerating AI from sci-fi to reality" (Nadella et al. 2017). Previously, Thomas Friedman had pointed out that the world is much flatter than it used to be as a result of the outsourcing of not only physical labor but also, with the adoption of new technology, white-collar labor (Friedman 2007). With such a quickly evolving landscape, it is important to note that the way companies adopt these technologies will have significant impacts on their communities, company structure and strategy, and ultimately their future. While robotic process automation (RPA) and AI are not necessarily novel concepts, downstream consumer demand, consultant-driven media blitz, and subsequent executive exploration give insight into where the B2B service sector is headed.

The Role of Digital Transformation

To take advantage and participate in the 4IR, many companies now find themselves in a position where they need to undergo what many are calling a "digital transformation." Digital transformation can be explained as the shift in work, jobs, and products through the use of technology in a company or the operational context of that company (Parviainen et al. 2017). Transformations driven by technological advancements have a history of being stalled as governments struggle to determine the policy to control them, companies lack resources to take advantage of them, or the labor force is not equipped to use them. Strong economic activity since 2008 has helped to sustain global economic growth, and corporate coffers had been full heading into the COVID-19 era. To evaluate the potential repercussions of 4IR, it is worthwhile to explore how some companies have already embraced these technologies and engrained them into their core business models.

Hub Economy Companies Leading the Way

Technological leadership has been spearheaded by a group of hub economy companies that do not operate by trying to compete on legacy products and services. These companies whose valuations now compare with some of the largest governments in the world use their vast networks in certain domains and then transfer those assets to redesign the competitive landscape of new domains and markets (Marco and Lakhani 2017). Hub economy companies like Amazon, Alibaba, and Alphabet are changing the way that value is created and gaining market share at a disproportionate and accelerated manner (Marco and Lakhani 2017). Grown out of the Third Industrial Revolution, these companies are exemplifying the potential of the 4IR as they innovate, grow exponentially, and leverage multiple forms of technology along the way.

Traditional B2B companies are struggling to adapt and catch up as their lower-level white-collar worker ranks are bogged down by complacency and low levels of retention. The sense of urgency has not set in, yet. Protests over Amazon opening up a new headquarter in New York (Feiner 2019) and an increasingly larger media profile surrounding the automation involved in their warehouses (Bose 2019) reflect the communities' cognition that hub economy companies' consolidation of market share may not be beneficial to all despite offering convenience and a lower price tag. For every million dollars in revenue that companies like Amazon take from traditional businesses like Walmart, it is estimated that four jobs are lost in the community (Kaplan 2015). Adopting cutting-edge technology, such as automation and AI, and venturing into different market sectors have traditionally not been the core focus of many logistics companies as prioritization has tended to focus on more tangible short-term customer-oriented results or supplier management. The underlying global supply chain is shifting beneath us, and there is evidence to suggest a change is coming that will necessitate a reprioritization.

Current State of the Logistics Industry

Effects of the Second and Third Industrial Revolutions on Logistics

Logistics was born out of the need for military groups to travel efficiently around the world, and the term was borrowed by international trading companies as they looked to deliver their goods (Bonacich and Wilson 2011). The industry was further developed by the advent of the steam engine as its propelled cargo on trains and boats at faster paces. For this chapter, the term logistics will represent third-party logistics (3PLs) or freight forwarding companies, who provide transportation, warehousing, and compliance support to the global trade network. Logistics is an industry built on relationships, information arbitrage, and strategic outsourcing from retailers and manufacturers of their transportation coordination needs to specialized companies that only do this in one form or another.

Logistics has transformed in terms of complexity as it moved from break-bulk shipping into the containerized and air freight modes of transportation and now into an increasingly complex logistics network that is being constructed around consumer demand. The Third Industrial Revolution brought about a foundational shift in the levels of transparency available via Internet-enabled computers that opened up new forms of competition for incumbent players as the amount of capital needed to be involved in the process decreased dramatically. Transparency led to more competition that result to lower margins on the supply side; however, at the same time, it also decreased the amount of manual labor needed that reduced costs (McKinsey&Company 2018).

Technology Is Easy; But People Are Hard

With $1.5 trillion being spent every year on logistics, new entrants are quickly entering the market, and incumbents are frantically trying to solidify their position (Gould 2018). As a critical control mechanism of global supply chain activities, global logistics service providers, of which freight forwarding companies are part of, are facing a new set of requirements from supplier and customers, and they are having to adjust their organizational competencies and align processes (Chen et al. 2019). Societal adoption of technology has traditionally varied even when the talent and infrastructure have been in place, and researchers propose that there are typically social impediments to technology transplantation (Mokyr 1996).

Technological adoption in large logistics companies has been slow as top-line demands, corporate bureaucracy, large geographic presence, and a lack of innovation culture stifle even the most well-resourced technological investments. Senior executives from top 20 logistics company have stated, "Digital platforms are just for the next generation", "Temporary labor automated process flows would never work in a country like China" and "Blockchain is a topic from 10 years ago" (Bao et al. 2018). Changing perceptions and stances on the impacts of potentially pervasive technology takes time and is often a seismic event to put things into perspective as shown by recent events during the outbreak of COVID-19.

The Role of Startups in the Logistics Community

Large multinational logistics companies are seeing a steady outflow of senior talent to niche technology-driven supply chain and logistics startups. Startups are primed for leveraging technology as they have significantly less overhead and less bureaucratic structures to overcome as they explore new service and product value propositions (Tipping and Kauschke 2016). Much of the logistics and supply chain disruption starting to happen has been led by agile companies, startups or not, who can utilize digital tools to bolster strategic investments in sales, marketing, and distribution, to more quickly and more effectively deliver value to their customer (Schwab 2015). Logistics companies have typically not attracted the most innovative, the most intellectual, nor the most technologically sophisticated talent, simply because they have not needed to as the industry required strong communication and relationship soft skills, some basic computer, and math hard skills and not much more to get where the industry is today.

Logistics Companies Under Pressure

Macolm McLean's invention of the standardized shipping container not only gave rise to some of the largest international transportation companies today but also helped to decimate the manufacturing communities of some of the largest consumer nations like the United States and Japan and those in Western Europe (Levinson 2016). With the recent influx of protectionism and the decreasing cost of physical automation, near-shoring manufacturing is making localization of production more of a reality, and those same nations above can bring manufacturing home again. Even though freight rates are a fraction of what they were in the days of break-bulk shipping, logistics costs are on the rise and are no longer considered a small percentage of the landed cost. When brands or trading companies evaluate their total landed cost of goods, logistics have traditionally represented a small amount, hence the lack of scrutinization and upskilling of in-house talent to handle these roles. Manufacturing costs are lowering due to the affordability of robotics, and this is causing transportation to be a much larger factor when determining the selling price of a product to the end consumer (Kelly 2017). Under more scrutiny, logistics companies must adapt by lowering their selling rates that can only be sustainable by lowering the inputs of their services and products.

Technology Investments by Logistics on the Rise

Increased pressure from customers, startups, economic forces, and suppliers is causing senior executives from many top logistics companies to more closely evaluate what role technology will play in reducing their costs (Riedl and Chan 2019). From sales to finance, white-collar positions in logistics are increasingly seeing that everyday technology available on their phones has not been represented at their work. Using investments in technology by logistics companies as a barometer, the industry is at a tipping point. The investment community has taken notice, and much-needed capital injections to digitize this traditional B2B industry are happening regularly.

From 2012 to 2017, there were over 3 billion dollars invested in logistics and shipping startups (Riedl and Chan 2019). In response to this, industry-leading logistics companies are responding in turn by earmarking huge investments in technology. C.H. Robinson, a leading global logistics company, just committed to investing one billion dollars over the next five years, citing technology as having a "profound impact on the supply chain marketplace" (Prairie 2019). Those that do not make the jump will be left behind as there is more and more consolidation in the logistics community through mergers and acquisitions as there is still no "hub economy" logistics player, yet. One of the leading contenders in terms of size and financial resources, Deutsche Post DHL Group's CEO Frank Appel, stated that "Digitalization will change our industry, it will change our company. . .We want to invest 2 billion [euros] in the next couple of years" (Frangoul 2019). Industry leaders are finally responding and opening up their checkbooks, but have they been too complacent and opened up the doors to new players?

New Entrants Threaten Status Quo

Where larger outsourced logistics companies have already offshored many of their back-office functions to shared service centers, there is now an effort to automate those same functions, and medium-sized logistics companies now are directly comparing outsourcing with automation. As the cost of these documentation, coordination, and billing tasks, which had previously been done in the more expensive labor consumer nations, decreases, so will the margin on the overall price arbitrage. These previously low- and mid-skill-level positions who had been dependent on labor inputs will be offshored or automated.

Aggressive competition and agile startups will seize market share by reducing their selling rates. This is hypothetically made possible in highly automated and technologically savvy companies that have little overhead and a low marginal cost. It is also worth noting that startups chasing market share and revenue, while ignoring profitability in hopes of raising seed capital, have the potential to impact things as well. As an example, Morgan Stanley's analysis of Uber Freight, a platform designed to facilitate the movement of cargo by matching truck drivers with cargo owners, shows that Uber Freight is giving 99% of its revenue back to its throughput carriers and keeping only a 1% margin for itself (Aouad 2019). This is a prime example of how a hub economy, and in this case a sharing economy company, is crossing industry segments and threatening the role of domestic trucking brokers just as it did with the taxi industry.

The impacts of technology on logistics from a product perspective are worth examining. The US trucking industry, estimated at $700 billion, still uses "phone, fax, and email as primary methods of connecting"; however companies like Transfix are building online marketplaces to remove that inefficiency (McAfee and Brynjolfsson 2017). Since 2017, there has been a boom in companies offering the same services from Uber Freight to Next Trucking. Many warehousing companies sit on excess space in which Flexe, a Seattle-based company, is looking to optimize by offering a platform for those companies to rent that space out (McAfee and Brynjolfsson 2017). The Neighbor application has taken that a step further, and now anyone can rent the extra space in their home (Holder 2019). These new products are causing major disruptions for logistics companies in the same way the Uber and Netflix have disrupted the transportation and television industries while also maintaining low overhead costs. These new entrants are pushing logistic companies to respond by increasing technological capabilities to offer similar products and customer experiences.

Disintermediation Threat Looming

The rise of Uber Freight and many others like it through the supply chain and logistics ecosystem should frighten logistics companies as the comparison with a travel agency is strikingly real. Disintermediation, defined as suppliers dealing directly with customers and eliminating the needless cost and interaction of an intermediary, happened in the travel agency as the number of sales went 80% online within seven years of its commercial and technological enablement (Dodu, 2008). Whether it is expedia.com or Uber Freight, there is a lesson to be learned. Freight forwarders must move up the value stream. Beneficial cargo owners (BCOs), otherwise known as the direct customers of logistics and freight forwarding companies, typically have limited knowledge on evaluating different carrier options and even less experience dealing with all the formalities involved in booking directly (Gasparik et al. 2017); therefore the value that a freight forwarder provides is in the knowledge that they have in terms of carrier and customs (Burkovskis 2008). The ultimate value that they offer is in being a single source of information and a centralized window to provide their clients with the services and relevant information needed to move a shipment (Gasparik et al. 2017).

The challenge remains on how to move that centralized point away from phone calls and emails to a more efficient medium that offers system connectivity and instant access. The first step in the digitalization journey for logistics companies is not abundantly clear as often they are heavily reliant on outside stakeholders, including customers, governments, and vendors. With $1.5 trillion of value at play in the logistics industry, there is an influx of new players, and incumbents are scrambling to figure out their strategy (Gould 2018).

Navigating a Digital Transformation

Budget Considerations for Technological Upgrades

Struggling with where to start and how to keep their business afloat, talent-strapped logistics companies must face the difficult discussion of where the money will come from to fund their digitalization journey. Logistics is a fixed cost heavy business in that there are many pass-through costs, which are generally much higher than the final margin, and therefore there are inherent risks if things go wrong. Profit in the logistics industry is quite low compared with the sharing economy companies mentioned above. Even within the logistics industry, there is a large gap between the courier/express/parcel (CEP) companies who could have double-digit EBIT margins compared with the logistics service providers whose profitability is usually between −1% and 8%(Tipping and Kauschke 2016). The necessity of logistics as it facilitates the movement of goods around the world and the inability for one company to control each step along that supply chain will ensure that logistics providers have a vital role for the foreseeable future. Logistics companies need to focus on projects that are within their control where they have the foundations in place, the budget to realistically invest, the talent to drive, and lastly, the permissions from shareholders to execute (see Figure 1.2).

Shipment coordination is necessary, and for a long time, it was a profitable business to buy space along the chain and resell it at a higher rate by offering an integrated seamless solution to the world's market. The ability to move goods efficiently in and out of international markets independent of the distance but dependent on

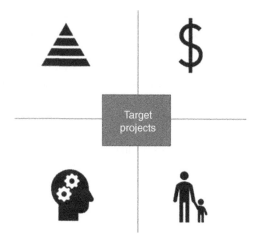

Figure 1.2 Target digitalization projects.

the speed, quality, and price of coordination determines the value of logistics (Tseng et al. 2005). In other words, logistics companies are not going to get an injection of additional margin from their customers to facilitate this digital transformation; this will need to be a capital expense that will be recouped through either reducing payroll, which often means cutting high-paid executive staff, or procuring from their suppliers at a lower rate.

Lessons Learned from Other B2B Industries

A multitude of traditional service industry businesses has been disrupted by the advancements of technologies associated with cloud computing, faster processing power, and the beginnings of AI through sophisticated algorithmic filtering processes. From international logistics to international communication, there will continue to be further economic growth, and new markets emerge as supply-side gains will enable efficiency that will lead to reduced costs (Schwab 2015). For example, there has been a large shift in the way that the publishing, taxi, hotel, finance, music, movie, and travel agency businesses all operate as they fight to compete against their zero marginal cost platform model competitors.

Automation Leading in Terms of Return on Investment

Outsourced manufacturing to the emerging and developing world may have peaked in terms of manufactured goods as tariffs and nontariff barriers (NTB) threaten the effectiveness of cross-border supply chains (Großer and Weinert 2019). If there is negative growth in the global economy, the 0.8–1.4% productivity gains from automation will decrease the polarization that has previously skewed toward low-skill, repetitive, and rote work (Manyika et al. 2017). Companies have to utilize their assets, internal capabilities, and cash to "transform their core businesses" in hopes of producing the same outputs with a lower cost or striving to produce new revenue generators, as the adoption of new technologies and entrants of a new form of competitors is encroaching on many economic silos that had been previously untouched (Marco and Lakhani 2017).

Up until the spread of COVID-19, shareholders had been content with positive returns and had not exerted pressure for automation as they considered the media backlash and subsequent revenue ramifications of subzero job creation. As competition grows fiercer through more transparent marketplaces in a somewhat commoditized and mature industry, potential economic constrictions loom, and increasingly nominal technical factors make automation more realistic. A combination of these factors will erode the societal and governmental protectionary boundaries that have kept workers at their current subsistence levels (Manyika et al. 2017). Low-skill workers tend to recover fairly well after recessions; however, middle-skill cognitive workers, like logistics clerks, tend to bounce back slower (Foote et al. 2015).

The pursuit and possibility of near-zero marginal costs as in the freight market are driving companies to reengineer their current business processes and make them look at automation. Companies are always trying to increase productivity to reduce costs. However, until recently, economists did not think it was possible that technology could push that marginal cost of "communication, energy, and transportation, as well as many other physical goods and services" as close to zero as seen in the "emerging sharing economy" (Rifkin 2015). This is more of a reality today as the cost of automating tasks is dropping at a rapid rate with new emerging technological advancements in computing power and machine learning.

There is a shift away from outsourcing manual labor in terms of both physical and digital tasks, which still have a significant operational expense, toward the capital expenditure of investing in automation. This is looking more appealing as labor costs go up and talent shortages of skilled workers continue to plague R&D leading countries like the United States and China. Logistics company leaders, as representatives of businesses leaders, will be at the forefront of technological change and will be able to take advantage of the efficiency gains offered by technology; however, they must focus on rethinking their business to see how automation can work alongside their workforce rather than think in terms of simply automating tasks using legacy processes (Bughin et al. 2016). See more on this topic in the RPA chapter by Sullivan, Simpson, and Wesley Li, Chapter 5.

Winners and Losers of a Digital Transformation

Utilizing robotic process optimization, upgrading to a cloud server, or setting up a customer-facing API may sound like great first steps along this journey. However, whether a company decides to use an outside consultant or drive this initiative from within their organization, there can be substantial roadblocks that hinder the digital transformation. The potential benefits of better technology must be weighed against its costs, including not only the cost of obtaining the technology but also the costs of learning how to use it and integrating it with other technologies already in use. There is also the risk of rushing into a new technology and having a more efficient technology come out just after that makes the investment that the company just made obsolete. Technological change will help some workers do their job more efficiently but potentially put others out of a job. In the end, it will be the market, profitability, and economic efficiency that have the final say on what technology will be used and how (Mokyr 1996).

The Necessity of Economies of Scale

Having a strong in-house IT team coupled with a CIO would certainly help companies along this path, as would a contract with a top-tier consulting firm. However perhaps the thing that will help companies the most is a clear written understanding of their global processes and procedures. Having a roadmap of processes helps to lead the way to be able to identify where commonly occurring tasks happen and facilitate a smooth transition to improving internal mechanisms. This has been traditionally not very well documented in logistics as much of the process has relied on email communication. Companies with business processes that necessitate a significant amount of person-to-person communication have held on to their legacy communication mechanisms because their stakeholders valued it and there was not a more or "equally efficient digital alternative" (McAfee and Brynjolfsson 2017). Logistics, which is representative of the B2B service industry, fits this description well. Even container ship operators, multibillion-dollar juggernauts who went through their evolution as they adopted the use of the standardized container in the 1960s and 1970s, are struggling to upgrade their communication methods (Tipping and

Kauschke 2016). To learn about how China is using its Belt and Road Initiative to lift its trading partners, see Wheeler's chapter on the Digital Silk Road.

While companies like Maersk have set up shared service centers in developing countries, the adoption of cutting-edge processes is staggeringly slow as can be seen by response times and incorrect invoicing still plaguing the industry. Processing hundreds of thousands or more of virtually identical transactions that had traditionally been handled, monitored, or audited by a human worker has been the best use case so far for process automation and implemented across the banking, insurance, telecommunication, and travel industries. This perfectly applies to logistics and transportation companies whose operations are at high risk of automation (Bughin et al. 2016). Chasing the top line has been long integrated into the mindset of logistics companies who are consistently looking for the next target customer to improve organizational health instead of looking toward process improvement.

Foundations of a Digital Transformation

Prioritization of Technology Exploration

Logistics companies must adapt and proactively look for ways to use technology to reinvent themselves (Tipping and Kauschke 2016). Consistently, logistics companies have been looking outward to new investments in assets or new client acquisition instead of building technological prowess in-house that would help to best prepare for a digital transformation. Unlike finance or manufacturing companies, commercially focused individuals who spend a large part of their time selling and catering toward their tier-one customer lead most logistics companies.

In addition to the financial and time constraints, companies are faced with an overload of exciting new technologies that they should or are being encouraged to evaluate. Logistics companies could invest in capturing data through IoT devices; developing sophisticated self-learning algorithms based on multilayer regressions to create some form of machine learning AI; eliminating emails, faxes, and spreadsheets by capturing data via new forms of ecosystems using blockchain or optical character recognition (OCR) technology; rolling out digital handshakes in the form of electronic data interchange (EDI); or allowing systems to communicate better with each other through application programming interfaces (APIs). They could even outsource their data management, integration, and software development. These are the conversations that have happened in board rooms of leading logistics companies around the world as they scramble to come up with a technology investment plan.

Connectivity Standardization in Logistics

It is feasible to imagine a network of systems that independently coordinate the facilitation of efficient movement of goods without human intervention, assuming all went according to plan. The novel concept of smart contracts powered by blockchain technology shows the potential for powering this system by eliminating trust and security concerns that plague the current system, but this has yet to reach mass adoption. For more details on this, see Ariguiz, Tran, Margheri, and Xu's chapter on smart contracts.

As if navigating the hundreds of three-letter acronyms used in logistics was not enough, when it comes to the IT portion of logistics, systemic standardization and incompatibility issues are holding the industry back. From a systems integration perspective, just as a company needs staff that speaks the same language, the global trade network also need systems that speak the same language. Information may need to flow through 12–15 different systems ranging from manufacturers in developing world countries to automated ports. An industry must have some standards in place, as these are the foundations of a company's ability to implement these technologies.

For a company who wants to ship goods and to get shipment information back into its ERP system, it has to connect to the freight forwarder who is coordinating the transportation, who in turn has to communicate with a trucker to pick up the goods, a customs broker, and a shipping line often through specialized messaging partners called. The freight forwarder also needs to link to its internal origin, in China, for example, and the destination office, in the United States, either of which could be a different agent company. With little coordination in terms of standardizing digital connectivity beyond EDI between core segments of the supply chain, there is a huge opportunity for companies like Chain.io who are offering a digital connectivity platform to translate all the different fields from different providers to the freight forwarding ecosystem. With no player having more than a 12% market share, it will take the efforts of entrepreneurs and startups to innovate solutions (Riedl and Chan 2019).

New Business Models Emerging

Now that the majority of logistics companies have adopted enterprise resource planning (ERP) systems as their core systems, such as from Oracle, SAP, or CargoWise One, there has been a rise in the standardization of some global processes that have enabled them to potentially use other technology. Native integrations with SaaS providers like Salesforce, QuickBooks, and Workday are allowing ERPs to exchange data seamlessly and instantaneously through APIs. This exchange creates a mountain of data that if correctly analyzed can be used to identify inefficiency, improve forecasts, and reduce labor costs. AI for logistics is particularly attractive in that there are bill of lading databases, shipment logs, and customs documents of millions of previously moved shipments that could be scrubbed to backtest algorithms.

AI, defined here as sophisticated algorithms that can parse information, has been shown to predict when a customer is going to buy something and when an aircraft engine needs servicing or alert a person that they are at risk of disease (Economist 2017). The hub economy companies have shown the amazing potential of data wrapped with algorithms to solve consumer problems. Technology and investment simply are not enough to enable AI to help give sophisticated business intelligence or better evaluate customer needs. To mine the data that drives AI, companies must have the infrastructure in terms of data management, the will and power to ensure data governance, and the talent to be able to identify, isolate, and cleanse data flows. Talent, as shown in Figure 1.3, is the last step in the foundations of a

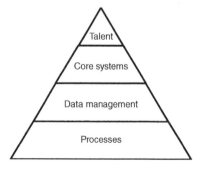

Figure 1.3 Foundations for digital transformation.

digital transformation. Traditionally programmers had set about training AI or a robot in rule-based "teach" patterns. However, with neural networks, raw data can be fed into the network, and the patterns are identified (Lee 2018). All of this is great but is useless unless an organization can feed the network huge amounts of data with clear algorithmic parameters focused on a narrow and specific goal (Lee 2018). In short, it takes a concerted effort by a talented, well-funded team with a clear set of business goals and strong organizational support in terms of technical and permissioned access to unlock the potential for AI.

As more logistics companies move away from working off spreadsheets and emails, they will be able to more effectively utilize technology due to the offering of Software as a Service (SaaS) that is discussed in Berry's chapter on the rise of cloud-based systems in logistics.

Participation in Platforms and Marketplaces

As logistics companies upgrade core systems and storage mechanisms and open up new forms of connectivity options, the reality of submitting rates to a marketplace or having an externally facing quotation portal becomes more of a reality. Platforms will continue to gain attention and capture market share from the traditional service providers. Platforms are discussed in Margand and Heuck, Chapter 16, and in Sullivan, Wong, and Tang, Chapter 23. These marketplaces not only can scale and connect suppliers and buyers with the click of a mouse bypassing the traditional sales cycle, but they also offer instant quotes and the ability to book immediately as opposed to the traditional email-and-wait model (Riedl and Chan 2019). Marketplaces are gaining attention especially as they enable a low-cost solution to direct selling entrants, like steamship lines Maersk and CMA, who now can open their digital sales channels and accelerate disintermediation.

Ironically, research has found that using the performance metric of return on assets (RoA), publicly traded companies in the United States have declined by at least 75% since 1965 (Hagel et al. 2016). This means that despite all the technological improvements that have happened in the past 50 years, companies still are not doing a great job of utilizing technology. Perhaps this could be explained by the lack of focus and the sheer complexity of a large organization. To start a digitalization journey, logistics companies need to invest in talent who have some understanding of logistics, the processes involved, and the integration bottlenecks between core systems that technology could potentially address.

Zoom Out/Zoom In Approach

Assuming that a company can get these foundations in place, leadership teams can then brainstorm on what they know or have learned about their companies' ability to adapt and look to hone their strategy on achieving more in an unpredictable world (Hagel and Brown 2018). The zoom out/zoom in approach to successfully navigating the waters of digital transformation explains how a company can explore not only the short-term stakeholder-driven targets but also the longer-term survival needs of the company (Hagel and Brown 2018). Here are a few takeaways that can be applied:

Zoom Out

1) Take leadership to an outside event or company to open their eyes and give them a glance at the future.

2) Bring in outside parties to consult and advise as they will have a fresh outlook on the company's position on what it will need to be successful in the next 10–20 years.
3) Look at how the customer base is evolving and identify unmet needs. Work backward to figure out what internal resources could be leveraged to add the most value.
4) Define what the IT department's purpose and goals are in hopes of providing a unique value proposition that will protect the company from competitors.

Zoom In

1) Examine areas where business meets technology that would be crucial to the success of the zoom out initiative and earmark resources to grow this.
2) Pick one near-term initiative that would bolster the company's core value proposition. Ideally, this would be in an area that would have a measurable ROI from improving operational efficiency or selling.
3) Eliminate poor-performing initiatives to free up resources to fund the above two points.

Executives will inherently tend to stray from above zoom out initiatives as urgent issues arise in the micro- and macroeconomic environments that they live in (Hagel and Brown 2018). Figure 1.4 presents a set of digitalization goals based on a zoom out/zoom in approach.

This approach is just one of many strategies that could enable a company to understand all of the emerging technologies and then start their digitalization journey that will involve a change in customer experience, the acceleration of how they achieve their digital goals, and the best way to allocate their resources (McKinsey&Company 2018).

The Time Is Now

Change is not coming; it is already here. Electrification opened up a vast amount of new possibilities but was only made possible by fossil fuels that power it, smart innovative businesses that monetize it and drive adoption, engineers that tinker and enhance it, and lastly

Figure 1.4 Digitalization goals.

Increase productivity velocity

Accelerate speed to market

Pragmatize digital transformation

Automate rote work

supportive governments that invest in the infrastructure to support it. 4IR technologies, such as AI, are opening up newfound opportunities and need many of the same inputs to make their potential a reality (Lee 2018). As noted above, there is a flood of startups and financial investors pouring into the supply chain and logistics market. Trade and shipment data is abundant in the databases of the logistics and shipping companies, as well as the governments who collect information when doing customs clearance. While traditionally governments are more reactive, efficiency gains driven by transportation that lead to more industrial outputs are being supported. The key element that the logistics industry may be missing is the talent to tie all these pieces together.

The 4IR is not meant to be flipping of a switch but more the tinkering and thoughtful collaboration of a network of companies, individuals, and industries that collaborate to create a new vision for the future. By creating and spreading innovative ways to coordinate the global movement of goods, logistics companies can reduce costs and reduce waste that will help to create a more sustainable way of life for all those involved. Due to customer demand, competitive risk, and socioeconomic factors, logistics companies have no choice but to embrace digital transformation. By taking steps to audit and optimize a company or even a department within a company in terms of its technological capabilities, talent pool, and processes susceptible to automation, individuals can help to future-proof the sustainability of their company.

Conclusion

Global economic stability and GDP growth have delayed the automation of white-collar jobs to a lesser extent than blue-collar jobs. Companies that had healthy balance sheets and content shareholders were less likely to have capital expenditures on expensive automated solutions as social ramifications were daunting and productivity gains and immediate ROI were not abundantly clear. More competition from new entrants and startups enabled by new systems integration solutions, as well as increasingly cheaper and more powerful cloud computing and processing speeds, will lower the barrier for hub economy companies to consider investing in the logistics industry. Profit per job derived from human input will decrease with higher labor costs, and increased transparency by marketplaces will continue to accelerate consolidation in the logistics market.

Talent shortages for documentation work will continue to be an issue in the logistics industry, and the industry will see the sun set on the era of information and pricing arbitrage. Investment in automating tasks whether through an API or RPA will outweigh the cost savings derived from outsourcing rote tasks to a lower-cost location. Increased costs due to human error are currently negated by the complexity and number of variables dealing with different geographies, commodities, and customer types; however, technology is allowing for a wider view across multiple systems. While the operations of many logistics providers have remained unchanged for decades, there is an abundance of opportunities for improving efficiency through the use of technology, but company culture, organizational prioritization, and a strategy for change management need to be addressed first. Logistics companies will be forced down the digital transformation path, and those who start on it sooner will have a better chance to survive.

Key Takeaways

- Technology has changed drastically since the start of the Third Industrial Revolution and can be difficult to keep up with. If unfamiliar with many of the emerging technologies presented in this chapter, then now is the time to start learning about them and considering how they could change the industry.
- Technology is easy; changing people is the hard part. Convincing staff to adopt new technologies and onboarding stakeholders to use them will be one of the biggest challenges of digital transformation. Resistance to change is widespread among logistics providers, and overcoming this will take refined soft skills from IT departments and other interested parties.
- The outcomes of digitalization will not be positive for everyone. There will likely be an increase in the adoption of automation in the coming years. For some employees, automation will free them of redundant tasks such as data entry, allowing them to focus on value-added strategic tasks. Other employees could see their jobs eliminated.
- Starting a digital transformation journey can be very difficult for logistics providers. The best place to start is by looking at the current state and trying to understand it holistically. Look at and identify current systems that can be improved; then decide what the organization can do on its own, and lean on suppliers and consultants to fill in the gaps.

References

Aouad, A. (2019). Uber freight is paying big to compete in the digital freight marketplace—99% of its revenue goes to carriers. *Business Insider*: 2019–2021. https://www.businessinsider.com/uber-freight-gives-nearly-all-revenue-to-freight-carriers-2019-6.

Bao, K., Sun, N., and Guo, E. (2018). *Status of the Logistics Industry*.

Bonacich, E. and Wilson, J. B. (2011) *Getting the Goods: Ports, Labor, and the Logistics Revolution*. Cornell University Press. https://books.google.com/books?id=R0EQk_0Ldd8C (accessed 15 March 2019).

Bose, N. (2019). *Amazon Dismisses Idea Automation Will Eliminate all its Warehouse Jobs Soon - Reuters*. Reuters https://www.reuters.com/article/us-amazon-com-warehouse/amazon-dismisses-idea-automation-will-eliminate-all-its-warehouse-jobs-soon-idUSKCN1S74B9.

Brynjolfsson, E. and McAfee, A. (2014). *The Second Machine Age: Work, Progress, and Prosperity in a Time of Brilliant Technologies*. W. W. Norton https://books.google.com/books?id=PMBUAgAAQBAJ.

Bughin, J., Manyika, J., and Woetzel, J. (2016). *MGI the Age of Analytics: Competing in a Data-Drivel World*, 126. http://www.mckinsey.com/mgi.

Burkovskis, R. (2008). Efficiency of freight forwarder's participation in the process of transportation. *Transport* 23 (3): 208–213. https://doi.org/10.3846/1648-4142. 2008.23.208-213, https://www.mckinsey.com/featured-insights/future-of-work/jobs-lost-jobs-gained-what-the-future-of-work-will-mean-for-jobs-skills-and-wages.

Chen, I.S.N., Fung, P.K.O., and Yuen, S.S.M. (2019). Dynamic capabilities of logistics service providers: antecedents and performance implications. *Asia Pacific Journal of Marketing and Logistics* 31 (4): 1058–1075. https://doi.org/10.1108/APJML-12-2017-0308.

Economist, T. (2017). Regulating the internet giants - the world's most valuable resource is no longer oil, but data | leaders | The Economist. *The Economist.* https://www.economist.com/leaders/2017/05/06/the-worlds-most-valuable-resource-is-no-longer-oil-but-data.

Feiner, L. (2019). Amazon to lease space in manhattan, less than a year after HQ2 fallout. *CNBC.* https://www.cnbc.com/2019/12/09/amazon-to-lease-space-in-manhattan-less-than-a-year-after-hq2-fallout.html (accessed 25 May 2020).

Foote, C.L., Reserve, F., and Ryan, R.W. (2015). *Labor- Market Polarization over the Business Cycle.* National Bureau of Economic Research (NBER) https://www.journals.uchicago.edu/doi/full/10.1086/680656.

Frangoul, A. (2019). *Digitalization "Will Change our Industry," Says Deutsche Post CEO.* CNBC https://www.cnbc.com/amp.

Friedman, T.L. (2007). *The World Is Flat [Further Updated and Expanded; Release 3.0]: A Brief History of the Twenty-First Century.* Farrar, Straus and Giroux https://books.google.com/books?id=-mv_ryTuvo0C.

Gasparik, J., Zitricky, V., Abramovic, B., David, A. (2017). *Role of CRM in supply chains using the process portal.pdf. 17th International Scientific Conference Business Logistics in Modern Management,* Osijek, Croatia (12–13 October 2017).

Gould, R. (2018). How robotics technology is modernizing third-party logistics. *Material Handling & Logistics* 73 (3): 18–20. http://search.ebscohost.com/login.aspx?direct=true&db=bsu&AN=130412754&site=ehost-live.

Großer, G. and Weinert, G. (2019). World economic outlook. *International Monetary Fund* 26 (1): 45–48. https://doi.org/10.1007/BF02928895.

Hagel, J. and Brown, J. (2018) '*Zoom out/Zoom in an Alternative Approach to Strategy in a World that Defies Prediction'.* https://www2.deloitte.com/us/en/insights/topics/strategy/alternative-approach-to-building-a-strategic-plan-businesses.html.

Hagel, J., Brown, J.S., Wooll, M., de Maar, A. (2016). *2016 Shift Index,* 40. Deloitte University Press.

Holder, S. (2019). *Rent Out Your Closet with an 'Airbnb for Storage' - CityLab.* CityLabs: Bloomberg https://www.citylab.com/life/2019/07/storage-unit-for-rent-shed-neighbor-app-side-hustle-airbnb/592630.

Kaplan, J. (2015). *Humans Need Not Apply: A Guide to Wealth and Work in the Age of Artificial Intelligence.* Yale University Press https://books.google.com/books?id=yatJCgAAQBAJ.

Kelly, K. (2017). *The Inevitable: Understanding the 12 Technological Forces that Will Shape Our Future.* Penguin Books https://books.google.com/books?id=zUzNDgAAQBAJ.

Lee, K.F. (2018). *AI Superpowers: China, Silicon Valley, and the New World Order.* HMH Books https://books.google.com/books?id=KdVHDwAAQBAJ.

Levinson, M. (2016). *The Box: How the Shipping Container Made the World Smaller and the World Economy Bigger.* Princeton University Press.

Manyika, J., Lund, S., Chui, M. et al. (2017). Jobs lost, jobs gained: workforce transitions in a time of automation. *McKinsey Global Institute*: 1–160. https://doi.org/10.1002/lary.20616.

Manyika, J., et al. (2017). A future that works: automation, employment, and productivity. *Mckinsey Global Institute,* 148. http://njit2.mrooms.net/pluginfile.php/688844/mod_resource/content/1/Executive Summary of McKinsey Report on Automation.pdf.

Marco, I. and Lakhani, K. (2017). Managing our hub Economystrategy, ethics. *Harvard Business Review*: 85–92. https://hbr.org/2017/09/managing-our-hub-economy.

Marr, B. (2018). The 4th industrial revolution is here - are you ready ? *Forbes*: 2018–2021. https://www.forbes.com/sites/bernardmarr/2018/08/13/the-4th-industrial-revolution-is-here-are-you-ready/#2e1a9f94628b.

McAfee, A. and Brynjolfsson, E. (2017). *Machine, Platform, Crowd: Harnessing Our Digital Future*. W. W. Norton https://books.google.com/books?id=zh1DDQAAQBAJ.

McKinsey&Company (2018). *Travel and Logistics: Data Drives the Race for Customers*, 40.

Mokyr, J. (2000). Innovation and its enemies: the economic and political roots of technological inertia. In: *A Not-so-dismal Science: A Broader View of Economies and Societies*. Oxford University Press Retrieved 24 September 2020. https://oxford.universitypressscholarship.com/view/10.1093/0198294905.001.0001/acprof-9780198294900-chapter-2.

Nadella, S., Shaw, G., Nichols, J.T., Gates, B. (2017). *Hit Refresh: The Quest to Rediscover Microsoft's Soul and Imagine a Better Future for Everyone*. Harper Business https://books.google.com/books?id=SeGMDAAAQBAJ.

Parviainen, P., Tihinen, M., Kääriäinen, J., Teppola, S. (2017). Tackling the digitalization challenge: how to benefit from digitalization in practice. *International Journal of Information Systems and Project Man agement* 5 (1): 63–77. http://www.sciencesphere.org/ijispm/archive/ijispm-0501.pdf#page=67.

Dodu, P. (2008). The internet, threat or tool for travel agencies?. http://steconomice.uoradea.ro/anale/volume/2008/v2-economy-and-business-administration/017.pdf.

Prairie, E. (2019). C.H. Robinson CEO BOB biesterfeld announces company's largest investment in innovation and technology. *CH Robinson*. https://www.chrobinson.com/en-us/newsroom/press-releases/2019/09-17-19_technology_investment.

Riedl, B.J. and Chan, T. (2019). The digital imperative in freight forwarding the traditional offline quotation and booking process is lengthy and. 1–13. https://www.bcg.com/publications/2018/digital-imperative-freight-forwarding.

Rifkin, J. (2015). Welcome to the third industrial revolution. *Wharton Magazine* (2015 January), pp. 1–20. https://www.bcg.com/publications/2018/digital-imperative-freight-forwarding

Schwab, K. (2015). The fourth industrial revolution | foreign affairs. *Foreign Affairs*: 1–5. https://www.foreignaffairs.com/articles/2015-12-12/fourth-industrial-revolution.

Simkin, M. (2016). Networking the revolutions. *Ágora* 51 (1): 67.

Tipping, A. and Kauschke, P. (2016). *Shifting patterns. The Future of the Logistics Industry, PWC*. https://doi.org/10.7748/ns.11.9.14.s28.

Tseng, Y., Yue, W., and Taylor, M. (2005). The role of transportation in logistics chain. *Eastern Asia Society for Transportation Studies* 5 (1): 1657–1672. https://doi.org/10.21522/tijmg.2015.05.01.art015.

Section II

Technologies

2

Technologies Driving Digital Transformation

Mac Sullivan

NNR Global Logistics, Dallas, TX, USA

The Fourth Industrial Revolution (4IR) and the past three industrial revolutions before it are defined by the technological innovations that occurred during each period of socio-economic change. Our exploration of 4IR therefore is not simply about the understanding the transformation of systems around us, but it is also about the opportunity to understand, discuss, and guide how these technologies will shape our future (Schwab and Davis 2018). In order to analyze the impacts of 4IR, the authors of this section will go through some of the most influential technologies that are defining our latest technological revolution.

Technologies like self-driving cars and virtual reality have often fascinated science fiction lovers; however here the authors will examine them in the context of how they will shape the future of the supply chain industry. These innovations are often presented as being solutions to some of the biggest issues plaguing businesses and consumers. However, there is a certain amount of hype that must be set aside. On the other hand, McAfee and Brynjolfsson (2014) famously wrote that "Many people are falling behind as technology races ahead," giving way to the idea that maybe there is not enough excitement to push business to think of emerging tech as viable solutions. In the following chapters, in order to keep us from falling behind or getting too far ahead of ourselves, the authors will peel back the layers of mystery and excitement that have grown around these new ideas and solutions and show some amazing use cases that are having a real impact on the future of work. Each chapter in this section focuses on a particular technology that has come about or been significantly improved by innovations that followed the Third Industrial Revolution. The authors present these technologies that are fueling 4IR through a pragmatic lens by explaining each technology, how it works, its application in logistics and supply chain, and the additional challenges that it may bring.

To start the section, Axel Neher discusses one of the defining topics of 4IR that is the Internet of Things (IoT). IoT enables traditionally offline, disconnected devices to gather and transmit data on a network. Neher discusses applications of IoT in logistics management and describes how it can be used to increase supply chain visibility and monitor goods within a container or on a truck as they are moving.

3D printing, otherwise known as additive manufacturing, is broken down in the Chapter 4 by Johannes Kern, who offers a background to the technology and use cases that are being seen in production. He discusses different methods of 3D printing and describes to us how giving end users a new role in the supply chain is having a disruptive effect.

White-collar workers throughout the supply chain ecosystem are not immune to the effects of automation; however, this shift in work has been less visible in mainstream media. In *The Role of Robotics Process Automation (RPA) in Logistics*, Mac Sullivan, Walter Simpson, and Wesley Li show that robotic process automation (RPA) is offering a real alternative to offshoring of clerical work. Underpinning the success of process automation and optimization is the need for logistics companies to map out how they currently operate and systematically break down complexity.

One of the hottest topics in the past five years across the financial world has been blockchain, which is known mostly as the foundation of cryptocurrency, but it has numerous other applications outside of this. Here, Nic Krapels breaks down the history and basics of blockchain for our readers to have a general context for why this technology has created such a stir. Krapels explores how innovative startups are taking the concepts and core functionality of blockchain and applying them to supply chain and logistics problems. The other end of the spectrum is a venture between Maersk and IBM who have created a blockchain product named TradeLens that is attempting to revolutionize segments of ocean freight transportation.

The courier, express, and parcel (CEP) delivery services have received a lot of attention in recent years with the rise of global e-commerce giants like Amazon or Alibaba. The fastest growing segment of this industry is in China where a highly developed and specialized market allows people to buy almost anything at the convenience of a fingertip. However, local logistics networks are still struggling to keep up with demand and are constrained by various operational challenges such as high costs, low efficiency, and a lack of integrated intermodal transport networks. Here, digitalization becomes key to increasing efficiency and transparency. In Chapter 7, Scott Wang and Johannes Kern explain the challenges in the logistics operations of key industries, describe contemporary digitalization solutions, discuss e-commerce in China and its CEP market, and highlight new challenges in the competitive environment that increase the burden on all market players. In addition, they share a case study about the successful application of artificial intelligence in a CEP company in China.

Thanks to the inventor and entrepreneur Elon Musk and his company Tesla, autonomous vehicles have been pushed to the forefront of discussion about emerging technology. Many other automotive manufacturers are also hard at work creating their own versions of driverless vehicles. The topic has received a lot of attention in the logistics industry due to the potential cost savings of using autonomous truck fleets as well as the concerns that it will lead to the mass unemployment of truck drivers. The technology also has applications within factories and warehouses where autonomously guided vehicles could handle the picking and storage of goods. In *Understanding the Impacts of Autonomous Vehicles in Logistics*, Lionel Willems writes about the feasibility and timeline of rolling out automated trucks and warehouses as well as the intricacies of adopting these solutions.

To conclude this section, industry expert John Berry gives an overview of a variety of topics related to the evolution of IT departments in the logistics companies. In this chapter, he exposes the opportunities and challenges driven by a transition to the cloud. Having less physical technology along with less hands-on information technology staff in-house has been driving the need for logistics providers to be brought into the cloud-powered workplace.

References

McAfee, A. and Brynjolfsson, E. (2014). *The Second Machine Age: Work, Progress, and Prosperity in a Time of Brilliant Technologies*. Zurich: W. W. Norton.

Schwab, K. and Davis, N. (2018). *Shaping the Future of the Fourth Industrial Revolution: A Guide to Building a Better World*. New York: Currency.

3

Logistics Management in an IoT World

Axel Neher

Bosch Rexroth, Shanghai, China

Introduction

The way of doing business has entered the new era of the Internet of Things (IoT). Many new technological innovations have boosted the ability to collect, store, and analyze data in unprecedented availability, speed, and volume. IoT is spreading to almost all sectors including manufacturing, logistics, agriculture, healthcare, and building management. At home, smart speakers are making it easier to get weather information or play music, smart thermostats allow us heat our home before we arrive back, and smart home security systems let us easily monitor what is happening inside and outside or talk to visitors. In companies, data can be collected and analyzed to better understand product performance, and manufacturers can optimize machine performance or replace components before they cause damage (Ranger 2020). Fundamental to this is the IoT that enables various physical devices, embedded with electronics, software, sensors, actuators, and network connectivity, to collect and exchange data (Oztemel and Gursev 2020). The physical devices represent "Things" like machines, material, or even people. IoT is therefore also referred to as cyber-physical systems (CPSs), merging the physical and digital world.[1]

In this chapter, the topic of how this "new" world of connected Things impacts logistics and supply chain management (SCM) will be discussed. In analogy to Industry 4.0, some call this Logistics 4.0 (Bousonville 2017; Strandhagen et al. 2017).

---

1 Based on how wide the range of connected devices is defined, some talk about Industry 4.0 (in Germany), Smart Manufacturing (in US), Industrial Internet of Things (IIoT), or IoT. Whereas IoT encompasses in general all applications based on connected smart devices, the Industrial Internet of Things (IIoT) and Industry 4.0 focus on the industrial applications within the broad umbrella of IoT. To illustrate the difference we can think about a smart watch to monitor the health condition as an example of the wide range of IoT compared with a smart device monitoring the condition of a machine in a factory or a component on the transport as an example for the IIoT (Serpanos and Wolf 2018).

As IoT is still under development, a broad range of technologies is discussed to describe what IoT is about (Strandhagen et al. 2017; Kern and Wolff 2019; Oztemel and Gursev 2020). This is a small selection:

- *Identification and sensor technology*
  To connect Things, the first step is to know about the Thing (what, where). So, one crucial aspect of IoT is the identification and sensor technology enabling the automated identification of an object (Thing) and collection of "sensed" real-time data (e.g. temperature, acceleration, humidity, vibration, and location).
- *Networking technology* (e.g. *5G, Wi-Fi 6*)
  Enabling to connect the Things in a fast and reliable way to interact (network) with each other. With more Things to be connected and thus more data to be transmitted, new powerful networking technologies are required to provide the capability of transmitting big data volume of many Things in parallel with a short latency.
- *Big data*
  As a result (or intention) of connecting everything with everything, a lot of data are created. New technologies provide the possibility to transmit (see networking technology), store, and analyze a huge amount of data – big data. Especially the topic of big data analytics and decisions based on a huge amount of data is seen as one big lever to improve speed and quality of management and pave the way for artificial intelligence (AI).
- *Artificial intelligence*
  Enabling automated and autonomous decision making and operations, based on self-learning systems without the intervention of humans.
- *Autonomous robots and vehicles (e.g. drones, automated guided vehicle [AGV])*
  Material handling and transportation are done by robots or AGVs that execute the defined tasks in an automated or autonomous way in interaction with other Things.
- *Augmented reality (AR) and virtual reality (VR)*
  Devices supporting decisions and processes by providing instructions via augmented and virtual reality technologies (e.g. headsets or smart glasses showing the worker on a small display in the glasses what to do next).

Logistics managers not only have to understand these new technologies, but also the overall concept of IoT and its implications for managing material and information flows in global supply chains.

One prevailing "new" quality of IoT is "connection." Therefore, the first new topic in logistics management is to think about how to "get connected". Acknowledging that the overall approach of IoT is to connect everything with everything, it is important to reflect what this will mean for logistics. What are the "Things" in the supply chain we want to connect to be able to operate faster and more efficiently?

One of the major goals of having connected Things is to gather big data. Logistics management must define which data is needed, e.g. data about the on-time delivery of products, quality of the product in transit, or security of facilities and vehicles, and how to make the best use of the data for their business customers. Hence, the second new topic in logistics management is to define how IoT is used to "get decisions" in a better way than before.

To connect Things in supply chains and to come to better decisions in logistics management, new management strategies, new mindsets, and new skills must be developed. Similar to other disruptive changes in the previous industrial revolutions, like mechanization, mass production, and computerization, IoT not only is a new way of using technology but also very much requires a change of thinking. The third topic for logistics in an IoT world is therefore to "get prepared" for this new way of logistics management. Besides just ensuring high data quality, the industry needs to reconsider organization, skills, and the whole ecosystem of logistics in an IoT world.

What follows is a discussion about how to get connected, how to get decisions, and how to get prepared. Regarding getting connected, relevant topics are sensors, identification, standards, and security. In the section about getting decisions, descriptive, diagnostic, predictive, and prescriptive ways of data analytics are introduced, and manual, automated, and autonomous decision making contrasted. Finally, key topics to get prepared such as data quality, organization, skills, and ecosystems are addressed. This is also summarized in Figure 3.1. The conclusion in the end highlights that logistics management in an IoT world will require managing new IoT ecosystem setups with new actors and new skill requirements.

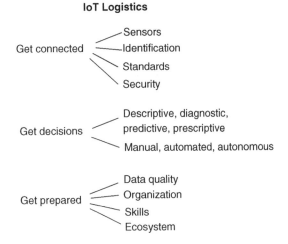

IoT Logistics

Get connected
- Sensors
- Identification
- Standards
- Security

Get decisions
- Descriptive, diagnostic, predictive, prescriptive
- Manual, automated, autonomous

Get prepared
- Data quality
- Organization
- Skills
- Ecosystem

Figure 3.1 IoT logistics.

Logistics Management in an IoT World

Get Connected

The defining element of IoT is connectivity. Therefore, the first aspects of setting up the supply chain and its management in an IoT style are to define what are the relevant Things that need to be connected, in what way data should be collected, and finally, which data needs to be collected. Simply collecting more and more data might just increase the amount of available data exponentially, but not necessarily lead to improved patterns of doing business.

When is a thing a Thing in the sense of logistics in an IoT world? Is a small component a Thing that needs to be monitored by sensors and always be online? Or does the Thing start with complex finished goods that are themselves composed of multiple components? Here, we need to distinguish between the user perspective and the logistics perspective. From the logistics perspective, i.e. the material and information flow point of view, everything is a Thing. From the classical user IoT point of view, the Thing is more related to machines and devices, such as finished products, components in a production line, transport vehicles, or whole warehouses. This chapter will examine Things from the logistics perspective.

Criteria for Defining the Right Things

Some examples of criteria based on which a Thing may be included in an IoT system or application would be as follows:

- Requests from customers (e.g. to stay informed about the placed order).
- The criticality of the Things to influence the supply chain performance (e.g. keeping a certain temperature or avoiding shocks during transport).
- Requirements from the business model (e.g. provide services).

Depending on the needed information, the appropriate sensor and identification technology has to be selected. For instance, to keep customers informed about their delivery, it might be sufficient to collect information once the Thing is passing certain checkpoints, like a warehouse, cross-dock, or the truck for the last mile, through simple scanning. Meanwhile, in case the Thing needs to be cooled or is sensitive to shocks, information on temperature or acceleration provided by sensors might always be necessary. Otherwise, the business model might require a continuous condition monitoring of a forklift or AGV through embedded sensors. In case deviations are detected, an alert can be triggered and then, for example, spare parts ordered.

Sensors and Identification

The level of monitoring and identification, thus the Thing itself, e.g. product, box, container, or transport device (see Figure 3.2), can vary according to the requirements. Although in a narrow sense the Thing is equal to the product, for example, for tracking a shipment from Europe to Asia by an ocean carrier, tracking the vessel itself during its way might be enough – knowing and trusting that the product was loaded onboard the ship. Thus, tracking information does not have to be sent from every single of the millions of products on a containership, but only from the ship itself. If the sensor data from individual products are requested, other solutions are required.

To define this level, the overall supply chain process and the intended business services have to be considered. Is a pure identification/localization[2] enough, or are sensors needed to get certain data, such as condition monitoring? It could be necessary to use sensors implemented in the product to monitor certain conditions. At the same time, although the product itself might have some sensors inside, to optimize transport, handling, or warehousing of a mix of sensor and non-sensor products, it might be needed to have additional sensors on container or transport device level. Furthermore, in case a sensor is needed, which conditions,

Levels of sensors and identification

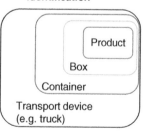

Figure 3.2 Levels of sensors and identification.

---

2 To identify/localize Things several methods are available, ranging from labels (e.g. barcode, QR codes) via RFID up to GPS or other ways.

such as temperature, humidity, or acceleration, do the sensor need to measure? Modern sensors are quite compact, small, and capable of measuring multiple of those.[3]

Means of Connection

Based on the application of the data, the means of connecting the Things can be done in an active or passive way. *Active* means that data is actively sent out from the Thing. This might be necessary if the condition of the Thing changes quickly, and the knowledge about these changes is decisive for the next step, such as the coordination of many transport devices in a warehouse to avoid collisions. To actively send out data, more technologies and especially energy supplies are necessary that might limit the device size and the usage lifetime and require additional services such as exchanging or charging of batteries. In contrast, if conditions are relatively stable or the intended application is purely identification, a passive mode might be sufficient. In a *passive* mode, the Thing reveals its data when it is triggered from the outside, just as when a box labeled with a barcode passes through a barcode reader.

Triggers

Data from the sensors to trigger specific actions or decisions enable new business models. For example, if the measured data of the sensors deviate from the target level, a truck driver gets an alert that the condition of the products on the truck needs to be checked or actions like adjusting the cooling function must be taken. Another use case can be machine monitoring. Based on information provided by the sensors on a machine, the next maintenance can be scheduled with minimum impact on downtime. A machine could even automatically order necessary spare parts preventively or once it breaks down. To gain full advantage of these actions, the logistics systems have to be prepared to quickly facilitate the transport of these necessary spare parts. In connected systems, the signal from the machine will already trigger the picking in the warehouse, prepare the necessary transport devices, and synchronize the material flow with the availability of the service technician.

Standards

As already well known in SCM, *standards* are important. With an increasing number of IoT devices involved, there is an exponentially large impact on the standards of the data format, the identification, and connecting technology that necessitates the use of middleware to connect different formats used throughout the whole supply chain. Everything is connected with everything, but based on which "language"? This issue can, for instance, be seen in the discussion about cables and connectors in the computer/mobile phone sector. It started with a lot of diverse single solutions and finally led to the USB solution as unified standard.

---

3 See, for example, the Nexeed Track and Trace devices of Bosch Manufacturing Solutions (Robert Bosch Manufacturing Solutions 2020).

One well-known and established standard for identification in the global retail business is the GS1 EPCIS standard and the Global Trade Item Number (GTIN). By scanning the GTIN number/barcode at defined "control points" in the supply chain, a track and trace function can be established from the source to the customer. Another common standard that is mainly used in the automobile industry is the VDA/Odette barcode label. Although such common identification systems exist, we can find additionally numerous labels of each logistics service provider for their specific operations or parts of the supply chain. The further development of the IoT in the future will show which identification and communication technology will prevail. The "Industrial Internet Consortium" (Object Management Group 2020) and the "Industry 4.0 platform" (Plattform Industrie 4.0 2018), for example, are working closely together to standardize and ensure interoperability in the industrial sector (Plattform Industrie 4.0 2019).

Security

With increasing amounts of data transferred between the connected Things and increasing relevance of and dependability on data to manage supply chains, the *security* of data transfer, data handling, and data storage needs to be looked at. Any connected smart Thing could be a potential target for cybercriminality as it could grant access to a company's network. In recent years, a lot of "leaks" stories in social media and the financial world have clearly shown the importance of data security. For a further extension of a connected supply chain, it will be crucial to ensure secure data exchange between the Things and the actors along the supply chain, like the customer, their supplier, logistics service provider, and customs agencies. Technologies like blockchain might provide a solution for this secure data transfer and secure identification.[4]

To ensure trust and security in digital systems, Bosch has launched an initiative and invited different supply chain partners to the first Digital Trust Forum in Berlin on 16 May 2019. Representatives from leading international associations and organizations, including the Institute of Electrical and Electronics Engineers (IEEE), DigitalEurope, European Telecommunications Standards Institute (ETSI), the Eclipse Foundation, Trustable Technology, Plattform Industrie 4.0, the Industrial Internet Consortium (IIC), and the Trusted IoT Alliance, came together to discuss this topic.

With more Things connected, and more data being transferred, the capability or bandwidth of the "connecting technology" to transfer large amounts of data in a fast way has become a huge business. If short reaction times are needed, but the latency of the data transfer to and from a "center of decision making" is too long, certain applications might not work. Further innovations in connecting technology, like 5G, are on the way to provide such solutions. Another approach to come up with fast decisions on the Thing level is to avoid data transfer to a central brain and to have built-in computing power and intelligence on the Thing level, which is referred to as "edge computing."[5] To share these decentral deci-

---

4 This topic will be discussed Chapter 6 in this volume.

5 With edge computing data produced by Internet of things (IoT), devices are processed decentralized closer to where it is created (the edge of the network, e.g. the machine) instead of sending all data to data centers or clouds (Shi et al. 2016).

sions and their results across the whole system and to enable big data learning for a continuous improvement process, a transfer of data from the Thing to a central data pool, a so-called data lake, is still needed. Fast reaction or decision can be achieved through local, decentral intelligence while the continuous improvement process is based on big data on a central level.

Get Decisions

From a logistics management point of view, one major purpose of connecting Things and transferring, handling, and storing data in the IoT is to get decisions in a better (time, cost, quality) way than before. In general, we can separate into three different types of decision making: manual, automated, or autonomous.

In the *manual decision* way, the collected data is analyzed or displayed by certain applications that help the decision maker to come to better decisions. The decision itself remains a task of the related person. For example, in a transport company, based on GPS tracking, all trucks are displayed on a screen showing the map of the road combined with traffic and weather information. In the case of a certain event such as a traffic jam, the "decider" must decide where and how to intervene to keep promised lead times.

In case the decision process can be described in a standardized way, *automated* solution finding and decision making can be possible. Based on predefined checking and decision finding processes, the "application" (App) in use executes automatically the predefined solution process or gives a proposal to a human decider. Logistics management has to think about which kind of decision situations (e.g. data given by a Thing) can occur in which process and how the decision processes can be defined in a standardized way, including standardized answers. This is also discussed in Chapter 5.

The next level of decision making is currently in development using AI. Such AI systems can *autonomously* find solutions and come to decisions. Using big data and a self-learning mode, the "machine" or application itself is smart enough to decide what to do next at a much higher speed and accuracy than humans. Logistics management has to determine which type of decision making is appropriate for which process based on which data from which Thing. Regardless of which type of decision making is applied, the analytics of the data as an input for the decision making can be done in mainly four different ways: descriptive, diagnostic, predictive, or prescriptive Figure 3.3 (Porter and Heppelmann 2015).

In a *descriptive* way, all selected information is gathered, and the monitored situation is described (e.g. condition, environment, process). Using the *diagnostic* way, the root causes of deviations between the set target value and the actual monitored value are analyzed. Hence, both the descriptive and diagnostic ways of analytics look back and explain what happened in the past. In a *predictive* way, the analyzed data is used to detect indications that signal impending events in the future. For example, if the system shows a certain number of Things in the inbound area of a warehouse while in parallel the internal data indicates that several workers are ill, it will most likely result in a delay or congestion in the goods receiving process – if no actions are taken. Lastly, the *prescriptive* form of data analytics strives for identifying measures to improve results or correct errors.

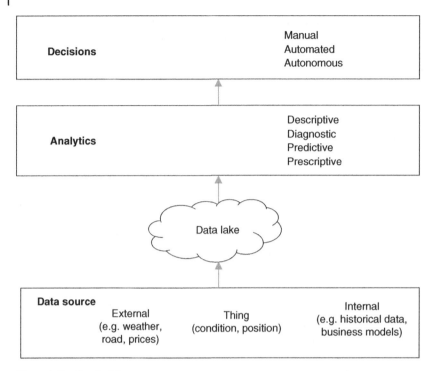

Figure 3.3 Get decisions.

Different applications need different ways of conducting data analytics. What they all have in common is that they get the data out of the pooled data location. In the first part of this chapter, we have discussed how the data get into the data lake and how to get improved decisions out of it. In the section that follows, we will discuss what needs to be considered to get prepared for logistics management in an IoT world.

Get Prepared

In logistics management in an IoT world, the management of material and information flows in supply chains still has the same overall goals as before: to provide the right Thing, in the right quantity, at the right time, to the right place (Pfohl 2016, 2018). As outlined above, the IoT world provides new tools to realize these targets more effectively and efficiently. But this new IoT approach also demands new skills and a new way of thinking in logistics management. Aside from the new IoT-related technologies, the prevailing characteristics of IoT and its management are the data.

Data Quality

Logistics management has always had to handle a lot of data to manage supply chains effectively, but with IoT, a new level of big data is reached. In current daily operations, we still see a lot of inefficiencies due to bad data quality. Material master data (e.g. weight, dimension, country of origin, packaging information, etc.) are missing or often incorrect that can lead to wrong decisions or interruption of operations. As disturbing as such problems of data quality are already today, data quality will be *the* crucial prerequisite for logistics management in an IoT world. In summary, getting more bad data in a faster way will not improve the process but instead have the opposite effect.

Organization

Most international companies still have a functional supply chain organizational structure. However, operating in an IoT world requires close cross-functional collaboration. Research and development (R&D), information technology (IT), engineering, logistics, purchasing, marketing, and sales will need to be aligned on how the "Thing" will be used and how to use the data lake. Especially the integration of and collaboration with the IT team have shown to be particularly important. A functional setup of data storage, data analytics, and data security might not be the most effective and efficient way of dealing with an increasing data volume. For this reason, some companies are already building up some dedicated "data groups" who only deal with these increased requests of data storage, analytics, and security in a cross-functional way.

Skills

People working in IoT logistics need to have the right skills to get Things connected and get the best out of big data (Heckel 2017). This involves know-how of IoT technology, logistics process understanding, data analysis, and systems. Companies who want to make full use of IoT have to develop such multiskilled logistics employees.[6] On a management level, this requires skills to "see" supply chains from an IoT point of view. This contains the abilities to handle the above-shown topics: "get connected," "get decisions," and "get prepared." On an operational level the following skills are needed:

- *Data analytics* – skills to handle and analyze big data and discover new insights/patterns of the IoT logistics processes.
- *Process mapping and description of reaction processes* – understand the actual processes and define/describe (for automated execution) potential reaction processes on detected deviations.
- *Ability to react faster* – to get full advantage of online data, fast reactions are necessary: Is the associate prepared for this? Does he have the authorization to do so?
- *Check reliability of IT/AI decisions* – Is the data correct? Do the automated/autonomous decision-making processes make the right decisions and show the expected results?
- *Human–machine interaction* – with more automation and robotics, especially in handling areas like goods receiving, warehousing, and loading/unloading, the interaction between robots and machines will only increase (Klumpp 2018). Is your staff prepared and trained how to work with a robot? In the case of small technical errors, can your staff troubleshoot the issue in terms of working with the software program of the robot, or can they do maintenance on the machine, such as exchanging batteries or conducting simple repairs?

---

6 Cf. the chapter by Sun in this volume.

Ecosystem

To set up and run an operation in an IoT world, new players are necessary. Besides the logistics service providers, producers of embedded systems and sensors, telecommunication companies, IoT platform providers and operators, application developers, certification companies, data analytics providers, and a lot more have to be coordinated and managed to put IoT solutions into practice. A whole IoT ecosystem (Papert and Pflaum 2017) needs to be managed (see Figure 3.4).

This stresses once more the strong need for close coordination with IT-related players and the "classical" material flow-related players. Today, the different players in a supply chain often use their own proprietary systems to handle their data and to manage the material flow (Vial 2019). In an IoT world, the sensors of the Thing will provide all data needed for the business. In the collaboration with internal and external partners, this brings up several questions:

1) Who owns the data?
2) Who gets access and controls permissions to the data lake for which kind of data?
3) Will these original data from the Things replace the different labels, barcodes, and other storage technology in use today?

The ownership of the data might change the role of the supply chain manager and the current logistics service provider. Some big OEM customers might be the driver of the implementation of IoT logistics solutions, some logistics service provider might see some new business opportunities, but the full success of IoT logistics will only come in a connected IoT ecosystem. One example of such an IoT logistics service provider and IoT ecosystem can be seen with Alibaba's Cainiao that operates a logistics network in China (Chou 2019). Cainiao is managing the supply chain and associated express delivery companies such as STO Express, Yunda Express, YTO Express, or ZTO Express (Wang 2018). Based on the big data of the parent company, the Cainiao network uses its own IoT platform for standardized communication and collaboration with the associated companies, which allows using IoT devices and robots in all of its warehouses and deliveries.

Conclusion

The target of logistics management in an IoT world is still the same as before: managing material and information flows. With smart devices (Things) connecting everything with everything, the way of doing to reach this target will change. Logistics management must address new questions related to connected Things. Data must be collected in order to be able to analyze it, so that decisions can be made. New IoT ecosystems consisting of a lot of new players must be managed.

Beside the Things, the major topic of logistics in an IoT world will be the big data generated by the Things. The major challenge will be how to use and share this data in an IoT ecosystem most efficiently and effectively. Standards, trust, and connectivity will play a crucial role.

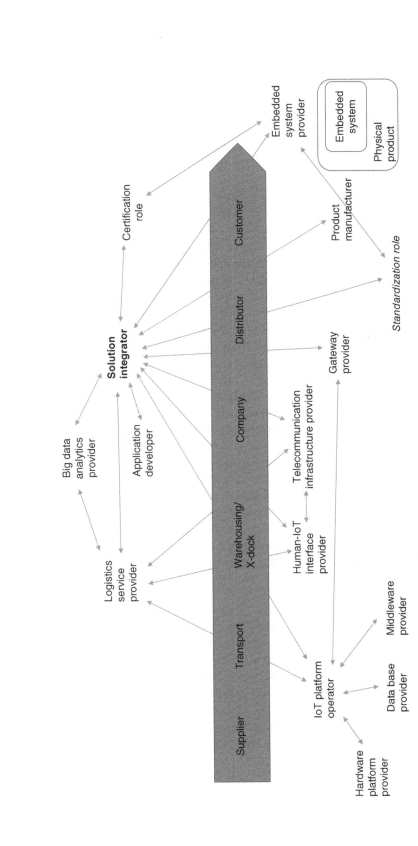

Figure 3.4 IoT ecosystem. *Source:* Based on Papert and Pflaum (2017). © John Wiley & Sons.

The connected world will offer new possibilities of doing business in terms of providing sensors for track and trace services, offering sensor-based preventive maintenance and repair services, and helping companies to reach their logistics goals in terms of cost, lead time, and quality in a better way. This also requires new skills and a mindset change. Logistical process know-how has to be combined with data analytics and technology. Lastly, cross-functional collaboration, especially with internal or external IT teams, will be a crucial success factor.

The use of more smart and connected Things will enable the supply chain to implement an increasing number of automated processes. Things in the supply chain triggers the automated goods receiving, putaway and transport, commissioning, and packing of boxes and pallets for the next customer order. All devices in the physical world will be represented as a "digital twin" in the digital world. In addition to the already known classical 3D models or computer simulations, the digital twin is a virtual model of the real Thing based on "sensed" real-life data. Design, operation, and optimization of logistical processes, infrastructure, or global supply chains can be done based on the digital twin either in real-time or as a simulation of different scenarios (Gesing and Kückelhaus 2019).

The implementation of IoT logistics and automation solutions requires substantial investments in hardware like RFID gates, barcode readers, AGVs, smart racks, and Wi-Fi, as well as the corresponding software and IT infrastructure.[7] For all investments in IoT, logistics management has to show a clear calculation on the expected return of investment (ROI). Reducing costs, most likely personnel, on the one hand, and increasing revenues through new and better services, on the other hand, will be the major driving factors in this calculation.

As in every transition phase from old technologies to new ones, there is a certain time period where it is necessary to handle both worlds. This also applies for IoT. Logistics management in an IoT world must ensure that the IoT ecosystem can handle IoT and non-IoT logistics at the same time although the realization of the future picture of IoT logistics must clearly be in focus. A vision for the IoT logistics of the future can be that Things will be transported, handled, and stored in an automated way by robots and controlled by self-organizing, decentralized, smart, AI-facilitated networks of Things.

Key Takeaways

- Internet of Things (IoT) is spreading and transforming almost all sectors including logistics, manufacturing, agriculture, healthcare, or building management. IoT enables physical devices, embedded with electronics, software, sensors, actuators, and network connectivity, to collect and exchange data.
- Logistics managers must understand IoT-related technologies such as sensors, networking technology (e.g. 5G), big data, AI, autonomous robots and vehicles (e.g. drones, AGV), augmented reality (AR) and virtual reality (VR), and the overall IoT concept as well as its implications for managing material and information flows in supply chains.

---

7 Cf. the barriers to technology adoption discussed Chapter 25 in this volume.

- As IoT is about "connection," logistics management needs to assess how to get connected. It is important to define which "Thing" in the supply chain, e.g. a product component, finished product, box, or container, needs to be monitored by sensors and always be online.
- Based on customer's requirements, criticality of the "Things" to influence the supply chain performance, and requirements from the business model, it must be decided which Thing should be connected to enable faster and more efficient operations.
- Connecting "Things" allows to gather big data that allows for better decision making in terms of time, cost, or quality than before. Logistics management must decide which data is needed, e.g. data about the on-time delivery of products, quality of the product in transit, or security of facilities and vehicles, and how decisions should be made.
- Decisions can be either manual, automated, or autonomous, with varying degrees of human intervention necessities and possibilities. In the highest level of automated decision making, AI systems will autonomously find solutions and come to decisions based on self-learned behavioral patterns.
- Logistics management in an IoT world demands new skills and a new way of thinking about how to provide the right Thing, in the right quantity, at the right time, to the right place. Such skills include data analytics, process mapping, ability to react faster, verification of AI decisions, and human–machine interaction.
- Finally, logistics management must work on improving the currently often low level of data quality to enable more automated decisions and closely collaborate not only with the "classical" material flow-related players but also IT-related ones.

References

Bousonville, T. (2017). *Logistik 4.0*. Wiesbaden: Springer Fachmedien Wiesbaden (essentials) https://doi.org/10.1007/978-3-658-13013-8.

Chou, C. (2019). How cainiao plans to digitize China's logistics industry. https://www.alizila.com/how-cainiao-plans-to-digitize-chinas-logistics-industry (accessed 17 February 2020).

Gesing, B. and Kückelhaus, M. (2019). Digital twins in logistics. https://www.dhl.com/content/dam/dhl/global/core/documents/pdf/glo-core-digital-twins-in-logistics.pdf (15 March 2020).

Heckel, M. (2017). Skills for industry 4.0. http://www.delivered.dhl.com/en/articles/2017/02/skills-for-industry-4-0.html (accessed 8 October 2018).

Kern, J. and Wolff, P. (2019). The digital transformation of the automotive supply chain–an empirical analysis with evidence from Germany and China. https://www.innovationpolicyplatform.org/www.innovationpolicyplatform.org/system/files/imce/AutomotiveSupplyChain_GermanyChina_TIPDigitalCaseStudy2019_1/index.pdf (accessed 7 May 2020).

Klumpp, M. (2018). Automation and artificial intelligence in business logistics systems. *International Journal of Logistics Research and Applications* 21 (3): 224–242.

Object Management Group. (2020). Industrial internet consortium. http://www.iiconsortium.org (accessed 14 January 2020).

Oztemel, E. and Gursev, S. (2020). Literature review of industry 4.0 and related technologies. *Journal of Intelligent Manufacturing* 31 (1): 127–182. https://doi.org/10.1007/s10845-018-1433-8.

Papert, M. and Pflaum, A. (2017). Development of an ecosystem model for the realization of internet of things (IoT) Services in Supply Chain Management. *Electronic Markets* 27 (2): 175–189. https://doi.org/10.1007/s12525-017-0251-8.

Pfohl, H.-C. (2016). *Logistikmanagement. 3. Aufl. 2*. Berlin, Heidelberg: Springer Berlin Heidelberg http://www.worldcat.org/oclc/967690663.

Pfohl, H.-C. (2018). *Logistiksysteme. 9. Aufl. 2*. Berlin, Heidelberg: Springer Vieweg. doi: 10.1007/978-3-642-04162-4, T4 - Betriebswirtschaftliche Grundlagen M4 - Citavi.

Plattform Industrie 4.0. (2018). https://www.plattform-i40.de/I40/Navigation/EN/ThePlatform/PlattformIndustrie40/plattform-industrie-40.html (accessed 14 September 2018).

Plattform Industrie 4.0 (2019) *Shaping Industrie 4.0. Autonomous, Interoperable and Sustainable*. Berlin. https://www.plattform-i40.de/PI40/Redaktion/EN/Downloads/Publikation/2019-progress-report.pdf?__blob=publicationFile&v=7 (accessed 2 March 2020).

Porter, M.E. and Heppelmann, J.E. (2015). How smart, connected products are transforming companies. *Harvard Business Review* 93 (10): 96–114.

Ranger, S. (2020). What is the IoT? Everything you need to know about the Internet of Things right now, ZDnet. https://www.zdnet.com/article/what-is-the-internet-of-things-everything-you-need-to-know-about-the-iot-right-now (accessed 7 May 2020).

Robert Bosch Manufacturing Solutions (2020). Nexeed track and trace starter kit. https://www.bosch-connected-industry.com/en/connected-logistics/nexeed-track-and-trace/nexeed-track-and-trace-starter-kit (accessed 7 May 2020).

Serpanos, D. and Wolf, M. (2018). Industrial internet of things. In: *Internet-of-Things (IoT) Systems*, 37–54. Cham: Springer International Publishing https://doi.org/10.1007/978-3-319-69715-4_5.

Shi, W., Cao, J., Zhang, Q. et al. (2016). Edge computing: vision and challenges. *IEEE Internet of Things Journal* 3 (5): 637–646. https://doi.org/10.1109/JIOT.2016.2579198.

Strandhagen, J.O., Vallandingham, L.R., Fragapane, G. et al. (2017). Logistics 4.0 and emerging sustainable business models. *Advances in Manufacturing* 5 (4): 359–369. https://doi.org/10.1007/s40436-017-0198-1.

Vial, G. (2019). Understanding digital transformation: a review and a research agenda. *Journal of Strategic Information Systems* https://doi.org/10.1016/j.jsis.2019.01.003.

Wang, L. (2018). Third-party logistics development in China. In: *Contemporary Logistics in China. Collaboration and Reciprocation* (eds. J. Xiao, S. Lee, B. Liu and J. Liu), 71–93. Singapore: Springer SV - 8.

4

Additive Manufacturing

Shaping the Supply Chain Revolution?

Johannes Kern

Tongji University, Shanghai, China

Introduction

Additive manufacturing (AM), casually also called 3D printing or rapid prototyping, is one of the key elements of the Fourth Industrial Revolution that is predicted to strongly impact supply chains (Verboeket and Krikke 2019; Fan et al. 2020). While in subtractive manufacturing, the typical way of how things are made today, material is removed in a controlled way, in AM, material is added based on a digital 3D drawing or model (Carr 2017). Depending on the specific technology used, complex and light designs can be created, parts developed more dynamic, and production lead times reduced (General Electric 2020). AM is envisioned to become widely spread as currently still existing constraints are progressively overcome (Schwab 2017). Already today, the technology is applied across a wide array of products. In aerospace, NASA has been experimenting with printed rocket injectors since 2013 (Holmström et al. 2016). In aviation, in 2014, a 3D-printed titanium bracket took to the skies on board of an Airbus jetliner for the first time (Airbus 2020). Also in automotive, the technology has been used already for a decade in F1 racecars. Recently, in its commercial vehicle segment, the OEM[1] Daimler started to fully integrate it into its development and production process (Sher 2020). The company also uses AM to print spare parts, where according to a spokesperson "production and delivery of a 3D-printed part takes only a few days as opposed to several months (while producing) considerably less waste (Garnsey 2020)." In the medical industry, customized prosthetics, implants, and anatomical models are widely manufactured with AM, and research about producing tissue and even organs ("bioprinting") is ongoing (Ventola 2014). Although such applications are noteworthy, it must be recognized

---

1 Original equipment manufacturer. In the automotive industry, the term is commonly used for car producing companies.

The Digital Transformation of Logistics: Demystifying Impacts of the Fourth Industrial Revolution,
First Edition. Edited by Mac Sullivan and Johannes Kern.
© 2021 by The Institute of Electrical and Electronics Engineers, Inc.
Published 2021 by John Wiley & Sons, Inc.

that AM is not suitable for every manufacturing need and that there are still barriers to overcome.

This chapter takes the form of six sections. First, the vision of the AM supply chain will be compared with a conventional supply chain. Then, the technology's advantages and remaining bottlenecks will be explained. This is followed by an overview of various AM technologies and materials. Typical application scenarios will be subsequently derived. Afterward, markets and trends will be accounted for. The chapter then ends with a conclusion that summarizes the status of the technology and its potential supply chain impact.

AM Supply Chain

In a conventional supply chain, suppliers from various places distributed all over the world deliver certain raw materials and components to a manufacturer that is often located in a low-cost location. This manufacturer then produces goods in large quantities, benefiting from "economies of scale," and ships them to a central hub like an airport or port. The goods will be transported by carriers with transit times ranging from one week (airplane) up to over a month (ocean carrier) and ultimately arriving at a wholesaler. From there, the parts are distributed to various retailers until they ultimately arrive at the final customer. Such a conventional supply chain is shown in Figure 4.1.

In contrast to a conventional supply chain, an AM supply chain is much more concise (Mills and Camek 2004). Suppliers only have to deliver basic materials instead of specific components, which allow the manufacturer to already store the right materials before any customer order occurs. Considering that only a few basic materials are required, the number of suppliers in the AM supply chain would be drastically reduced. This would allow for a stronger focus on few collaboration partners that deliver high-quality materials and the most suitable software program (Kothman and Faber 2016; Chekurov et al. 2018). The actual production process could then happen at a decentralized location close to the customer, such as at an AM manufacturer, a wholesaler, or even a retailer. This would change the conventional flow of goods, reducing upstream transportation to basic materials and downstream transportation to locally produced finished goods (Thiesse et al. 2015; Ford and Despeisse 2016). As advanced AM machines might be able to manufacture a wide range of products, physical inventory holding could be replaced by "digital inventory" holding in the form of digital CAD[2] models. This would result in less stock, less material handling, and less packaging (Ruffo et al. 2007).

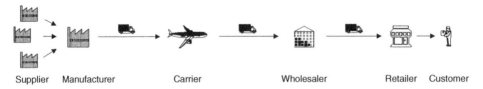

Supplier   Manufacturer   Carrier   Wholesaler   Retailer   Customer

Figure 4.1   Conventional supply chain.

---

2 Computer-aided design.

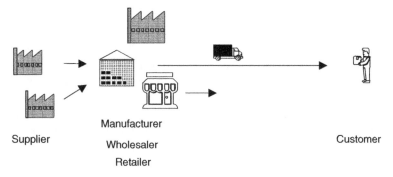

Figure 4.2 Additive manufacturing supply chain.

Distributed manufacturing can occur close to customers, disregarding that they might be situated in locations that are hard to reach or where transportation is difficult, e.g. due to political risks. Some scholars even argue that customers themselves could have AM machines at home, cutting out all middlemen of a conventional supply chain. In such per-sonalized manufacturing setup, production and consumption would occur at the same place (Rayna and Striukova 2016). In this supply chain, "economies of one" benefits would allow a decentralized production of custom-made products or spare parts. However, such produc-tion setup with numerous locations close to customers might also result in underutilized AM machines. Research suggests that to overcome the risk of such inefficiencies, machine sharing could be used, and new AM service providers emerge (Chekurov et al. 2018).[3] Figure 4.2 shows the AM supply chain. While this supply chain appears to be more straight-forward and beneficial, it also has a variety of disadvantages that will be discussed in the next section.

Evaluation

This section will review the advantages and currently existing bottlenecks of AM.

Advantages

As described in the previous section, adopting AM would create shorter, smaller, more local-ized supply chains (Gebler et al. 2014). This would reduce costs for holding inventory, as less safety stock is required, obsolescence rates can be reduced, and work in process would basi-cally not exist (Atzeni and Salmi 2012; Chiu and Lin 2016; Holmström et al. 2016). Also cost for production, especially for small parts, is reported to decrease (The Economist 2012; Petrick and Simpson 2013). As costs for assembly, setup, energy, material handling, person-nel, and packaging are lower, the overall costs could be reduced when compared with the conventional manufacturing (Walter et al. 2004). Therefore, small production batches, with

---

3 An alternative approach is specialized platforms such as Shapeways or Materialise (Wolff and Kern 2019) that offer "printing on demand" services to consumers (Rayna et al. 2015).

customized designs, which also can be quickly changed, become feasible and economical (Holmström et al. 2010). Moreover, investment costs for assets such as tools, machines, or facilities could be decreased (Atzeni and Salmi 2012; Khajavi et al. 2014). And as processes are largely computer-controlled, the required level of operator expertise could be lowered (Lipson and Kurman 2013). Employing AM reduces the time to market, especially for complex shapes, leading to increased responsiveness in the supply chain (Mellor et al. 2014; Ryan et al. 2017). Consequently, also demand fluctuations could be better managed (Chiu and Lin 2016). AM further increases flexibility, as a wide range of products can be manufactured with one process and the product mix be extended (Weller et al. 2015; Steenhuis and Pretorius 2017). In addition, sustainability benefits are predicted as resources could be used more efficiently due to improved products and production processes (Ford and Despeisse 2016). Researchers also assume that through reduced repair and refurbishment and more sustainable "socioeconomic patterns" in the form of stronger person–product affinities and a closer relationships between producers and consumers, the product life could be extended (Kohtala 2015; De la Torre et al. 2016; Ford and Despeisse 2016). Other environmental and efficiency benefits result from reduced wastage, where reports claim that it can be reduced by 90% compared with conventional manufacturing, and a more energy efficient production process[4] (Singamneni et al. 2019). Another advantage of AM is that novel, complex structures, such as free-form enclosed structures and channels, and lattices are achievable (Ford and Despeisse 2016). For example, due to stringent certification criteria, aircraft designers typically have very little design freedom, and optimizations beyond the normal would result in complex geometries that are not possible to produce with conventional methods. As AM removes such limitations, new opportunities for optimized designs are opened up (Singamneni et al. 2019). Finally, AM fosters innovation (Dwivedi et al. 2017). Customers can be involved in the innovation process, providing ideas via crowdsourcing or cocreating via open innovation platforms[5] (Rayna et al. 2015). However, there are still various product-, processing-, and regulation-related bottlenecks to overcome before AM can truly revolutionize manufacturing and disrupt supply chains (Verboeket and Krikke 2019).

Bottlenecks

In a study about AM spare part production in consumer electronics, Chekurov and Salmi highlight that only certain components without strict surface requirements can be produced. They concluded that, for example, internal parts inside consumer electronics are compatible with AM where the look and feel is not so critical (Chekurov and Salmi 2017). Other researchers also report issues related to the accuracy of the product, which is depending on the AM machine device mechanism, material, and resolution (Kothman and Faber 2016; Moore et al. 2016). Experts further cite achieving the desired part strength

---

4 A subject matter expert in a study by Mellor, Hao, and Zhang relativizes this, suggesting that "in terms of material usage . . . particularly with the metals it's very good but you have to go through a high energy process of turning it into powder in the first place, do you really gain a benefit there? It's marginal to be fair. The actual machines are they really efficient at building stuff now? No . . . In terms of efficiency of use of energy they are not very good at all (Mellor et al. 2014)."

5 For an overview about the main services offered by 3D printing platforms cf. (Rayna et al. 2015).

and durability with a current set of material and AM technologies as primary barrier to implement AM (Chiu and Lin 2016; Dwivedi et al. 2017). In addition, problems with process predictability and repeatability result in increased costs due to build failure and quality issues (Baumers et al. 2016). Neely also points out that there might be product safety-related constraints, considering that while in conventional manufacturing products are tested and certified and factories inspected, in AM the main appeal is the ability to manufacture in dispersed locations (Neely 2016).[6] Aside from these product-related bottlenecks, also production-related ones exist. For instance, a main production-related restriction is the high cost for implementing AM. This includes high material costs, high machine/equipment costs, high costs for technology acquisition, and high maintenance costs (Baldwin and Lin 2002; Dwivedi et al. 2017). Production speed is criticized for being still too slow, and throughput rates for being too low (Gebler et al. 2014; Baumers et al. 2016; Khorram Niaki and Nonino 2017). This is expedited considering that post-processing of parts is often required, typically caused by stair-stepping effects that arise from incrementally placing one layer on top of another or because finishing layers are needed (Ford and Despeisse 2016). Also, based on case study research, Mellor, Hao, and Zhang identified that machine suppliers partly implement restrictions such as which specific powders can be processed – that they typically also offer – or locking down process parameters that hinder R&D practices in the form of process parameters optimization (the machine suppliers offer R&D services that fill this artificially generated gap) (Mellor et al. 2014). Finally, regulation-related bottlenecks prevail. One big concern is the ambiguous intellectual property (IP) situation. With AM machines becoming more ubiquitous, the traditional forms of IP protection will be significantly challenged (Kurfess and Cass 2014). IP protection in AM is particularly critical considering that while in conventional manufacturing, the copying of a design can be readily traced to a source, in AM there is no need for a specific infrastructure, which renders it difficult to prevent unauthorized replications (Brown et al. 2016). Other challenges include the certification for components (e.g. spare parts) and further liability-related legal issues, including warranty (Ford and Despeisse 2016; Holmström et al. 2016). This is aggravated by the fact that traditional destructive and nondestructive tests that assess critical product characteristics might not be applicable for parts that are produced with variation by an AM machine or only produced once (Petrick and Simpson 2013). Petrick and Simpson also point out that validation of complex internal geometries equally remains an issue (2013). Considering these limitations, it also remains elusive how processes can be certified (Sirichakwal and Conner 2016). In addition, relying on AM increases the risk for knowledge leaks and product piracy as data is transferred openly and products are possibly manufactured on decentralized manufacturing stations (Bogers et al. 2016; Chekurov et al. 2018). Finally, there is also a military dimension to AM. As guns and high-capacity magazines for assault weapons could be 3D printed, terrorists could reduce their reliance on supply chains for

---

6 She suggests that this could be overcome by regulating software instead of physical objects. As AM machines manufacture objects according to specifications provided in a design plan, product safety could be addressed by controlling the sharing of created plans and preventing unsafe ones from being distributed or sold (Neely 2016).

weapons, increasing the risk for the general population (Garrett 2014). Policy makers could respond with regulation, potentially harming AM business models and value chains.

That some of the advantages and bottlenecks outlined above are contradicting is partly caused by the different technologies and materials that are used under the umbrella of AM. This will be discussed in the next section.

Technologies and Materials

Various technologies and materials are discussed in the context of AM. This section will introduce them, stating their advantages and disadvantages.

Technologies

While the first commercial AM printers that were developed as early as 1986 relied on stereolithography (SLA) technology, where light exposure hardens a photopolymeric resin, over time, more technologies emerged so that now seven different categories of AM technologies can be distinguished[7] (Silva et al. 1999; Wohlers 2012; Ngo et al. 2018). The three most popular ones – SLA, fused deposition modeling (FDM), and selective laser sintering (SLS) – will be subsequently described.

During an *SLA* manufacturing process, a concentrated beam of ultraviolet (UV) light or a laser is focused onto the surface of a container filled with a liquid photopolymer. The beam is focused and creates each layer of the target product through cross-linking or degrading of the polymer (Dassault Systèmes 2020). Advantages of this technique are high accuracy, a smooth surface finish, and a range of functional applications. However, the resulting product is sensitive to long exposure to UV light. It is most suitable for functional prototyping, patterns, molds, and tooling, dental applications, jewelry prototyping and casting, and model making (Formlabs 2020b). In *FDM*, a continuous filament of a thermoplastic material is used as base material. The filament is fed from a coil, through a moving heated printer extruder head. The molten material is forced out of the extruder's nozzle and deposited onto a platform. Once the first layer is completed, the extruder and the platform are parted away in one step, and the second layer can then be directly deposited onto the growing item. A variety of materials can be processed. Most popular are thermoplastics, but also paste-like materials such as ceramics, concrete, and chocolate can be used (Dassault Systèmes 2020). Advantages of FDM is its speed and that there are low-cost consumer machines and materials available. However, its drawbacks are low accuracy, low level of details, and limited design compatibility. Therefore, it is most suitable for low-cost rapid prototyping and basic proof-of-concept models (Formlabs 2020b). For *SLS*, lasers are used to sinter powdered material layer by layer to create a solid structure. The final product, rendered enveloped in loose powder, is then cleaned with brushes and pressurized air. The main materials used in the SLS AM process include

---

7 The seven categories are photopolymerization (including the technologies VAT, SLA, DLP, CDLP), powder bed fusion (DMLS, SLS, SLM, MJF, EBM), material extrusion (FDM), material jetting (MJ, NPJ, DOD), binder jetting (BJ), sheet lamination (LOM, SL), and directed energy deposition (DED, LENS, EBAM) (Wohlers 2012; Dassault Systèmes 2020).

polyamide, alumide, and rubber-like materials (Dassault Systèmes 2020). The technology's advantages are design freedoms and that there is no need for support structures. However, there are issues regarding rough surface finishes and limited material options. It is therefore best applied in functional prototyping and for short-run, bridge, or custom manufacturing (Formlabs 2020b). These technologies are compared in Table 4.1.

Materials

Due to the fast development in AM technologies, an increasingly diverse range of materials from chocolate to advanced multifunctional materials can be used (Petrick and Simpson 2013). Most common are polymers, such as thermoplastic polymers (e.g. polyamide, polycarbonate) and thermosetting powders (e.g. polystyrene or photopolymer resins), which have been developed for aerospace, automotive, sports, medical, architectural, and toy industries (Ngo et al. 2018). Benefits of using polymers and composites are that they allow for fast prototyping, are cost-effective, and permit complex structures as well as mass customization. However, challenges are weak mechanical properties, a limited selection of polymers and reinforcements, and anisotropic mechanical properties[8] (especially in fiber-reinforced composites) (Ngo et al. 2018). Using metals and alloys in AM, preferably titanium, nickel, and iron alloys (particularly stainless steels), is especially interesting for the aerospace, medical, and dental industries (Fisher 2020). Benefits of metals and alloys are the possibilities for multifunctional optimization, mass customization, reduced material waste, fewer assembly components, and the possibility to repair damaged or worn metal parts. Meanwhile, there is only a limited selection of alloys available, there are dimensional inaccuracies, and the surface finish is poor. Furthermore, post-processing such as machining, heat treatment, or chemical etching might also be required (Ngo et al. 2018). In comparison with these metals and polymers, ceramics are difficult to process, particularly for complex shapes as they cannot be cast or machined easily (Eckel et al. 2016). Advantages of ceramics are that the porosity of lattices can be controlled and that complex structures and scaffolds for human body organs can be printed. Also, reduced fabrication time is needed, and composition and microstructure can be better controlled. Challenges are that there is a limited selection of 3D-printable ceramics, dimensional inaccuracies, and poor surface finish and that post-processing such as sintering might be required (Ngo et al. 2018). These materials are compared in Table 4.2.

For the AM process, the raw material must be formed into a compatible state such as powder, sheet, wire, or liquid. For the most stringent service applications, AM parts are usually post-processed to improve the microstructure, to reduce porosity, and to finish surfaces (Bourell et al. 2017). Currently, there are still material-related hurdles in AM, such as high costs of materials, poor material characteristics, variable quality across shipments due to a lack of standardization, and a limited variety of available materials (Casalino et al. 2015; Joshi and Sheikh 2015; Lewandowski and Seifi 2016; Hitzler et al. 2017a, b). However, these barriers could be overcome, and the share of printable materials be increased in the future. A substitution of original materials must be taken into account, which potentially leads to inclines in costs and varying mechanical properties. Also, the printability of some

---

8 Physical and mechanical properties differ with the part orientation.

Table 4.1 Comparison of AM technologies.

	Stereolithography (SLA)	Fused deposition modeling (FDM)	Selective laser sintering (SLS)
Type	Vat polymerization	Material extrusion	Powder bed fusion
Advantages	• Very high dimensional accuracy/ intricate details • Very smooth surface finish • Specialty materials available (e.g. clear, flexible, castable resins)	• Cost-effective way of producing custom thermoplastic parts/ prototypes; low-cost consumer machines and materials available • High production speed • Wide range of thermoplastic materials available	• Good, isotropic mechanical properties • No support required (designs with complex geometries can be easily produced) • Excellent manufacturing capabilities for small to medium batch production
Disadvantages	• Brittle (not suitable for functional prototypes) • Mechanical properties/visual appearance degrades with long exposure to UV light • Support structures always required; post-processing needed to remove visual marks	• Low dimensional accuracy and resolution (not suitable for parts with intricate details) • Likely to have visible layer lines, requires post-processing for smooth finish • Anisotropic (directionally dependent) • Low accuracy, low level of details	• Grainy surface finish, internal porosity (may require post-processing, if smooth surface or water tightness needed) • Cannot print large flat surfaces and small holes accurately
Applications	Functional prototyping, patterns, molds, and tooling, dental applications, jewelry prototyping and casting, and model making	Low-cost rapid prototyping, basic proof-of-concept models	Functional prototyping, short-run, bridge, or custom manufacturing

Source: Based on Dassault Systèmes (2020), Formlabs (2020b), and Varotsis (2020a, b, c). © John Wiley & Sons.

Table 4.2 Comparison of main materials used in AM.

	Polymers and composites	Metals and alloys	Ceramics
Advantages	• Fast prototyping • Cost-effective • Complex structures • Mass customization	• Multifunctional optimization • Mass customization • Reduced material waste • Fewer assembly components • Possibility to repair damaged or worn metal parts	• Controlling porosity of lattices • Printing complex structures and scaffolds for human body organs • Reduced fabrication time • A better control on composition and microstructure
Disadvantages	• Weak mechanical properties • Limited selection of polymers and reinforcements • Anisotropic mechanical properties (especially in fiber-reinforced composites)	• Limited selection of alloys • Dimensional inaccuracy and poor surface finish • Post-processing may be required (machining, heat treatment, or chemical etching)	• Limited selection of 3D-printable ceramics • Dimensional inaccuracy and poor surface finish • Post-processing (e.g. sintering) may be required
Applications	• Aerospace • Automotive • Sports • Medical • Architecture • Toys • Biomedical	• Aerospace • Automotive • Military • Biomedical	• Biomedical • Aerospace • Automotive • Chemical industries

Source: Based on Ngo et al. (2018). © John Wiley & Sons.

materials could be already demonstrated in a research environment, but they are not yet available for industrial usage (Kretzschmar et al. 2018).

Given this wide range of technologies and materials, AM will highly likely impact production processes, which will be discussed in the next sections.

Application Scenarios

AM will surely affect supply chains and the whole economy, although there are opposing views to what extent. For instance, Richard D'Aveni, professor of strategy at Dartmouth College's Tuck School of Business, refers to it as "3-D Printing Revolution," explaining that "Industrial 3-D printing is at a tipping point, about to go mainstream in a big way" (d'Aveni 2015). Others also refer to it as "gold rush" that will dramatically change "supply chains, firm strategies, competition, and industrial geographies" (Sasson and Johnson 2016). Contrary to these optimists, some experts are more skeptical, considering AM only as complementary to conventional manufacturing that will be applied in very specific scenarios (Verboeket and Krikke 2019). AM is particularly suitable for products that have complex shapes. In conventional manufacturing, the more complicated the shape of an object, the higher the manufacturing costs. However, in AM, fabricating an intricate, complex shape does not require more time or cost than a simple block (Lipson and Kurman 2013). This also allows for the manufacturing of "generative design parts." While in traditional design, humans are drafting the concrete end product, generative design relies on computational algorithms, which generate designs based on certain input parameters such as purpose, size, strength, weight, or cost constraints. By simulating physics, sometimes inspired by biological systems, thousands of design options are virtually explored until the best, optimal design solution is found for a particular problem (Xponentialworks 2020). For example, the industrial control and automation company Festo developed a gripping arm that, modeled based on the complex kinematics of a bird's beak and designed considering the forces acting on the component, possesses a unique force-to-weight ratio (Calignano et al. 2017). AM additionally offers a chance for parts that are characterized by a high risk of obsolescence and high shortage costs, such as spare parts (Holmström and Partanen 2014). John T. Lee, responsible for managing print store ABC Imaging's 3D modeling and rapid prototyping services, explained that he sees a "future where people and architects and engineers – rather than sending an order to a warehouse to get a spare part – will download a CAD file and have it printed in their neighborhood print shop." He also added that "right now, we're not that far away from that model already. All day long, our bicycle couriers come in and out of here to deliver printed parts to our customers" (Lipson and Kurman 2013). Products in short supply can also be quickly and flexibly manufactured, thanks to AM (Pérès and Noyes 2006). For instance, at the height of the COVID-19 pandemic in the United States, additive manufactures produced urgently needed and difficult to obtain protective equipment material such as face shields. The chief product officer of the AM company Formlabs stated "I can't even tell you how many hospitals, and various other health institutions and health-care providers and governments, have asked us to help out with the situation" (Page 2020). AM is also suitable to produce production tools such fixtures, jigs, gauges, molds, or dies. Tools produced via AM offer a new approach to improve productivity through better cycle times, machine performance, and tool changeovers (Ford and

Despeisse 2016; Materialise 2020). In addition, without geometric constraints, design efforts can be concentrated on part functionality and assembly. Therefore, AM can be used to produce assembly parts that are optimized for low part count and fabrication in an assembled state. This is especially valid as conventional manufacturing constraints such as material sizes or coordinate systems and symmetric axis for machining do not have to be considered, so that parts can be created with doors and attached interlocking hinges at the same time (Atzeni and Salmi 2012). AM is also predestined to manufacture single unit items. As AM removes the overhead costs needed for retraining operators and retooling machines, units with the lot size one, such as individual gears, discontinued and antique parts, custom-fit dental protheses, hearing aids, or even components for the Mars rover can be produced at affordable costs (Lipson and Kurman 2013). This goes along with the possibility to manufacture customized consumer products. Traditionally, executives must balance being as customer driven as possible, including inventing new programs and procedures to meet every customer's request, while not adding too many unnecessary cost and complexity to operations (Gilmore et al. 1997). As companies can overcome this dilemma with AM, fully customized consumer products such as bicycle frames (UK bicycle frame manufacturer Reynolds), razors (shaving and razor brand Gillette), or shoes (sportswear manufacturer Adidas) emerged (Vialva 2019).

Given these application scenarios and the considerable research undertaken in the field, the AM market is growing, and new technology trends are emerging. This will be described in the following section.

Market and Trends

Market

From the 1990s on, the technology evolved from rapid prototyping, focused on formal and functional prototypes, over rapid manufacturing, with a focus on final parts, to AM, where the target is mass production and hybrid manufacturing (Monzón et al. 2019). Each stage was accompanied by a varying degree of frequently overly optimistic extravagant publicity and promotion. For instance, the 3D printing technology developer and manufacturer Formlabs explained that "while AM technologies have been around since the 1980s, the industry went through its most striking hype cycle during the early 2010s, when promoters claimed that the technology would find broad usage in consumer applications and reorder businesses from The Home Depot to UPS. Since the breathless hype subsided a few years ago, professional 3D printing technologies have been rapidly maturing in many concrete ways" (Formlabs 2020a).

With this gradual maturing of the technology, the AM market is also continuously increasing (Petrick and Simpson 2013; Schwab 2017). According to the research firm Wohlers Associates, the global AM market grew from some 1 billion USD in 2009 to 9.8 billion USD in 2018, as shown in Figure 4.3. North America accounts for 40%, Europe for 28%, and Asia Pacific for 27% (UPS 2016; Fan et al. 2020). The largest contributors to the AM revenue are the consumer electronics and the automotive sector with each 20%, followed by the medical device industry at 15% (UPS 2016).

In this figure, products include "AM systems, materials, and aftermarket products, such as software and lasers." Services include revenues generated from parts produced on AM

USD billion

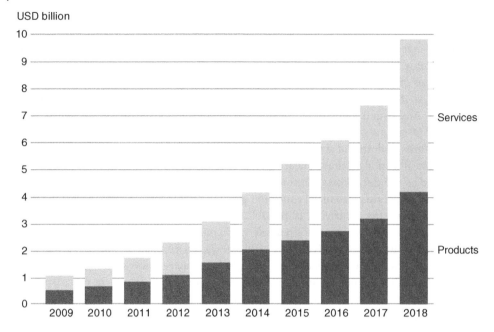

Figure 4.3   Global AM revenues. *Source*: Based on Fan et al. (2020). © John Wiley & Sons.

systems by service providers and system manufacturers, system maintenance contracts, training, seminars, conferences, expositions, advertising, publications, contract research, and consulting services (Fan et al. 2020).

Trends

The research firm Gartner considers AM an important technology trend that accelerates product development, improves operations, and can create finished goods (Basiliere and Shanler 2019). The firm therefore tracks related trends and publishes it annually in its "hype cycle," a graph where the maturity and adoption of a particular technology are charted (3ders 2017). In the 2019 revision, AM in supply chain is considered as sliding into the Trough of Disillusionment,[9] where "interest wanes as experiments and

_____

9 Gartner explains all phases as follows. *Innovation Trigger*: A potential technology breakthrough kicks things off. Early proof-of-concept stories and media interest trigger significant publicity. Often no usable products exist and commercial viability is unproven. *Peak of Inflated Expectations*: Early publicity produces a number of success stories – often accompanied by scores of failures. Some companies take action; many do not. *Trough of Disillusionment*: Interest wanes as experiments and implementations fail to deliver. Producers of the technology shake out or fail. Investments continue only if the surviving providers improve their products to the satisfaction of early adopters. *Slope of Enlightenment*: More instances of how the technology can benefit the enterprise start to crystallize and become more widely understood. Second- and third-generation products appear from technology providers. More enterprises fund pilots; conservative companies remain cautious. *Plateau of Productivity*: Mainstream adoption starts to take off. Criteria for assessing provider viability are more clearly defined. The technology's broad market applicability and relevance are clearly paying off (Gartner 2020).

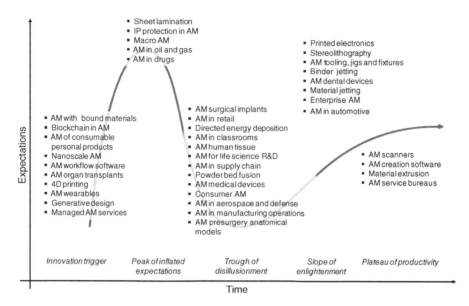

Expectations

- Sheet lamination
- IP protection in AM
- Macro AM
- AM in oil and gas
- AM in drugs

- Printed electronics
- Stereolithography
- AM tooling, jigs and fixtures
- Binder jetting
- AM dental devices
- Material jetting
- Enterprise AM
- AM in automotive

- AM with bound materials
- Blockchain in AM
- AM of consumable personal products
- Nanoscale AM
- AM workflow software
- AM organ transplants
- 4D printing
- AM wearables
- Generative design
- Managed AM services

- AM surgical implants
- AM in retail
- Directed energy deposition
- AM in classrooms
- AM human tissue
- AM for life science R&D
- AM in supply chain
- Powder bed fusion
- AM medical devices
- Consumer AM
- AM in aerospace and defense
- AM in manufacturing operations
- AM presurgery anatomical models

- AM scanners
- AM creation software
- Material extrusion
- AM service bureaus

| Innovation trigger | Peak of inflated expectations | Trough of disillusionment | Slope of enlightenment | Plateau of productivity |

Time

Figure 4.4 Hype cycle for AM. *Source*: Based on Fenn et al. (2018) and Basiliere and Shanler (2019). © John Wiley & Sons.

implementations fail to deliver. Producers of the technology shake out or fail. Investments continue only if the surviving providers improve their products to the satisfaction of early adopters" (Basiliere and Shanler 2019; Gartner 2020). This is shown in Figure 4.4.

Other technologies are just at the Innovation Trigger state, where early proof-of-concept stories and media interest trigger significant publicity although no usable products exist and commercial viability is unproven (Gartner 2020). For example, in nanoscale AM, parts can be produced with a depth resolution of 175 nm (Saha et al. 2019). This would allow for the production of flexible electronics, electrochemical interfaces, micro-optics, and other micro- or nanostructures (Boissonneault 2019). Another technology at this maturity stage is 4D printing, a new process that entails multi-material prints with the capability to transform over time. Here, manufactured structures are programmably active and can transform independently, instead of being simple static objects (Tibbits 2014). Furthermore, AM organ transplants (also referred to as 3D bioprinting) could open a new world of possibilities for the medical field. Instead of taking the risk of the body rejecting a transplanted organ or waiting until a suitable donor hopefully appears at some point in time, this technology would allow a patient to have an organ fabricated specifically to them to replace their faulty ones (Ng et al. 2019).

Conclusion

This chapter set out to discuss AM, one of the key elements of the Fourth Industrial Revolution (Verboeket and Krikke 2019). The technology has come a long way since it was first developed in the 1980s and today includes applications across a wide range of industries, from aerospace to aviation and from automotive to the medical industry. From a supply chain point of view, AM is very appealing as it allows for more concise, shorter, and localized physical and information flows (Gebler et al. 2014). With production happening more decentralized and closer to the place of consumption, supplier relationships, transportation patterns, inventory policies, and packaging processes are drastically altered (Mills and Camek 2004; Ruffo et al. 2007; Thiesse et al. 2015; Ford and Despeisse 2016). AM has a variety of advantages related to costs, flexibility, sustainability, and innovation (Walter et al. 2004; Petrick and Simpson 2013). It allows to economically produce small batches with customized designs and complex structures (Holmström et al. 2010; Ford and Despeisse 2016). However, still various product-, processing-, and regulation-related bottlenecks exist that hinder a true supply chain disruption due to AM (Verboeket and Krikke 2019). The advantages and bottlenecks are also depending on the technology and material used. Therefore, currently mainly SLA is used for functional prototyping, patterns, molds, and tooling, FDM for low-cost rapid prototyping and basic proof-of-concept models, and SLS for functional prototyping or custom manufacturing. The most common materials, polymers and composites, metals and alloys, and ceramics, are all used in the aerospace, automotive, and biomedical industries, while ceramics are also common in chemical industries. Although there is little doubt that AM will affect production processes and the economy, it is unclear to what extent. While some researchers call AM a revolution, ready to go mainstream in a big way and dramatically changing supply chains, others consider AM only complementary to conventional manufacturing and believe that it will be only applied in very specific scenarios (d'Aveni 2015; Sasson and Johnson 2016; Verboeket and Krikke 2019). The technology is particularly suitable for products that have highly complex shapes, spare parts, products in short supply, tools, and parts that are developed through a generative design process (Pérès and Noyes 2006; Lipson and Kurman 2013; Holmström and Partanen 2014; Xponentialworks 2020). Despite of the often unmerited hype that surrounded this technology for the past years, the global AM market indeed reached 9.8 billion USD in 2018 and is expected to further grow considerably (Petrick and Simpson 2013; Schwab 2017; Fan et al. 2020). With the gradual maturing of the technology, even trends such as nanoscale AM, 4D printing, or AM organ transplants that today sound like science fiction might become reality one day.

In the future, it is likely that companies will have to decide whether they are competing based on *economies of scale* with low cost and high volumes or based on *economies of one* with end-user customization (Petrick and Simpson 2013). Similar to the development of the office printer, where one or different documents can be pushed out at about the same costs for each item, the development of AM shrank the traditionally prohibitive cost of customization to zero (The Economist 2012). And as Lipson and Kurman explained, historically "bursts of innovation happen when an emerging technology removes a once prohibitive barrier of cost, distance, or time" (Lipson and Kurman 2013). In the spirit of the Fourth Industrial Revolution, AM definitely has the potential to truly revolutionize the way we develop, manufacture, and provide products to customers.

Key Takeaways

- Additive manufacturing (also called 3D printing or rapid prototyping), where a product is built by adding instead of subtracting material, is one of the key elements of the Fourth Industrial Revolution.
- The additive manufacturing supply chain is much more concise, shorter, and localized compared with a conventional supply chain.
- AM allows for decentralized production close to the place of consumption, thus drastically altering supplier relationships, transportation patterns, inventory policies, or packaging processes.
- The technology has advantages in terms of costs, flexibility, sustainability, and innovation and enables an economic production of small batches – also with customized designs and complex structures.
- Product-, processing-, and regulation-related bottlenecks are currently existing and preventing a breakthrough of the technology and consequently a disruption of supply chains.
- Currently the AM technologies – stereolithography, fused deposition modeling, and selective laser sintering – are mainly employed. Used materials are polymers and composites, metals and alloys, and ceramics.
- AM is predicted to affect production processes and the whole economy, but opinions vary to what extent.
- Application scenarios for AM are especially in the production of products with highly complex shapes, spare parts, products in short supply, tools, and parts that are developed through a generative design process.
- The global AM market reached 9.8 billion USD in 2018 and is expected to further grow considerably.
- With the maturing of the technology, companies in the future might have to decide whether they are competing based on economies of scale (low cost and high volumes) or economies of one (end-user customization).

References

3ders (2017). Gartner's 2017 3D printing hype cycle. http://www.3ders.org/articles/20170804-gartners-2017-3d-printing-hype-cycle.html. (accessed 17 May 2020).

Airbus (2020). 3D printing, corporate website. https://www.airbus.com/public-affairs/brussels/our-topics/innovation/3d-printing.html (accessed 17 May 2020).

Atzeni, E. and Salmi, A. (2012). Economics of additive manufacturing for end-usable metal parts. *The International Journal of Advanced Manufacturing Technology* 62 (9–12): 1147–1155. https://doi.org/10.1007/s00170-011-3878-1.

Baldwin, J. and Lin, Z. (2002). Impediments to advanced technology adoption for Canadian manufacturers. *Research Policy* 31 (1): 1–18. https://doi.org/10.1016/S0048-7333(01)00110-X.

Basiliere, P. and Shanler, M. (2019). Hype cycle for 3D printing, 2019. https://www.gartner.com/en/documents/3947508/hype-cycle-for-3d-printing-2019 (accessed 17 May 2020).

Baumers, M., Dickens, P., Tuck, C. et al. (2016). The cost of additive manufacturing: machine productivity, economies of scale and technology-push. *Technological Forecasting and Social Change* 102: 193–201. https://doi.org/10.1016/j.techfore.2015.02.015.

Bogers, M., Hadar, R., and Bilberg, A. (2016). Additive manufacturing for consumer-centric business models: implications for supply chains in consumer goods manufacturing. *Technological Forecasting and Social Change* 102: 225–239. https://doi.org/10.1016/j.techfore.2015.07.024.

Boissonneault, T. (2019). Researchers Pioneer 1000x faster nanoscale 3D printing method, 3D printing media network. https://www.3dprintingmedia.network/1000x-faster-nanoscale-3d-printing/ (accessed 17 May 2020).

Bourell, D., Kruth, J.P., Leu, M. et al. (2017). Materials for additive manufacturing. *CIRP Annals* 66 (2): 659–681. https://doi.org/10.1016/j.cirp.2017.05.009.

Brown, A., Yampolskiy, M., Gatlin, J., et al. (2016). Legal aspects of protecting intellectual property in additive manufacturing. In: *Critical Infrastructure Protection X. ICCIP 2016. IFIP Advances in Information and Communication Technology* (eds. M. Rice and S. Shenoi), 63–79. Springer https://doi.org/10.1007/978-3-319-48737-3_4.

Calignano, F., Manfredi, D., Ambrosio, E.P. et al. (2017). Overview on additive manufacturing technologies. *Proceedings of the IEEE* 105 (4): 593–612. https://doi.org/10.1109/JPROC.2016.2625098.

Carr, S. (2017). What is additive manufacturing? https://www.energy.gov/eere/articles/what-additive-manufacturing (accessed 17 May 2020).

Calignano, F., Manfredi, D., Ambrosio, E.P. et al. (2015). Experimental investigation and statistical optimisation of the selective laser melting process of a maraging steel. *Optics & Laser Technology* 65: 151–158. https://doi.org/10.1016/j.optlastec.2014.07.021.

Chekurov, S. and Salmi, M. (2017). Additive manufacturing in offsite repair of consumer electronics. *Physics Procedia* 89: 23–30. https://doi.org/10.1016/j.phpro.2017.08.009.

Chekurov, S., Metsä-Kortelainen, S., Salmi, M. et al. (2018). The perceived value of additively manufactured digital spare parts in industry: an empirical investigation. *International Journal of Production Economics* 205: 87–97. https://doi.org/10.1016/j.ijpe.2018.09.008.

Chiu, M.-C. and Lin, Y.-H. (2016). Simulation based method considering design for additive manufacturing and supply chain. *Industrial Management & Data Systems* 116 (2): 322–348. https://doi.org/10.1108/IMDS-07-2015-0266.

Dassault Systèmes (2020). Introduction to 3D printing: additive processes. https://make.3dexperience.3ds.com/processes/photopolymerization (accessed 17 May 2020).

d'Aveni, R. (2015). The 3-D printing revolution. *Harvard Business Review* 93 (5): 40–48.

De la Torre, N., Espinosa, M.M., and Domínguez, M. (2016). Rapid prototyping in humanitarian aid to manufacture last mile vehicles spare parts: an implementation plan. *Human Factors and Ergonomics in Manufacturing & Service Industries* 26 (5): 533–540. https://doi.org/10.1002/hfm.20672.

Dwivedi, G., Srivastava, S.K., and Srivastava, R.K. (2017). Analysis of barriers to implement additive manufacturing technology in the Indian automotive sector. *International Journal of Physical Distribution and Logistics Management* 47 (10): 972–991. https://doi.org/10.1108/IJPDLM-07-2017-0222.

Eckel, Z.C., Zhou, C., Martin, J.H. et al. (2016). Additive manufacturing of polymer-derived ceramics. *Science*. American Association for the Advancement of Science 351 (6268): 58–62.

Fan, Z., Sotelo, J., and Sundareswaran, V. (2020). 3D printing: a guide for decision-makers. http://www3.weforum.org/docs/WEF_Impacts_3D_Printing_on_Trade_Supply_Chains_Toolkit.pdf (accessed 17 May 2020).

Fenn, J., Raskino, M., and Burton, B. (2018). Understanding Gartner's Hype Cycles. *Gartner Research Report*, May, pp. 1–35.

Fisher, D.J. (2020). *Additive Manufacturing of Metals*. Materials Research Forum LLC (Materials Research Foundations) https://books.google.ae/books?id=qKXMDwAAQBAJ.

Ford, S. and Despeisse, M. (2016). Additive manufacturing and sustainability: an exploratory study of the advantages and challenges. *Journal of Cleaner Production* 137: 1573–1587. https://doi.org/10.1016/j.jclepro.2016.04.150.

Formlabs (2020a). Additive manufacturing: industry trends and outlook, corporate website. https://formlabs.com/uk/blog/additive-manufacturing/ (accessed 17 May 2020).

Formlabs (2020b). Compare FDM, SLA, and SLS 3D printing technologies. https://formlabs.com/blog/fdm-vs-sla-vs-sls-how-to-choose-the-right-3d-printing-technology/ (accessed 17 May 2020).

Garnsey, S. (2020). Daimler buses to start 3D printing of spare parts, automotive logistics. https://www.automotivelogistics.media/supply-chain-management/daimler-buses-to-start-3d-printing-of-spare-parts/39832.article (accessed 17 May 2020).

Garrett, B. (2014). 3D printing: new economic paradigms and strategic shifts. *Global Policy* 5 (1): 70–75. https://doi.org/10.1111/1758-5899.12119.

Gartner (2020). Gartner hype cycle. https://www.gartner.com/en/research/methodologies/gartner-hype-cycle (accessed 17 May 2020).

Gebler, M., Schoot Uiterkamp, A.J.M., and Visser, C. (2014). A global sustainability perspective on 3D printing technologies. *Energy Policy* 74: 158–167. https://doi.org/10.1016/j.enpol.2014.08.033.

General Electric (2020). What is additive manufacturing? Corporate website. https://www.ge.com/additive/additive-manufacturing (accessed 17 May 2020).

Gilmore, J.H. and Pine, B.J. (1997). The four faces of mass customization. *Harvard Business Review* 75 (1): 91–102.

Hitzler, L., Hirsch, J., Heine, B. et al. (2017a). On the anisotropic mechanical properties of selective laser-melted stainless steel. *Materials* 10 (10): 1136. https://doi.org/10.3390/ma10101136.

Hitzler, L., Hirsch, J., Merkel, M. et al. (2017b). Position dependent surface quality in selective laser melting. *Materialwissenschaft und Werkstofftechnik* 48 (5): 327–334. https://doi.org/10.1002/mawe.201600742.

Holmström, J. and Partanen, J. (2014). Digital manufacturing-driven transformations of service supply chains for complex products. *Supply Chain Management: An International Journal* 19 (4): 421–430. https://doi.org/10.1108/SCM-10-2013-0387.

Holmström, J., Partanen, J., Tuomi, J. et al. (2010). Rapid manufacturing in the spare parts supply chain. *Journal of Manufacturing Technology Management* 21 (6): 687–697. https://doi.org/10.1108/17410381011063996.

Holmström, J., Holweg, M., Khajavi, S.H. et al. (2016). The direct digital manufacturing (r)evolution: definition of a research agenda. *Operations Management Research* 9 (1–2): 1–10. https://doi.org/10.1007/s12063-016-0106-z.

Joshi, S.C. and Sheikh, A.A. (2015). 3D printing in aerospace and its long-term sustainability. *Virtual and Physical Prototyping* 10 (4): 175–185. https://doi.org/10.1080/17452759.201 5.1111519.

Khajavi, S.H., Partanen, J., and Holmström, J. (2014). Additive manufacturing in the spare parts supply chain. *Computers in Industry* 65 (1): 50–63. https://doi.org/10.1016/j. compind.2013.07.008.

Khorram Niaki, M. and Nonino, F. (2017). Additive manufacturing management: a review and future research agenda. *International Journal of Production Research* 55 (5): 1419–1439. https://doi.org/10.1080/00207543.2016.1229064.

Kohtala, C. (2015). Addressing sustainability in research on distributed production: an integrated literature review. *Journal of Cleaner Production* 106: 654–668. https://doi. org/10.1016/j.jclepro.2014.09.039.

Kothman, I. and Faber, N. (2016). How 3D printing technology changes the rules of the game. *Journal of Manufacturing Technology Management* 27 (7): 932–943. https://doi.org/10.1108/ JMTM-01-2016-0010.

Kretzschmar, N., Chekurov, S., Salmi, M. et al. (2018). Evaluating the readiness level of additively manufactured digital spare parts: an industrial perspective. *Applied Sciences* 8 (10): 1837. https://doi.org/10.3390/app8101837.

Kurfess, T. and Cass, W.J. (2014). Rethinking additive manufacturing and intellectual property protection. *Research-Technology Management* 57 (5): 35–42. https://doi.org/10.543 7/08956308X5705256.

Lewandowski, J.J. and Seifi, M. (2016). Metal additive manufacturing: a review of mechanical properties. *Annual Review of Materials Research* 46 (1): 151–186. https://doi.org/10.1146/ annurev-matsci-070115-032024.

Lipson, H. and Kurman, M. (2013). *Fabricated: The New World of 3D Printing*. Indianapolis: Wiley.

Materialise (2020). Applications: 3D printed production tools, corporate website. https://www. materialise.com/en/manufacturing/3d-printed-production-tools (accessed 17 May 2020).

Mellor, S., Hao, L., and Zhang, D. (2014). Additive manufacturing: a framework for implementation. *International Journal of Production Economics* 149: 194–201. https://doi. org/10.1016/j.ijpe.2013.07.008.

Mills, J.F. and Camek, V. (2004). The risks, threats and opportunities of disintermediation. *International Journal of Physical Distribution and Logistics Management* 34 (9): 714–727. https://doi.org/10.1108/09600030410567487.

Monzón, M., Paz, R., Ortega, Z. et al. (2019). Knowledge transfer and standards needs in additive manufacturing. In: *Additive Manufacturing – Developments in Training and Education* (eds. E. Pei, M. Monzón and A. Bernard), 1–13. Cham: Springer International Publishing https://doi.org/10.1007/978-3-319-76084-1_1.

Moore, S.S., O'Sullivan, K.J., and Verdecchia, F. (2016). Shrinking the supply chain for implantable coronary stent devices. *Annals of Biomedical Engineering* 44 (2): 497–507. https://doi.org/10.1007/s10439-015-1471-8.

Neely, E.L. (2016). The risks of revolution: ethical dilemmas in 3D printing from a US perspective. *Science and Engineering Ethics* 22 (5): 1285–1297. https://doi.org/10.1007/ s11948-015-9707-4.

Ng, W.L., Chua, C.K., and Shen, Y.-F. (2019). Print me an organ! Why we are not there yet. *Progress in Polymer Science* 97: 101145. https://doi.org/10.1016/j.progpolymsci.2019.101145.

Ngo, T.D., Kashani, A., Imbalzano, G. et al. (2018). Additive manufacturing (3D printing): a review of materials, methods, applications and challenges. *Composites Part B: Engineering* 143: 172–196. https://doi.org/10.1016/j.compositesb.2018.02.012.

Page, J. (2020). How 3D printing could save lives in the coronavirus outbreak. *Technology Review* https://www.technologyreview.com/2020/03/27/950240/3d-printing-coronavirus-covid-19-medical-supplies-devices/ (accessed 17 May 2020).

Pérès, F. and Noyes, D. (2006). Envisioning e-logistics developments: making spare parts in situ and on demand. *Computers in Industry* 57 (6): 490–503. https://doi.org/10.1016/j.compind.2006.02.010.

Petrick, I.J. and Simpson, T.W. (2013). 3D printing disrupts manufacturing: how economies of one create new rules of competition. *Research-Technology Management* 56 (6): 12–16. https://doi.org/10.5437/08956308X5606193.

Rayna, T. and Striukova, L. (2016). From rapid prototyping to home fabrication: how 3D printing is changing business model innovation. *Technological Forecasting and Social Change* 102: 214–224. https://doi.org/10.1016/j.techfore.2015.07.023.

Rayna, T., Striukova, L., and Darlington, J. (2015). Co-creation and user innovation: the role of online 3D printing platforms. *Journal of Engineering and Technology Management* 37: 90–102. https://doi.org/10.1016/j.jengtecman.2015.07.002.

Ruffo, M., Tuck, C., and Hague, R. (2007). Make or buy analysis for rapid manufacturing. *Rapid Prototyping Journal* 13 (1): 23–29. https://doi.org/10.1108/13552540710719181.

Ryan, M.J., Eyers, D.R., Potter, A.T. et al. (2017). 3D printing the future: scenarios for supply chains reviewed. *International Journal of Physical Distribution and Logistics Management* 47 (10): 992–1014. https://doi.org/10.1108/IJPDLM-12-2016-0359.

Saha, S.K., Wang, D., Nguyen, V.H. et al. (2019). Scalable submicrometer additive manufacturing. *Science* 366 (6461): 105–109. https://doi.org/10.1126/science.aax8760.

Sasson, A. and Johnson, J.C. (2016). The 3D printing order: variability, supercenters and supply chain reconfigurations. *International Journal of Physical Distribution and Logistics Management* 46 (1): 82–94. https://doi.org/10.1108/IJPDLM-10-2015-0257.

Schwab, K. (2017). *The Fourth Industrial Revolution*. Currency.

Sher, D. (2020). How major automakers use AM for production today, Part 3: Daimler Benz additive manufacturing, 3D printing media network. https://www.3dprintingmedia.network/daimler-benz-additive-manufacturing/ (visited on the 23 June 2020) (accessed 17 May 2020).

Silva, J.V.L., Yamanaka, M.C., and Saura, C.E. (1999). Rapid prototyping: concepts, applications, and potential utilization in Brazil. *Proceedings of the 15th International Conference in CAD/CAM Robotics and Factories of the Future*, Águas de Lindóia, Sp, Brazil (18–20 August 1999).

Singamneni, S., Yifan, L.V., Hewitt, A. et al. (2019). Additive manufacturing for the aircraft industry: a review. *Journal of Aeronautics and Aerospace Engineering* 8 (214): 2.

Sirichakwal, I. and Conner, B. (2016). Implications of additive manufacturing for spare parts inventory. *3D Printing and Additive Manufacturing* 3 (1): 56–63. https://doi.org/10.1089/3dp.2015.0035.

Steenhuis, H.-J. and Pretorius, L. (2017). The additive manufacturing innovation: a range of implications. *Journal of Manufacturing Technology Management* 28 (1): 122–143. https://doi.org/10.1108/JMTM-06-2016-0081.

The Economist (2012). A third industrial revolution. https://www.economist.com/special-report/2012/04/21/a-third-industrial-revolution (accessed 17 May 2020).

Thiesse, F., Wirth, M., Kemper, H.-G. et al. (2015). Economic implications of additive manufacturing and the contribution of MIS. *Business and Information Systems Engineering* 57 (2): 139–148. https://doi.org/10.1007/s12599-015-0374-4.

Tibbits, S. (2014). 4D printing: multi-material shape change. *Architectural Design* 84 (1): 116–121. https://doi.org/10.1002/ad.1710.

UPS (2016). 3D printing: the next revolution in industrial manufacturing. https://www.ups.com/assets/resources/media/en_US/3D_Printing_executive_summary.pdf (accessed 17 May 2020).

Varotsis, A.B. (2020a). Introduction to FDM 3D printing, 3D hubs. https://www.3dhubs.com/knowledge-base/introduction-fdm-3d-printing/ (accessed 17 May 2020).

Varotsis, A.B. (2020b). Introduction to SLA 3D printing, 3D hubs. https://www.3dhubs.com/knowledge-base/introduction-sla-3d-printing/ (accessed 17 May 2020).

Varotsis, A.B. (2020c). Introduction to SLS 3D printing, 3D hubs. https://www.3dhubs.com/knowledge-base/introduction-sls-3d-printing/ (accessed 17 May 2020).

Ventola, C.L. (2014). Medical applications for 3D printing: current and projected uses. *P&T: A Peer-Reviewed Journal for Formulary Management* 39 (10): 704–711. https://pubmed.ncbi.nlm.nih.gov/25336867.

Verboeket, V. and Krikke, H. (2019). The disruptive impact of additive manufacturing on supply chains: a literature study, conceptual framework and research agenda. *Computers in Industry* 111: 91–107. https://doi.org/10.1016/j.compind.2019.07.003.

Vialva, T. (2019). The best 3D printed consumer products, 3D printing industry. https://3dprintingindustry.com/news/the-best-3d-printed-consumer-products-148352/ (accessed 17 May 2020).

Walter, M., Holmström, J., Tuomi, H., Yrjölä, H. (2004). Rapid manufacturing and its impact on supply chain management. *Proceedings of the Logistics Research Network Annual Conference*, 9–10, Dublin, Ireland (9–10 September 2004).

Weller, C., Kleer, R., and Piller, F.T. (2015). Economic implications of 3D printing: market structure models in light of additive manufacturing revisited. *International Journal of Production Economics* 164: 43–56. https://doi.org/10.1016/j.ijpe.2015.02.020.

Wohlers, T.T. (2012). *Wohlers Report 2012: Additive Manufacturing and 3D Printing State of the Industry: Annual Worldwide Progress Report*. Wohlers Associates.

Wolff, P. and Kern, J. (2019). Materialise: teaching along the silk road. In: *Management Practices in Asia*, 203–215. Cham: Springer International Publishing https://doi.org/10.1007/978-3-030-19662-2_15.

Xponentialworks (2020). The next generation of generative design, corporate website. https://xponentialworks.com/the-next-generation-of-generative-design/ (accessed 17 May 2020).

5

The Role of Robotic Process Automation (RPA) in Logistics

Mac Sullivan[1], Walter Simpson[1], and Wesley Li[2]

[1] *NNR Global Logistics, Dallas, TX, USA*
[2] *Konica Minolta, Tokyo, Japan*

Introduction

Companies Under Pressure

With margins shrinking across the board for many freight forwarders, executives are looking for ways to either increase top-line revenues, for instance, through value-added services, or reduce their operating costs. Outsourcing of manufacturing has matured with the help of cheaper transportation costs and the improvements in international communication, for example, through the Internet. This led the world's largest brands to have access to cheap manufacturing labor. In turn, these developing countries had more resources to devote to education, which then also opened other service avenues like call centers and information technology (IT) development centers. The delegation of such IT-enabled business processes to an external service provider is called business process outsourcing (BPO) (Mani et al. 2010).

Many of the large freight forwarders, otherwise known as logistics service providers, started to take advantage of this lower service labor by creating BPO centers of excellence. Instead of paying a person in Germany or the United States $10–20 an hour to create a bill of lading or track a container, they could do this same process in India, Malaysia, or the Philippines for a fraction of the price. However, many small- to medium-sized logistics companies did not have the scale or worked in a too decentralized, loosely knit environment that deterred them from outsourcing their documentation work up to this point. As especially these companies now look to reduce their operating costs, automation and digital workflow optimization are quickly becoming a real alternative to outsourcing to a cheaper labor cost country.

The Digital Transformation of Logistics: Demystifying Impacts of the Fourth Industrial Revolution,
First Edition. Edited by Mac Sullivan and Johannes Kern.

RPA as a Solution

Automation has been placed at the forefront of the digitalization trend that is sweeping across the business community. There are several forms of automation, but in this case, we will be evaluating robotic process automation (RPA) as an emerging technology that is garnering substantial attention in the logistics community. According to the Institute for Robotic Process Automation, "Robotic process automation is the application of technology that allows employees in a company to configure computer software or a 'robot' to capture and interpret existing applications for processing a transaction, manipulating data, triggering responses and communicating with other digital systems" (Institute for Robotic Process Automation 2020). Research into the application of RPA shows that this technology is enabling automation in areas that had in the past shown too expensive to do so (Barnett 2015). The cost of virtual RPA robot worker is between 10 and 19% of a local full-time employee (FTE) and roughly 33 and 50% of an FTE in an outsourced location (Prangnell and Wright 2015; Slaby 2012; Willcocks et al. 2015). Considering such price points, it seems that there is a potential for substantial savings over the way companies are currently operating. Therefore, in this chapter, we will investigate the use case of automating white-collar documentation work using RPA.

To address the subject of RPA, an explanation of what the technology is and how it fits into the context of an international logistics company is initially needed. A logistics company typically has large departments of clerks who are creating documents like bills of lading or invoices. These documents are often created by directly copying parsed information from documents received from clients and pasted into new templates. By mimicking the actions of these clerks, it is feasible that a properly trained automated technology could take over this task. Specifically, an RPA bot works in the presentation layer of a system and acts like a human using the same inputs that a mouse and keyboard would by clicking and typing (Slaby 2012). An RPA bot does not necessarily live within a source system or software like Microsoft Word or SAP's enterprise resource planning (ERP); it lives on top of infrastructure like a Windows desktop (Aguirre and Rodriguez 2017). This allows the bot to cross multiple software and reduces the need for integrations and the risk of disrupting system logic. By not living in the source code or database, RPA bots are much more like a human worker and do not require the user to have a lot of coding skills as opposed to traditional system-to-system automation (Asatiani and Penttinen 2016; Slaby 2012).

Evaluating Heavyweight IT to Lightweight IT Automation

When evaluating any project, it is important to figure out which resources you have and which resources make the most sense to invest in for your future. The role of IT departments within logistics companies is going through a fundamental shift as codeless and plug-and-play software-as-a-service (SaaS) solutions and technologies become more prevalent in the workplace. The idea of heavyweight IT, which is defined as a knowledge domain driven by IT professionals, who used digital technology, and made possible by software engineering, is giving way to lightweight IT, which is a socio-technical knowledge domain that is enabled by solution-driven user demand, mass adoption of digital technology, and innovative processes (Bygstad 2016). Traditional IT experts tend to gravitate back to heavyweight IT back-end system integration using electronic data interchange (EDI) or application programming interface (API) integrations, whereas lightweight IT looks to graphical user interface (GUI) automation like API (Bygstad 2016; Lacity and Willcocks 2016). Heavyweight IT automation is reliant on little variation in systems architecture, while lightweight IT automation is more reliant on interfaces having little change (Penttinen et al. 2018). If looking to enable non-IT professionals to be able to automate more processes, lightweight IT automation like RPA is worth exploring (Bygstad 2016). Regardless of where your organization is today, it is important to audit your internal capabilities in terms of skills and perhaps more important to think about where customer demand is going to take you with respect to the talent needed to provide that.

Achieving Operational Excellence

Process Improvement on the Rise

Information technology offers businesses a few different paths for improving their processes through automation (Penttinen et al. 2018). The classic business strategy dilemma is to either be led by external forces such as customers who push requests and demand from the "outside" and affect those processes "in" an organization. This is called the outside-in approach. The alternative is to enhance "inside" capabilities and resources to be more efficient. Over the past several decades, we have seen the growth and decline of straight-through processing (STP) (van der Aalst and van Hee 2002; ter Hofstede et al. 2010) as a type of workflow management (WFM) that evolved into business process management (BPM). BPM can be explained as process automation or used to refer to the daily exaction of processes managed through software (van der Aalst 2014; Harmon 2010). BPM became expensive as it took the outside-in approach where new systems had to be developed from scratch and extensive integration with outside systems was needed (van der Aalst et al. 2018). Modern trends point toward a need for an increased amount of connectivity, which leads us to believe that BPM and heavyweight IT are going to continue to increase in cost and complexity (Sommerville et al. 2010). BPM use cases have been unclear and sparse through the past decade (Le Clair 2017), and RPA presents an opportunity to enable the "outside-in" strategy where the core legacy systems can remain and the human labor is augmented or replaced by "virtual agents" (van der Aalst et al. 2018).

It is important to note that BPM can be interpreted dramatically different between companies; however, in general, it is referring to the strategy of a cross-functional attempt to increase productivity through systematic governance and management of processes. BPM does not look at isolated processes in a vacuum but ideally takes a more holistic strategic look from an organization level to highlight, reengineer, and ideally automate processes and services (Jeston 2014). There are plenty of companies where the BPM team has the bandwidth and internal power to be able to implement an RPA solution.

Outsourcing Versus Automation

Improving operational excellence to drive down internal costs takes a concerted effort and has typically fallen behind increasing revenue in the logistics industry. Attempting to lower costs through tightening operational expenses by looking to leverage technology to promote efficiency is nothing new and there is a lot of room for improvement left as many companies still have inefficient processes (Penttinen et al. 2018). Since the First Industrial Revolution, companies have been streamlining processes and segmenting work. However, the definition of work itself is changing as it did during that period of time when workers moved from the farm to the factory. The role and creation of a logistics service provider is an area where manufacturers and trading companies have been increasingly outsourcing their logistics department since the 1990s (Sohail and Sohal 2003). The obvious value that an automated solution would offer is that robots are not affected by the mental, physical, and environmental constraints of a human worker (Adams 2002). Robots are a sustainable solution if properly funded and implemented.

Funding is an area in which automation has been falling short in the past as many companies did not have the scale to outlay large capital expenditures and later recoup this investment. Things may be changing though as the cost to automate is consistently decreasing and technologies like RPA are becoming more available. A 2016 research report by Deloitte states that the cost of running a bot was most often cheaper than offshoring (Frank 2015). Slaby (2012) echoes this sentiment stating that RPA is a threat to BPO companies who have a singular value proposition of providing a lower cost labor force. If this statement holds up, there would be much higher utilization of RPA bots to perform tasks that are currently executed by numerous human labor in outsourcing locations. Defining processes and creating the proper documentation to hand over a task are very similar whether dealing with a human or a bot. Whoever will be doing the task in a new scenario, there is a need to create extensive contingency plans that requires a large amount of time to ensure that all scenarios have been carefully thought out. This is where tasks that have very little variability are preferred in outsourcing or automating as the number of scenarios would be less. Once again, the decision to step back from the daily activities of troubleshooting and temporarily putting stakeholder demands aside to dedicate resources to process mapping becomes a formidable challenge for logistics executives to address.

Center of Excellence as a Leader of RPA

As operations in developing countries like China, India, the Philippines, or Malaysia grew for many manufacturing companies and the governments of these countries started favorable IT investment incentives, logistics companies also began to set up outsourcing centers. These centers are heavily reliant on labor cost arbitrage as the key decreased input would be manpower. Considering that the abovementioned countries become more competitive in terms of available skills, while labor costs are increasing, automation and specifically RPA are opening up the argument that BPO itself could become redundant (Casale 2014). To do that, RPA must mitigate not only the cost component but also technical complexities that may not be apparently ready. Economies of scale are a consideration in evaluating RPA as well as BPO practices as the initial cost of developing or training a bot can be more than training a human to the same task. BPO providers who have the skillset of process mining, mapping, and relocating might be the perfect leaders of implementing RPA, especially for their existing clients (Hallikainen et al. 2018). When considering a project team to look at RPA, involving your BPO team may be a path to consider.

RPA: Hype or Realistic Solution

The Facts

As with many of the Fourth Industrial Revolution technologies mentioned in this book, RPA has become a trending topic in the world of supply chain and logistics, and there are many speculations about wide adoption in the following years. The Chartered Institute of Procurement and Supply predicted that by 2019, there would be 72% of all companies using RPA to reduce costs and transaction times and increase levels of productivity and compliance (Deckard 2018). However, just as with other discussed technologies both present and past like Blockchain and RFID, companies are finding it much more difficult to implement it than originally expected. Gartner reports that RPA tools sit at the "peak of inflated expectations" in its hype cycle (Kerremans 2018). Fersht and Snowdan (2017) report that the RPA software market and RPA services expanded by 42% from 2017 to 2018 and are predicted to grow by 94% by 2021. Growth in RPA is not necessarily a question of when but more a question of how fast we will see it come to fruition.

The hype surrounding RPA has been around for the past five years though, and it warrants looking at past predictions to see where we are now. Transparency Market Research (2015) claimed that the IT RPA market would reach $4.98 billion by 2020. Kenneth Research (2019) reported that by 2026, the RPA market would be $8.8 billion. However, Grand View Research (2018) shows that in 2018, the total RPA market size was around $600 million well short of the progress that had been anticipated. Although until this point RPA has been looking more like hype than reality, we believe that this technology is worth further investigation.

Rote, Repetitive Tasks Ripe for Automation

If you are still on board to continue examining a task that you would prefer not to do yourself or have your local staff to handle, there are still a few more steps to identify whether it is the right candidate for RPA. Traditional process automation in the BPM sense where systems are configured to interact with each other requires many of the same cases of a task to be done in a short period to justify the costly investment. RPA, on the other hand, offers a cheaper and quicker implementation to target tasks that do have repetition, but a small amount of variation spread out over a longer time but still have enough scale to consider automation (van der Aalst et al. 2018). Insurance and credit card companies have utilized process automation as they had a large pool of claims and payments that were often being handled in very similar ways.

Jesuthasan and Boudreau (2018) explain that there are three characteristics to categorize the components of jobs to identify if they are suitable for automation. The first characteristic of the task is looking at whether it is repetitive or variable. Repetitive tasks are those that are done the same way every time, such as credit application, which pulls a report using the same customer data on each application, as opposed to a variable task, like human resource consultants whose work will vary greatly for each customer. The second characteristic is whether a task is independent or interactive. Independent work can be done by one person without interacting with others like generating accounting statements as opposed to collaborative work that requires communication skills like would be needed by an employee working for a call center. The last characteristic of a task ripe for automation is whether it is physical or mental work. This distinction is somewhat obvious: physical work requires dexterity and strength, whereas mental work requires cognitive ability. The tasks that are best fit for RPA and other forms of automation are those that are repetitive, independent, and mental.

Fung (2014) defines his criteria for potential RPA tasks as follows: (i) having low cognitive needs in terms of subjectivity or interpretation needs, (ii) being large in volume, (iii) needing to move between different applications, (iv) having small amounts of variability and exceptions, and (v) a task that has demonstrated human entry errors that have caused issues in the past. Since most RPA bots lack the cognitive capabilities of AI and machine learning (ML) algorithms, it won't be able to handle tasks with a large amount of variance, and due to it being software, it won't be able to complete any physical work (Jesuthsan and Boudreau 2019). An RPA bot is programmed to perform actions on the computer in the same way that a human would by navigating interfaces through clicking and typing. The bot is not smart in the sense that it knows what information it needs to pull or what button it needs to click; it simply knows where the button and information are and when the interaction should happen. Since the bot functions using location data to navigate elements of the interface, any changes to the interface or the appearance of the page will cripple the bot's functionality.

Process Considerations of Implementing RPA

While the initial capital expenditure in having the bot created by one of the 20+ RPA companies is continuing to drop, the return on investment (ROI) for that investment is still unclear for small- to medium-sized companies who do not have the scale to take advantage of automating these rote and repetitive tasks. Companies need to carefully identify certain processes that have a rule-based structure and are draining a considerable amount of resources (Lowers et al. 2016). Variability in the types of processes that logistics companies could automate and the 12–15 players included in almost every international shipment make it very difficult to map out how many of the tasks need to be accomplished. System errors, changes in forms, variability in documentation types, etc. all create the need for RPA tools to have the ability to "learn" like a human would. However, the tools are not developed to that level yet (van der Aalst et al. 2018). According to an article published by Inbound Logistics, automation not only can help processes to run more smoothly but also enables companies to monitor processes over time in hopes of that continuous improvement goal that many large logistics companies so proudly advertise. What may not be abundantly clear is that the "monitoring" of tasks in terms of an RPA bot has a cost associated with it, as you have to have trained staff or outside consultants available to ensure the continuity of the process being done correctly. Not only is there a cost to monitor bots, but also the more significant investment is also in process mining, process mapping, and training the bot to do the work.

Motivating Example: Konica Minolta Using RPA

The Japanese tech giant Konica Minolta Inc. adopted RPA in 2018 through the RPA provider Automation Anywhere and has since seen massive returns through working hours saved. In 2018, the company saved 19 000 full-time employee (FTE) hours through 55 automation deployments with higher aims in the coming years. Achieving these kinds of returns can be tough though. An IT director at Konica Minolta explained that "of course, like every other company, we have faced many obstacles while going through the journey of RPA before becoming successful. For example, in Konica Minolta's Asia Pacific team, we [had to] create an RPA governance structure, training framework for the region, internal RPA application process, disaster recovery plan. [Afterwards, in the local team], such as Konica Minolta Hong Kong, [had] to go through a series of learning workshops." This sheds some light on the complexity that comes with rolling out RPA in a large multinational company without even scratching the surface of the planning and preparation that must be done.

To successfully roll out RPA, Konica Minolta had to increase its RPA training program to include nearly five times the number of employees that were in training when they decided to adopt throughout 2018. Some of those trained would become RPA champions in different branches who oversaw further training, identifying processes with potential for automation and creating an RPA culture within their offices. This was a massive shift for parts of their workforce going from knowing little about RPA to being leaders of its development within their company. In addition to the careful planning and testing that comes before RPA is rolled out, upskilling and gaining buy-in from workers represents another huge challenge. However, navigating these complexities can be seen to have large returns when done correctly. After their successful initial RPA rollout, Konica Minolta aimed to nearly double the number of hours saved in the following year, and they show no intention of slowing down. The IT director shared, "The RPA journey in Konica Minolta has provided an invaluable insight into how automation can change a company's culture and the ways its employees work. However, we will not stop here, and we have already planned RPA 2.0, which is allowing staff to interact with the RPA bot in real-time. This is for sure an exciting time for all of us." With returns like Konica Minolta has seen, it certainly shows that RPA can be rolled out if the investment in training is there.

RPA is best understood when compared to a Microsoft Excel macro. In an Excel macro, you can automate cell activities by using VBA coding and Excel functions. RPA adds an advantage in that it can automate tasks across many different applications with simple programming language and function calls. As shown in the Konica Minolta case, one of the first tasks needed is to identify the RPA provider. The company ran through a series of RPA software provider comparisons and finally selected the application from Automation Anywhere, a leading provider in this space.

Konica Minolta's RPA Roadmap

Konica Minolta implemented RPA using the following roadmap:

1) Train a country RPA developer.
2) Identify the highly repetitive business processes that can make use of RPA from hundreds of processes and prioritize for which to develop a solution first.
3) Form a business technology communication unit to perform change management education.
4) Convert existing workflow (normal human can read) to become RPA workflow (RPA developer can read).
5) Create metrics to measure ROI in terms of calculated full-time employee (FTE) labor hours.
6) Communicate results to management in global and regional offices.
7) Investigate and integrate RPA into each of the different business units.
8) Expand RPA locally by training more local RPA champions who can help to spread knowledge and increase the number of RPA processes.
9) Multiply the power of RPA by combining it with Excel macros.
10) Distribute messages and hold further training to create an RPA culture within offices.
11) Continue to invest in upskilling the team and get them certified.

Processes that Konica Minolta has built RPA bots for include updating shipping schedules and generating customer consumption reports. More examples can be found in the chart below that demonstrates how Konica Minolta deployed bots across various department functions (Figures 5.1 and 5.2).

In 2018, over 55 automatic bots were developed and deployed, with a similar number in the work pipeline. Konica Minolta could save around 19 000 hours in 2018 and expects to save 33 000 hours in the fiscal year 2019. The goal here was not to replace staff through RPA but to add more business value to their customers. The company's RPA journey gives invaluable insights to how automation can change a company's culture and the ways its employees work.

Though Konica Minolta is not a logistics company, this roadmap would look extremely similar if the implementation was happening in one. Identifying highly repetitive tasks, converting workflow to one that an RPA developer can read, forming a team to manage adoption and training, and creating metrics to quantify ROI would be standard steps in any company's RPA journey whether they provide logistics services or technology solutions. The positive results and hours saved that Konica Minolta have experienced since deploying their first bots therefore could be shared by logistics companies since many of the functions that the bots handle will be the same regardless of the industry that they serve. For example, Konica Minolta's use of bots for creating market data reports, generating quality assessment reports, and managing inventory could very easily be applied to a logistics provider as these functions will generally look quite similar across different organizations. In addition to this, Konica Minolta used RPA specifically to update their shipping schedules, a task that is done daily in many logistics companies. Later on in this section, we will look at an example of how DHL have seen equally impressive results to Konica Minolta's during their own RPA journey.

Deploying RPA in various operations throughout the value chain

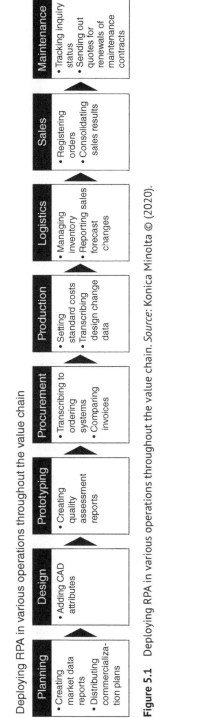

Figure 5.1 Deploying RPA in various operations throughout the value chain. *Source:* Konica Minolta © (2020).

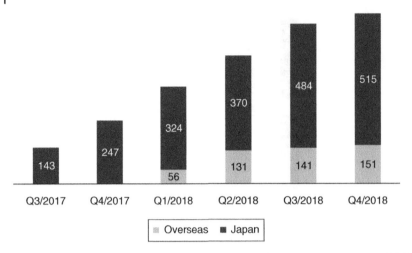

Figure 5.2 Internal RPA training course participants. *Source*: Konica Minolta © (2020).

Use Cases for RPA in Logistics

Track and Trace

Logistics service providers are seeing customer expectations increasingly rise in areas where they simply do not have the margins to throw extra staff at the solution. E-commerce and SaaS solutions are changing the visibility expectations of buyers. When you can price shop and buy a Boeing 747 online and track penny items turn by turn as they approach your house, the expectations in the business-to-business sector shift as well (Sheetz 2017). In international shipping, this has traditionally been a huge issue as carriers, forwarders, port authorities, rail operators, customs agencies, trucking companies, warehouses, and their customers often all have different levels of digital communication capabilities. In this scenario where a buyer of a container of goods in the United States wants to track his cargo from China to his warehouse door outside of Chicago, Illinois, they most often look to their freight forwarder as an aggregator of data and communication window. The freight forwarder typically does not directly own or control any of the carriers along the way, so they must find ways of grabbing event milestones to give updates to their customers. They have a few different options when it comes to retrieving such milestones. First, they could have direct integration with each of these carriers via EDI, although it requires a significant investment in mapping event fields from the carrier's system back to the forwarder's system. Second, they could have their staff locally go to each carrier website and retrieve the information as it is updated and key it back into their tracking database manually. Third, they could identify low-cost labor in an offshore location to update and key in the data. Fourth, they could pay a third-party aggregator for the milestones who often utilizes a combination of cheap labor and EDI connections. Lastly, they could train or hire someone to train an RPA bot that could go out and do the work of their local or offshore staff.

As previously mentioned, the key is to have a highly repeatable task. The task is mental and independent and does not have huge volumes happening at the same time if used by a small- to medium-sized company. Typically, the forwarder and carriers are not updating their system's interfaces frequently, so training the bot to navigate to different fields on a carrier's website and the forwarder's ERP system is feasible. Managers need to consider whether or not their systems will be changing soon in terms of the interface or the architecture as that will help to also decide whether to use RPA (Penttinen et al. 2018). If, for example, your company just rolled out the logistics platform CargoWise and there are no updates to your ERP coming soon, then it might make sense to start an RPA pilot. A difficulty here would be in the formatting of identifications used by different carriers and modes of shipping; however, this task is surmountable. The key is figuring out which trigger would initiate the bots tracking of events and dictating how often it would need to go out to find those events.

RPA Adoption at DHL

RPA adoption in logistics has shown many of the same positive results that have been seen in other industries. Notably, DHL, one of the largest global logistics providers, were early adopters of the technology and have seen massive returns through time saved by RPA bots. DHL Supply Chain in North America has offset over 5000 hours of manual work with RPA, and globally, there were over 30 000 hours saved per year during the pilot programs that they rolled out (Selko 2019). "Before RPA many basic, day-to-day tasks would need to wait to be completed until capacity allowed. Now they are done quickly and with a high level of accuracy, freeing our employees to spend less time doing busy work and more time doing the strategic value-added work they enjoy," said Jim Monkmeyer, president, transportation, DHL Supply Chain, North America (Selko 2019). DHL's RPA ambitions did not end with these savings though as they look to double the number of hours saved by the end of 2020. DHL's bots have been deployed primarily for assembling documents to create pay on delivery (POD) as well as scheduling deliveries and tracking milestones. In the future, they plan to use RPA combined with AI to automate more complicated tasks.

Navigating Your RPA Journey

Who Should Own This RPA Journey?

If you decided to go ahead and start the process of using RPA in your company, the first task would be to assign individuals to lead the project. A chief information officer (CIO) should have the knowledge and internal leverage to be able to coordinate between RPA developers and business teams. The CIO should be able to facilitate the cohabitation of bot and worker in terms of translating internal work rules into conditional statements that the bot could follow (Hallikainen et al. 2018). This is opposed to a director of operations or IT who could potentially only coordinate one side of the equation. In highly global organizations, there is also a question around the amount of leverage regional executives would have in influencing and capturing processes outside the scope of their entity. In this way,

RPA should be enacted in a controlled ecosystem at a local level or by a top-down approach from executives with similar skillsets to a CIO with the same amount of leverage to enact change. Regardless of whether an organization decides to build out its RPA program internally or have a consultant to support, the organization's IT department must be engaged and educated to facilitate negotiations and build the business case (Hallikainen et al. 2018). Having buy-in from a cross-functional leader and the support of IT will be key to the success of your project.

Process Mining and Process Mapping

Once you have decided who will lead your project, the next step is to start to understand which tasks could use an RPA bot to help create efficiency. To do this, leading RPA consultants recommend that you first do an audit of your processes through process mining and then map current processes. In some cases, there is a software that can sit on your computer that will mine the clicks and keyboard strokes of your staff to learn the most common behavior and map the steps for you. By observing the human handling complex cases, the RPA system can learn. There is also an obvious link with process mining (van der Aalst 2016; Kerremans 2018). For example, RPA vendor UiPath and process mining vendor Celonis collaborate to automatically visualize and select processes with the highest automation potential and subsequently build, test, and deploy RPA agents driven by the discovered process models.

UiPath and other RPA providers often offer process recorders that could speed up the time it takes to capture and map out processes manually by having the software record actions, such as mouse clicks and keyboard entries, of a human user (Frank 2015). RPA bots also can interact with multiple systems simultaneously and can even work with humans and indicate a need for further assistance (Hallikainen et al. 2018). Whether you decide to mine and map your processes before you choose your provider is probably best dictated after some preliminary budgets on how much you are willing to invest with your RPA provider.

Choosing the Right RPA Provider

With the people and processes now in place, you must pick which RPA provider to partner with. Blue Prism, a leading RPA provider that has been given credit for inventing the acronym RPA, was launched in 2001, and subsequently, there have been over 45 providers that have entered the market as of 2017 (Lacity and Willcocks 2018). As a buyer, this leaves a daunting task of trying to figure out which provider is right for you. Many of the providers and the products they offer that could be considered RPA are drastically different in terms of the upfront and monthly costs as well as the amount of human input needed (Hindle et al. 2018). Automation Anywhere, Blue Prism, and UiPath are widely regarded as the top three RPA providers and offer different levels of engagement depending on how much training you want to do yourself. Consultants often specialize in one of the above and are even willing to give you a free proof of concept. Other leading RPA providers include AutomationEdge, Cognizant, Conduent, Kofax, Kryon Systems, Pegasystems, and Softomotive.

Change Management Considerations

UiPath stresses the importance of establishing a center of excellence, a group of IT-savvy professionals who have the bandwidth to learn and champion this project, in any company that has decided to implement RPA. Within the center of excellence, they identify several roles to help support adoption, champion the setup and maintenance of bots, and analyze bot performance. One of these roles is the RPA change manager who should be tasked with overseeing the onboarding of stakeholders in using RPA technology. UiPath emphasizes that one of the most effective tools for these change managers is clear and open communication. By keeping stakeholders informed on the RPA adoption process and how they will be affected by it, the change manager will be much more successful in leading employees through the change. Managing this change is vital since any stakeholder that hears about bots being used to automate parts of their tasks will likely become fearful that they are being replaced. This fear would be rooted in their lack of knowledge of what the role of RPA is. This is where a change manager can step in and reassure the employee that the bot is not an AI solution meant to replace them, but more of a tool that can eliminate repetitive, rules-based tasks, thus allowing the employee to focus more on value-added process. If this point is not communicated well, employees will become fearful of the technology and unwilling to learn about or undergo training to use it, thus affecting the change manager's ability to manage the change. Clear and open communication to stakeholders is critical during the RPA journey to ensure stakeholders that the solution is not meant to replace them (UiPath, n.d.).

RPA Implementation Announcement

Whenever the prospect of automating tasks is brought into a business, concerns will arise among employees that they could potentially be laid off after being replaced by automation. Any company hoping to adopt RPA will likely have to deal with these concerns. How this is handled is extremely important since employee onboarding is a necessity for RPA to be successful. Opus Capital is a venture capital firm that manages over $1 billion in assets and was able to successfully implement RPA to handle the processing of new employee relationships and changes in employee payment details (Hallikainen et al. 2018). As news of RPA implementation moved through the payroll department, concerns grew among the employees. One employee stated, "Yes, I had these thoughts that . . . a robot is coming here to sit down there and do the typing, and then I would lose my job" (Hallikainen et al. 2018). The supervisors of the pilot dealt with these concerns by emphasizing to the employees that these bots were not meant to be replacements. Their functionality is limited to repetitive tasks, and there is still a great need in the organization for human workers who are better equipped to deal with cognitive tasks due to their adaptability which the bots lack. The supervisors framed RPA adoption as a way to free up employees from repetitive tasks. A supervisor at Opus Capital was quoted saying "[The robot] will free time for other type of work [by humans] that a robot could not do . . . it will [therefore] bring a positive change to everyone's workload" (Hallikainen et al. 2018). In addition to making this clear early in the process, it is also important to try to get employees involved in learning about the technology so they can be advocates themselves.

The abovementioned is a best-case scenario as there is always the possibility that some workers, whose jobs are entirely replaceable by the RPA bots, will no longer be needed. In a case study done by the KTH Institute of Technology, it was found that 13% of the clients of RPA providers laid off employees as a direct result of RPA implementation and another 13% offered retirement to some workers during the adoption process (Sandholm 2019). In addition to layoffs and retirement, 19% of these companies slowed their recruitment initiatives after adoption (Sandholm 2019). In an ideal world, RPA would not lead to any kind of reduction in full-time employees, which also is typically the case. However, it is undeniable that for certain positions, RPA will lead to layoffs and repurposing of staff. Employee concerns thus seem somewhat justified, but it must be kept in mind that if an RPA bot can replace them, their job could be anyways outsourced to a BPO center.

As a company goes through a digital transformation, assessment of staff is critical. Staff with the technical skill to work with RPA or staff with analytical, critical thinking, or creativity skills should be identified early in the process to avoid talent waste. It is advised that senior management and human resources align with RPA project leaders to identify the potential risks of the project and come up with the right messages for the workforce. Finding a balance of what information to share and when to share it could mean the difference in retaining key staff.

Liberated Knowledge Workers

Once RPA has been implemented and is running smoothly, employees will have been freed of some of their rote and repetitive work tasks. Employees will be able to concentrate their efforts on higher-value tasks that require cognitive and interpretive skills like analysis, innovation, and problem-solving (Slaby 2012). Usually, the rearrangement of employees' time takes the form of either redeployment within their unit or redeployment outside of their unit or sometimes a combination of the two (Lacity and Willcocks 2018). Having a plan in place of what workers will do with the time gained from implementing RPA is a key consideration for any company thinking about using automation. There may be some situations though where there isn't any work for the employee to fill their extra time with, or maybe their position can be entirely replaced by RPA. While employees could be laid off, it comes with repercussions about how layoffs in favor of automation impact your company's image as well as how internal stakeholders will react.

Interacting with RPA

Training the Bot

Once a task has been selected to be automated using RPA, the bot or bots must be trained to know how to do the task. A "bot" or "robot" in the context of RPA is the same as a unique software instance (Lacity and Willcocks 2016). RPA bots are not like AI and ML algorithms, i.e. they do not "learn" through iterations of large datasets in the same way that algorithms do. Instead, the bots must be given specific instructions to complete their task. Creating these instructions in a way that the bot can understand and follow is referred to as training, which is also what the same process is called in the field of AI. It is therefore important to remember that RPA training is not the same as AI training (Willcocks et al. 2015).

Various methods are used to configure a bot and making it carry out the intended process, such as screen recording, mapping out interactions on a GUI with process maps, or writing scripts (Willcocks et al. 2015). If programming instructions into the bot's code are required, technical knowledge is needed and will usually be done by an RPA provider with the necessary experience. On the other hand, recording screen activity and external inputs from a keyboard and mouse and then having the bot copy the actions in a way like recording a macro in Microsoft Excel are much simpler. It can be done by users with no technical coding knowledge, and many RPA providers have created easy tools for recording and replicating actions. The recording software of RPA provider UiPath even replicates your recorded actions into a flowchart that can be edited later using intuitive commands. In conclusion, depending on the skills and knowledge of the user and bot designer, training can be done with or without coding skills.

Next Evolution of RPA Training

Some RPA providers have started to bridge the gap between RPA and AI as they have designed ML algorithms to handle training. The software will observe a user performing a task over and over until it has enough data to "learn" how the task is done. However, the ML method is only employed to generate the set of instructions that the bot will follow. To achieve more widespread adoption, RPA and ML need to become "smarter." The promise is that with the use of ML techniques, more complex and less defined tasks can be supported. Companies like Microsoft or Apple are already recording your frequently used activities and promoting their products, Power Automate and Shortcuts, respectively, as workflow automation tools. For now, the bot learns instructions and then will simply follow them. It will not continue to learn and improve as it works, like an AI or ML algorithm would. This method makes training much easier since the ML software that handles training can run in the background while the task is done normally by a human worker. The longer the software watches the actions and the more variants of the execution of the tasks that the bot sees, the better it will learn to do the task. All of this happens without any human interaction aside from turning on the recording software. Depending on the technological capabilities of the RPA adopter as well as the RPA provider and the nature of the task being automated, different approaches to training will be used. It is important for companies to carefully review and select a training method as the bot will only be as effective as its training allows.

Conclusion

RPA technology is a key example of the democratization that the Fourth Industrial Revolution is bringing to non-IT professionals, which will be able to do more with their existing systems at a reasonable cost. The twenty-first century should have been showing much more of what Peter Drucker called knowledge work where nonroutine, mental work is driving the economic prosperity and value creation of the global business world

(Drucker 2012). Utilizing automation will be paramount to free up our knowledge work from menial, rote tasks and to help the logistics industry look to innovate in other areas. The use cases for RPA as representing lightweight IT are rising, as there are more IT proficient non-IT professionals that would be enabled to quickly make cheap, customized applications for automating tasks in an "innovation arena" (Bygstad 2016). As logistics companies around the globe are struggling to identify initial pilots into the digitalization space, we recommend that you take a hard look at whether RPA might be that first project.

Key Takeaways

- Workflow automation will threaten the offshoring of white-collar jobs as it becomes easier to use and more affordable.
- Internal communication regarding automation plans needs to be carefully considered and designed as the results of not informing staff about them can lead to unrest in the workplace. Training a bot can be as simple as pressing a button and having the software record the actions of a user at their computer.
- Process mining and mapping are a critical step in identifying tasks that are ripe for automation and process improvement.
- Consideration of using RPA should be driven by the number of tasks that you have that are repetitive, independent of human input, require no physical input, low subjectivity needs, and substantial in terms of the number of tasks being done by internal staff.

References

van der Aalst, W.M.P. (2014). Process Mining in the Large: A Tutorial. In: *Business Intelligence: Third European Summer School, eBISS 2013, Dagstuhl Castle, Germany, July 7-12, 2013, Tutorial Lectures* (ed. E. Zimányi). 33–76. Cham: Springer International Publishing. Doi: 10.1007/978-3-319-05461-2_2.

van der Aalst, W.M.P. (2016). *Process Mining: Data Science in Action*. Heidelberg: Springer.

van der Aalst, W.M.P. and van Hee, K.M. (2002). *Workflow Management: Models, Methods, and Systems*. Cambridge: MIT Press.

van der Aalst, W.M.P., Bichler, M., and Heinzl, A. (2018). Robotic process automation. *Business and Information Systems Engineering* 60 (4): 269–272. https://doi.org/10.1007/s12599-018-0542-4.

Adams, J.A. (2002). Critical Considerations for Human–Robot Interface Development. *AAAI Technical Report. FS-02-03*.

Aguirre, S. and Rodriguez, A. (2017). Automation of a business process using robotic process automation (RPA): a case study. In: *Applied Computer Sciences in Engineering. WEA 2017. Communications in Computer and Information Science*, vol. 742 (eds. J. Figueroa-García, E. López-Santana, J. Villa-Ramírez and R. Ferro-Escobar), 65–71. Cham: Springer.

Asatiani, A. and Penttinen, E. (2016). Turning robotic process automation into commercial success – case opuscapita. *Journal of Information Technology Teaching Cases* 6 (2): 67–74.

Barnett, G. (2015). Robotic process automation: adding to the process transformation toolkit. https://www.blueprism.com/wpapers/robotic-processautomation-adding-process-transformation-toolkit (accessed 20 September 2017).

Bygstad, B. (2016). Generative innovation: a comparison of lightweight and heavyweight IT. *Journal of Information Technology* 32 (2): 180–193.

Casale, F. (2014). *Bots in the Back Office, The Coming Wave of Digital Labor. KPMG*, November 2015. https://assets.kpmg/content/dam/kpmg/pdf/2016/02/bots-in-the-back-office.pdf.

Deckard, M. (2018). Benefits and impact of RPA for supply chain management. https://www.uipath.com/blog/rpa-the-glue-for-supply-chain-management (accessed 19 February 2020).

Drucker, P. (2012). *Management Challenges for the 21st Century*. Routledge.

Fersht, P. and Snowdan, J. (2017). The robotic process automation market will reach $443 million this year. https://www.hfsresearch.com/cognitive-computing/rpa-marketsize-hfs_061017/ (accessed 21 January 2020).

Frank, C. (2015). Introduction to robotic process automation. *Institute for Robotic Process and Automation*: 35.

Fung, H.P. (2014). Criteria, use cases and effects of Information Technology Process Automation (ITPA). *Advances in Robotic and Automation* 3(3): 1–11.

Hallikainen, P., Bekkhus, R., and Pan, S.L. (2018). RPA presents threats and opportunities for BPO providers 12. *The University of Sydney Business School*, 2018.

Harmon, P. (2010). *Business Process Change: A Guide for Business Managers and BPM and Six Sigma Professionals*. Elsevier.

Hindle, J., Lacity, M., Willcocks, L., and Khan, S. (2018). *Robotic Process Automation: Benchmarking the Client Experience*. Charleston, SC: Knowledge Capital Partners.

ter Hofstede, A.H.M., van der Aalst, W.M.P., Adams, M., and Russell, N. (2010). *Modern Business Process Automation: YAWL and its Support Environment*. Heidelberg: Springer.

Institute for Robotic Process Automation (2020). What is robotic process automation? https://irpaai.com/what-is-robotic-process-automation/ (accessed 2 March 2020).

Jeston, J. (2014). *Business Process Management*. Routledge.

Jesuthasan, R. and Boudreau, J. (2018). *Reinventing Jobs: A 4-Step Approach for Applying Automation to Work*. United States: Harvard Business Review Press.

Kerremans M (2018). Gartner Market Guide for Process Mining. *Report G00353970. Gartner*

Konica Minolta (2020). Integrated Report 2019. https://www.konicaminolta.com/shared/changeable/investors/include/ir_library/ar/ar2019/pdf/konica_minolta_ar2019_e_04.pdf (accessed 21 March 2020).

Lacity, M. and Willcocks, L.P. (2018). *Robotic Process and Cognitive Automation: The Next Phase*. Ashford: SB Publishing. ISBN: 9780995682016.

Lacity, M. and Willcocks, L.P. (2018). *Robotic Process and Cognitive Automation: The Next Phase*. SB Publishing.

Le Clair, C. (2017). *The Forrester Wave: Robotic Process Automation: The 12 Providers that Matter Most and How They Stack Up*. Cambridge: Forrester.

Lowers, P. and Cannata, F. (2016). *Automate this – The Business Leader's Guide to Robotic and Intelligent Automation*, 1–22. Deloitte LLP.

Mani, D., Barua, A., and Whinston, A. (2010). An empirical analysis of the impact of information capabilities design on business process outsourcing performance. *MIS Quarterly* 34 (1): 39. https://doi.org/10.2307/20721414.

Penttinen, E., Kasslin, H. and Asatiani, A. (2018). How to choose between robotic process automation and back-end system automation? Association for Information Systems AIS Electronic Library (AISeL), p. 15. https://aisel.aisnet.org/ecis2018_rp/66 (accessed 1 April 2020).

Prangnell, N. and Wright, D. (2015). The Robots Are Coming. *A Deloitte Insight Report.*

Robotic Process Automation (RPA) (2020). Market size, share & trends analysis report by type (software, service), by application (BFSI, retail), by organization, by service, by deployment, and segment forecasts, 2020–2027. https://www.grandviewresearch.com/industry-analysis/robotic-process-automation-rpa-market (accessed 21 April 2020).

Sandholm, I. (2019). Successfully utilising RPA. http://urn.kb.se/resolve?urn=urn:nbn:se:kth:diva-264128 (accessed 20 April 2020)

Selko, A.M. (2019). The digitization of transportation|material handling and logistics. *MH&L News.* https://www.mhlnews.com/technology-automation/article/22055914/the-digitization-of-transportation (accessed 25 April 2020).

Sheetz, M. (2017). Online special: two Boeing 747 jumbo jets sold on Alibaba auction site. *CNBC.* https://www.cnbc.com/2017/11/22/boeing-747-sold-in-online-auction-on-alibabas-taobao.html (accessed 18 April 2020).

Slaby, J.R. (2012). *Robotic Automation Emerges as a Threat to Traditional Low-Cost Outsourcing.* HfS Research Ltd.

Sohail, M.S. and Sohal, A.S. (2003). The use of third party logistics services: a Malaysian perspective. *Technovation* 23: 401–408.

Willcocks, L.P., Lacity, M., and Craig, A. (2015). The IT function and robotic process automation. *The London School of Economics and Political Science.*

6

Blockchain Will Animate Tomorrow's Integrated Global Logistics Systems

Nicholas Krapels

SKEMA Business School, East China Normal University, Shanghai, China

Introduction

With the globalization of supply chains across nearly every industry, international logistics were already struggling to accommodate required levels of throughput necessary for global growth. Then, the Sino–US trade war began, stretching recently established supply chains to the breaking point. Some logisticians have already called for the transformation of the very concept of a supply chain, which connotes linear step-by-step processes, into a "supply network," an abstraction that immediately brings to mind a more robust and adaptive system. While supply chains provide efficiency and cost-effectiveness at the expense of simplicity, supply networks would seem much more capable of creating a flexible and complexity-optimized organizational scheme better suited to the next phase of global logistics (Aelker et al. 2013; Mari et al. 2015).

The great enabler of this evolution in the industry structure of global logistics is likely to be blockchain technology. The minutiae of each and every interaction between counterparties will be recorded on to shared ledgers and analyzed for optimization potential. Whereas the supply chain industry has historically been monetized via opacity and information asymmetry, the leveraging of privileged and proprietary information is woefully inefficient. These bottlenecks explain why global blockchain spending is expected to grow at an annual rate of 76% until 2022 (International Data Corporation 2019). A subset of that broad industry, the still nascent blockchain supply chain market, is expected to outpace even that robust figure, growing at an annual rate of 80.2% until 2025. At that time, the spending on blockchain enhancements and tools for the supply chain industry alone is expected to be $10 billion per year (Big Market Research 2019). While isolating the blockchain employment opportunities available in the supply chain is difficult, a recent study by job listing site Indeed indicated that blockchain-related job postings have continued to surge despite the price volatility of bitcoin (Cavin 2019) (Figure 6.1).

The Digital Transformation of Logistics: Demystifying Impacts of the Fourth Industrial Revolution, First Edition. Edited by Mac Sullivan and Johannes Kern.

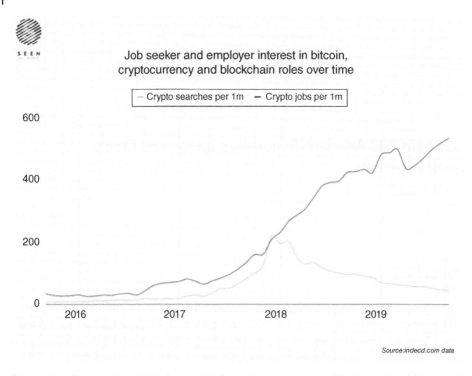

Job seeker and employer interest in bitcoin, cryptocurrency and blockchain roles over time

Crypto searches per 1m — Crypto jobs per 1m

Source:indecd.com data

Figure 6.1 The growth in blockchain-related employment opportunities is insulated from retail investor interest in the price movements of blockchain asset markets for bitcoin and other cryptocurrencies (Cavin 2019). *Source*: Indeed, Inc.

Clearly, blockchain as a means to transform the global supply chain is a trend that will increase in importance in the next few years. Supply chain and logistics firms will, at the very least, need to consider onboarding software engineers that can assist with the integration of blockchain tooling into existing enterprise resource planning (ERP) systems like SAP, front-end, and user interface (UI) engineers to make the tooling user-friendly, and blockchain consultants to ease with the transition of this digital transformation. Once basic blockchain technology has more or less been embedded into a firm's management systems, firms may want to consider hiring smart contract developers to develop on-chain solutions that fully leverage this new blockchain-enhanced infrastructure. As we will see in our discussion of the TradeLens platform, many of the heavily manual processes – negotiating container rates, checking bills of lading, navigating customs, settling up fees, etc. – that currently dominate the global logistics industry can be automated away within a robust blockchain-based platform ecosystem tailormade for the supply chain.

This chapter anticipates this transformative use case for blockchain technology and begins with a brief description of the origins of this twenty-first century innovation, the essential properties of blockchain, and an overview of contemporary blockchain applications for the global logistics industry. This background is essential to understanding the role of blockchain in the emerging complex adaptive systems that will define the efficient supply networks of the future.

The Origins of Blockchain Technology

Blockchain can be portrayed as a black box, a complex technology reserved for young people, IT professionals, and libertarians. However, a simple explanation of the technology is to call it a vastly more secure type of database. Looking at previous iterations of digital technology that have underpinned radical alterations to society provides a better perspective on the capacity of blockchain to transform the global logistics industry.

In 1969, the Department of Defense sponsored a project, ARPANET, which started the internet. Around the same time, a group of bankers wary of processing an increasing avalanche of paper checks in the United States created a trade association to study the feasibility of electronic payments. The Special Committee on Paperless Entries (SCOPE) established the guidelines for a national network of electronic payment systems (Homrighausen 1973). By 1974, the need to harmonize multiple regional protocols led to the establishment of the National Automated Clearing House Association (NACHA) to efficiently process batches of digital payments for payroll, retirement accounts, dividends, insurance payments, and other relatively low-value nonurgent debits. It is not a stretch, therefore, to claim that the pursuit of secure spendable digital currency has enchanted government bureaucrats, economists, and financiers since the advent of the computer.

While the Automated Clearing House (ACH) Network gained prominence in the United States, the Society for Worldwide Interbank Financial Telecommunication (SWIFT) Network was developed in Europe as a Belgian-based cooperative to securely send messages, including payment orders, through a global system of connected financial institutions. Back then, technological development prodded the finance industry into action by offering a means to almost infinitely scale transaction processing. Today, the ACH electronic payment network processes over $40 trillion annually. The SWIFT Network in 2019 facilitated 8.45 billion FIN messages (SWIFT 2019).

After the introduction of these SWIFT and ACH networks, the electronification of financial markets blossomed. The Nasdaq Stock Market and Bloomberg terminals allowed money to be increasingly abstracted. Indeed, of the approximately $60 trillion worth of currency in the world today, less than one-tenth of it exists in the form of physical coins and paper notes.[1] Once most money was electronic, the main question became how do counterparties that do not know each other trust that the money they are constantly sending each other is, in fact, real. The 11 000-plus financial institutions that participate in the SWIFT network participate in a trusted network based on preexisting business relationships that they have between each other. If Arthur State Bank in the American state of South Carolina wants to accept credit from de Volksbank in the Netherlands, the two entities might not have a preexisting relationship. However, they do likely have common partners in their business network, even if they exist only as third- or fourth-degree connections. Through these shared business connections, the SWIFT network can route payments in a trusted, albeit usually slow, fashion. Individual consumers do not have access to this network. Banks exploit this

---

1 The coronavirus scare of 2020 will surely lead this number to decline even further, as paper notes and metal coins were identified as a prominent contributor to the spread of the virus throughout the world. The use of digital money applications soared after the first wave of lockdowns in the United States (Scola 2020).

dichotomy and earn fees from their customers based on their ability to access global payment networks.

The financial services industry rode this wave of innovative digital transformation and comparatively rapid enterprise connectivity to a twenty percent share of the American economy (Witko 2016). At the same time, the supply chain industry remained mired in the habit of filling out forms in triplicate and having to wait for customs officials to finish their extended lunch break. Although there are many relevant criticisms of these financial networks for their frequent errors, inefficiency, and centralization, obviously, the digitization of paperwork in the banking industry was a massive success.

In response to excess banking profitability, starting in the 1990s, software engineers, libertarians, and privacy advocates launched a series of nongovernment attempts to digitize currency and, by extension, liberalize the payments industry in an attempt to return it to the hands of individuals. DigiCash, E-Gold, and Liberty Reserve were some of the more prominent projects in this vein. However, they all shared the same flaw. They relied on a single intermediary to process payments. Although this structure solved the "lack of trust" issue by imbuing trust within a centralized entity, it provided a uniform attack surface that proved far too easy for financial regulators to exploit. For similar reasons that peer-to-peer music file-sharing platform Napster got shut down for hosting links to MP3s on their own servers, these early digital currency startups got prosecuted or forced to close their doors due to the centralized nature of one or another aspect of their business models. But Bitcoin, via the breakthrough technology of the blockchain, mimicked what Napster's successors like BitTorrent and Gnutella figured out. The storage of data, even the most critical files, can be completely distributed across many servers that voluntarily opt in to the network. That single realization prompted an explosion of innovation in the software world and heavily influenced the development of Bitcoin.

For a class of politically motivated software engineers known as cypherpunks, the 2007–2009 Great Recession reignited the urgency to create a decentralized digital currency. In 2008, an entity writing under the pseudonym of Satoshi Nakamoto circulated a white paper on an obscure cryptography mailing list announcing the first virtual currency without a central point of failure and hence no way to regulate it – bitcoin. Its underlying technology was developed to provide the ultimate in cybersecurity, which also conveniently provided censorship resistance and an intriguing new trait called trustlessness.

Blockchain was born.

An important distinction must be made between bitcoin, the currency; Bitcoin's blockchain, the distributed ledger that underpins the bitcoin currency; Bitcoin, the software that governs the network on which the bitcoin ledger exists; and blockchain, a form of technology that secures a distributed network, which may or may not also include a decentralized digital currency.

Since the claim is often made that blockchain will revolutionize this or that industry, it is useful to know the exact process that blockchain employs to secure data in a distributed database. This unique innovation provides the foundation for the digital transformation of almost any industry, including the global supply chain and logistics industry.

In databases that are not secured by blockchain technology, a centralized entity controls the data that gets written to it. They can even amend data after the fact without a trace of the alteration of that amended data. As a result, companies hardly have reason to trust the data provided to them by other companies. Even if they grant access to their databases to other companies, there is no way to protect potentially market-sensitive data. Blockchain solves these conflicts by enabling the trustless sharing of data as well as the cryptographic hashing of data so that unwanted eyes can access a shared database without having access to all the data in it.

While the next section fully describes the implications of trustlessness, this section will conclude with a description of the process of a single transaction being permanently recorded on to a blockchain. Once a transaction occurs, that transaction data is included in a pool of other recent transactions. Validator nodes on the network bundle those transactions into blocks, usually by prioritizing those transactions according to the ones with the highest transaction fees included with the transaction data. Once a validator has successfully solved a cryptographic puzzle, they are allowed to propose their block of transactions to be the next block appended to their distributed database, referred to in the original bitcoin whitepaper as a "chain of blocks." This cryptographic solution, referred to as "proof of work," allows all the network validators to achieve consensus on the current state of the system by knowing which validator will append the next block. As this process is repeated, the accepted blocks are confirmed as they are covered over layer by layer with other blocks, much like viscous amber slowly covers over ancient insects and eventually entombs them in the fossil record. Transactions that are added to the "block of chains" in this manner are, after a certain number of confirmations, tamperproof (Nakamoto 2008). Please refer to Figure 6.2 for a visual explication of this process.

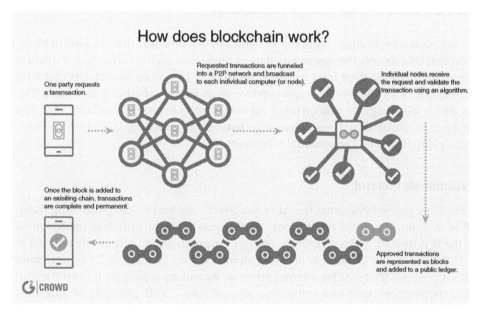

Figure 6.2 A blockchain is a public ledger of transactions maintained on a distributed peer-to-peer database wherein the network validators have agreed on a uniform cryptographic process that determines which blocks of transactions are added to the public ledger. As this process is repeated, more transaction blocks are appended to the ledger, irrevocably locking the transaction data onto the blockchain. *Source*: Graphic courtesy of Walker (2018). © G2.com, Inc.

Although this highly technical process may seem bland, as *The Economist* wrote, in one of the first widely disseminated articles about blockchain technology, back in 2015:

> The notion of shared public ledgers may not sound revolutionary or sexy. Neither did double-entry book-keeping or joint-stock companies. Yet, like them, the blockchain is an apparently mundane process that has the potential to transform how people and businesses co-operate.

That capacity for "mundane" revolution is the perfect backdrop for a discussion of the potential for blockchain to reshape global logistics. As the management of the supply chain relates to three basic flows – financing, information, and product – it is instructive to observe how financial and informational flows of an early digital network evolved to become the backbone of the global financial services system. Curiously, digital supply chain management systems took quite a bit of time to begin following a similar trajectory. While in theory the paperwork for cross-border transfers of intangible goods should not be that different from tangible products, in practice, much of the paperwork for international trade still relies on heavily manual processes. The reason why this overreliance remains embedded into the logistics industry is, of course, trust – or rather, the lack thereof between counterparties – the very aspect of business interaction that Satoshi Nakamoto intended blockchain to render unnecessary.

The Potential of Blockchain Technology

Blockchain has applications far beyond finance. In the same way that the Bitcoin blockchain indelibly records the transaction data of the bitcoin digital currency in a trustless manner, without the need for a trusted third party (TTP), other blockchains that have nothing to do with bitcoin can be spun up to support all sorts of other types of data. The process for recording that data on other networks occurs in a similar fashion. That data can be both financial and nonfinancial in nature. Programming languages can even interact with that data. The potential is enormous.

Revolution via Protocol

In the same year – 1974 – that the ACH and SWIFT networks were positioning themselves to dominate global remittances, the Internet protocol suite was implemented for the first time on ARPANET. These pairs of robust protocols, collectively referred to as TCP/IP, form the foundation of the modern Internet. TCP stands for transmission control protocol. IP stands for Internet protocol. By making it possible for two computers to communicate with each other via a network, the TCP/IP protocols provide most of the functionality needed to facilitate the World Wide Web within the base layers of the Internet. The application layer, where the Hypertext Transfer Protocol (HTTP) exists to provide the necessary infrastructure for websites, rests upon this base layer. Though Tim Berners–Lee did not introduce HTTP until 1991, seventeen years after ARPANET was spun up, the "phenomenally flexible and scalable networking strategy" (Cirani et al. 2019) of the TCP/IP protocols is widely credited for the wild success of the Internet today.

In computing, the Oxford English Dictionary defines a *protocol* as a "set of rules governing the exchange or transmission of data between devices." When two or more protocols work efficiently together, such as in the case of TCP/IP, or when a team or institution or firm has worked on a collection of similar protocols, they are referred to as a *protocol suite*. A complete set of network protocol layers that work together to provide complex networking and application capabilities is called a *protocol stack*. The primary conceptual model that guides computer engineering in this regard is the open systems interconnection (OSI) model, which delineates seven layers in a standard communication protocol stack. The TCP/IP model simplifies this abstraction to four layers.

Each protocol is meticulously constructed, typically through a series of working groups of computer scientists, to address one particular purpose. In the long run, this layered approach leads to the development of a robust open-source toolkit that facilitates the increasingly rapid construction of new protocols and different network architectures. By incorporating this type of modularization into a network design aesthetic, rapid iterations on new technological advancements are possible. Indeed, when a cohort of Internet experts convened to write "A Brief History of the Internet" in 2009, they agreed that "a key to the rapid growth of the Internet has been the free and open access to the basic documents, especially the specifications of the protocols" (Leiner et al. 2009).

One interesting aspect to note about the first three decades of the modern internet is that, for whatever reason, the large majority of the market value created accrued to the enterprises that innovated upon the application layer. The base layer protocols toiled away as profitless academic contributions. According to the "fat protocol" theory, some believe that blockchain will upend this order (Monegro 2016). At the very least, it should lead to a better distribution of value across the entire protocol stack. Indeed, blockchain provides the foundation for a whole new world of innovation yet to come.

Essential Properties of Blockchain

The first instances of the public internet, Web 1.0, were unidirectional. Users consumed content. As the Internet moved to mobile devices, this relationship became bidirectional. The sharing economy and social media, the crowning achievements of Web 2.0, are characterized by the interaction between people and platforms. Although *New York Times* technology reporter John Markoff coined the term Web 3.0 in 2006, its impact will be felt far into the next two decades of developing the digital economy. Even back then, tech mavens foresaw a "semantic Web" that functioned as "a layer of meaning on top of the existing Web that would make it less of a catalog and more of a guide – and even provide the foundation for systems that can reason in a human fashion" (Markoff 2006). Web 3.0's integrated and omnidirectional approach to people, machines, their respective data, and the connections between all these network participants is the next step in the evolution of the human relationship with technology. Society is still fumbling around trying to realize the full benefits of this gargantuan leap in technological ability. Blockchain provides the essential base layer of Web 3.0. Without it, all that the "semantic Web" envisions about the complex interaction of datasets, algorithms, application programming interfaces (API), data oracles,

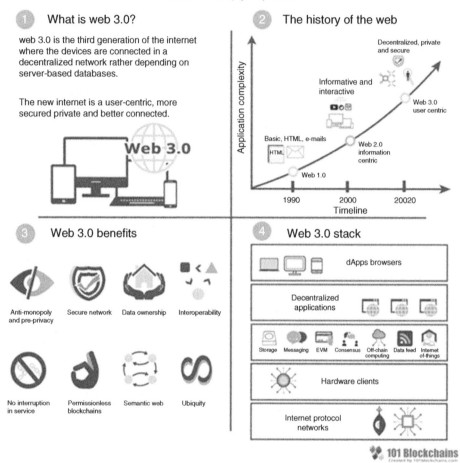

Figure 6.3 Web 3.0 is the next-generation decentralized "semantic Web" that will hopefully distribute value more equitably among all network participants, enable sophisticated automation, and offer robust privacy (Singh 2018). *Source*: Graphic courtesy of Aviv Lichtigstein of 101 Blockchains. Retrieved from: Singh (2018).

decentralized governance structures, artificial intelligence (AI), Internet of Things (IoT), and more are not possible. Chapter 3 discusses IoT applications in logistics (Figure 6.3).

Blockchains are often characterized as offering trustless transactions wherein two counterparties can enter into a binding agreement without having to rely on a TTP. To a certain extent, that is true, since one can be bound by code rather than law and as long as one uses a second-generation Turing complete[2] blockchain technology like Ethereum. The first generation of blockchain technology, Bitcoin and its many derivatives, does not offer this layer of functionality. It only offers a secure payment layer without an advanced computational

---

2 "Turing complete" is a term that describes programming languages that are computationally comprehensive and can thus simulate other real-world computations using data as inputs.

layer. However, many of the second and third generations of blockchain infrastructure systems use programming languages that offer the implementation of "smart contracts" on such a layer. These contracts are bits of code that are embedded in the blockchain and transform negotiated terms into systematic instructions or outline the performance measurements required to trigger a response in a binding agreement. The code automatically executes according to the terms set forth therein. Nick Szabo, who coined the term "smart contract" in 1996, defines them as "a set of promises, specified in digital form, including protocols within which the parties perform on these promises" (Szabo 1996). Chapter 12 will elaborate more on the applications of smart contracts and their application in logistics. Again, we are reminded of the importance of protocols and the time it takes to standardize them, in forging each successive layer of the digital economy.

Today, trustless transactions are more of an ideal than a reality. But the adoption of blockchain technology would change this situation. Companies that benefit from the lack of trust in the system dominate the supply chain industry. Their entire business model is based on their ability to function as one of the few trusted intermediaries in a global ecosystem that is incredibly difficult to wholly trust. These industry leaders provide value to their customers and are rewarded handsomely for the trust their reputation confers on the entire industry. Naturally, those companies are loathe to relinquish this portion of their competitive advantage. However, blockchain offers already trusted actors in the logistics industry an opportunity to enhance their leadership position on the issue of trust, or else risk being disrupted by a nimbler supply chain startup exclusively focused on blockchain applications. Once integrated fully into supply chain management systems, blockchain has the potential to undermine decades of industry leadership by removing the necessity for trust in transactions altogether.

A more practical characterization of how blockchain technology enhances trust between counterparties is to describe it as *trust minimizing*. A blockchain-enhanced database network, through a series of programming constructs and economic incentives, distributes the trust throughout the entire system rather than concentrating it in a single TTP, which is a vast improvement in information security (Szabo 2001). Because users can trust the system itself to behave as it was designed, the amount of trust any single user must place on any individual potential counterparty is thus minimized. For this reason, a blockchain is sometimes referred to as a "trust machine" (The Economist 2015).

Blockchain technology derives this feature of trustlessness from its four essential characteristics (Burniske and Tatar 2018). First, data stored in a blockchain structure is *distributed*. Having multiple touchpoints to access the data structure enables transparency and global trust. Redundancy across multiple machines also provides data security and loss protection. Second, all blockchain transactions are *cryptographic*. Computers on the network do not have to trust one another. They only need to verify the mathematical proof that underpins the cryptography encrypting the transaction data, which prevents system users without corresponding encryption keys from having access to the data. Third, the combination of a distributed database with transactions that must be cryptographically verified creates an *immutable* database. Whereas centralized data structures are prone to manipulation by a single malicious actor, blockchain data structures literally cannot be overwritten. This feature is provided by the nature of the technology itself wherein "blocks" of transactions are bundled together and "chained" to previous blocks. New blocks can only be appended to old blocks utilizing a nonce.[3] Over time, transaction data gets buried by layers of other transaction data. Recalling the previously used amber analogy, the ancient transaction data is forever preserved within the blockchain.

Fourth, blockchain-based data structures use a unique method, *securely verified by all*, to achieve consensus on the state of a system. The consensus mechanism for first- and second-generation blockchain systems like Bitcoin, Litecoin, and Ethereum is known as proof of work (PoW). Most of the newer third-generation systems are based on proof of stake (PoS), which better accommodates transaction speed and scalability. The difference between these two methods lies in where the infrastructure providers of system processing power invest their resources. PoW mandates that resources be invested in computing machines; PoS requires those resources to be invested in the native cryptocurrency token of that particular system. Multi-agent computer systems require a method to achieve consensus – to agree upon the current state of the distributed system. Otherwise, without some amount of at-risk investment dedicated to a particular blockchain, block producers would be incentivized to create their unique chain of blocks, which would render the entire system useless.

Blockchains are quite a contrast to the centralized datacenters that provide most of the infrastructure for the modern internet today, which serve as data honey pots that motivate headline hacks like those carried out against Yahoo, Equifax, and Marriot. Those hacks have compromised the private data of billions of users. Blockchain-based data structures that are distributed, cryptographic, immutable, and securely verified by all via a robust consensus mechanism, whether that be PoW or PoS, are exponentially more secure than the current infrastructure. As such, the supply chain industry should expect blockchain-based data structures to serve as the base layer for the industry-wide systems that will manage the complexity of interactions in Web 3.0.

Blockchain Applications in Logistics

After a review of the basics of blockchain technology, we can now dive into some of the different forms of that technology. When the anonymous creator of Bitcoin first implemented blockchain technology, it was as a distributed ledger accessible to everyone. A system that generates cryptocurrency with each new block of transactions to create economic incentives that secure the network, like Bitcoin and Ethereum, is known as a *public blockchain*. The coins on those public blockchains represent access tokens to the network. If all the users that transact within the system are trusted actors, then a *private blockchain* is an acceptable solution. This section will present public and private blockchain applications specific to the logistics industry.

Public Blockchain Applications in Logistics

Public blockchains are useful for business elements that face the consumer. Given the tainted milk scandal in China and *Escherichia coli* outbreaks in the United States, retail customers all over the world are concerned with supply chain provenance. Many

---

3 A nonce, or "a number only used once," is the number that bitcoin miners must find first before they can create a new block of transactions and thus be handsomely rewarded with newly minted cryptocurrency.

blockchain projects raised money during the cryptocurrency bubble of late 2017 and early 2018 by selling utility tokens to a voracious, highly speculative public. Since these capital raises, the projects have been trying to embed those utility tokens into a networked process that supports the value of the token, with varying degrees of success. In theory, *utility tokens* should have value, beyond their capacity to facilitate crowdfunding of course, because they offer access to the network on which they are stored. In practice, that value connection between protocol token and network protocol is incredibly difficult to achieve and requires skilled coders with experience working with blockchain technology to realize that goal.

As a result, all of the consumer-facing blockchain supply chain projects have done poorly in terms of investor return, even while their technology remains an interesting option for the logistics industry. Hungarian outfit TE-FOOD, operating its TFD token on its own public permissioned blockchain, focuses on farm-to-table food traceability. OriginTrail, which received Walmart's Food Safety Innovation Spark Award, uses its Ethereum-based TRAC token to enable trusted data exchange that fosters verifiable traceability. Modum, an innovative cold supply chain solution, used its Ethereum MOD token to raise $13 million but never could quite figure out how to also use the token to ensure the data integrity of its IoT temperature sensors. Ambrosus sold an AMB token with the ambitious vision of using it to power an end-to-end data integrity verification system to track, store, and transmit assembly-line and sensor data about food, pharmaceuticals, and chemicals. Tael raised money under the name of WABI, which its native utility token still uses as its ticker symbol, in order to cultivate an anti-counterfeiting system for critical e-commerce products in the Chinese market. Liqease, a supply chain finance solution that digitizes working capital assets, maintains a payment token on the Binance Chain called Traxia.

By far, the most successful public blockchain logistics project, at least in terms of marketing and global traction, is VeChain. Since the cryptocurrency bubble, VeChain's VET token has consistently remained a top-50 cryptoasset in terms of valuation. They have verifiably partnered with high-level groups such as PwC, Walmart China, and the French government on projects that track and trace the provenance of global food and beverage products. Through the use of QR codes to verify product authenticity, open APIs to ensure real-time data transmission, embedded NFC and RFID chips to expedite customs checks, environmental sensors to safeguard against spoilage, and secure identity devices that all interact with the VeChain public blockchain, VeChain technology could effectively scale the currently underutilized potential of the IoT industry (VeChain Whitepaper 2.0 2020). See Chapter 3 for more details on IoT and its application in the logistics industry.

Private Blockchain Applications in Logistics

In a private blockchain system, network participants do not need to be economically incentivized by the value of their cryptographic tokens in order to maximize the value of the network. Because these entities are not anonymous and presumably already participate precisely because they benefit from the network, crypto-economic incentivization structures are not as essential as they are in a fully trustless system. A private blockchain is a trusted system and can sometimes be referred to as an *enterprise blockchain, consortium blockchain,* or *permissioned blockchain* because, in such systems, there are barriers to becoming a block producer or even observer of the encrypted data.

Since most logistics interactions involve business-to-business (B2B) transactions, public blockchain projects are only a small part of the entire blockchain logistics landscape. For instance, although Tael tracks their loyalty points system on Ethereum, a public blockchain, they maintain their track-and-trace solution on a private blockchain built with Hyperledger Fabric. IBM is the main contributor to the Fabric codebase, which is built upon the Hyperledger project spearheaded by the Linux Foundation in December 2015. There are over 250 companies that have used the IBM Blockchain Platform to experiment with blockchain technical solutions in a familiar software-as-a-service (SaaS) environment to streamline the process for bringing Hyperledger Fabric tools to their existing network and application development.

The IBM Blockchain Platform is used in several successful applications. First, twelve European banks used the tooling to create the trade finance platform we trade, which aims to create an "ecosystem of trust for global trade" that an executive for a Finnish supplier in the fluid handling industry stated that it is so "surprisingly easy [that] I didn't need to know anything about blockchain" (IBM Blockchain 2019). Second, IBM Food Trust is a blockchain system arising from research initiated by Walmart and Tsinghua University to combat outbreaks due to tainted food supply. The system provides almost instant access to food supply chain information that is trusted, searchable, and tamperproof. The coalition, launched in August 2017, continues to add corporate members and SKUs because tracing the origin of food products back to their source now only takes 2.2 seconds (IBM Blockchain 2018). A third successful implementation of the IBM Blockchain Platform is a responsible sourcing blockchain network, which provides mine-to-market transparency for the automobile industry, created by RCS Global Group.

Enterprise blockchain developers interested in working with Hyperledger Fabric do not have to directly engage with IBM consultants. Although much of the code is contributed by IBM, Fabric is still an open-source project. This combination of code quality and flexibility attracted airspace supplier Honeywell to the codebase when they examined blockchain platforms on which to build GoDirect Trade, a blockchain-based marketplace founded in 2018 that services the $4 billion aircraft parts industry. General Manager Lisa Butters expects the innovative platform to capture a 25% market share by 2022 because the product is a trailblazer in a moribund industry (del Castillo 2020). In a recent interview with *Coin Telegraph*, she encapsulates the modern corporate perception of blockchain technology:

> If anyone argues about the fact that this is a permission-based network that is supposed to be decentralized then they are killing the dream of enterprise blockchain before it starts. There is no way you will get Fortune 500 companies participating in blockchain networks and sharing data if there are not permissions around that. You need some constraints for enterprises to operate in. (Haig, 2020)

By far, the most discussed Hyperledger Fabric project is an IBM–Maersk collaboration announced in March 2017. By December 2018, using Hyperledger Fabric, that joint initiative had quickly evolved into an actual working product with the grand vision to digitize global trade.

The result is an evolving open platform solution called TradeLens that provides a comprehensive environment for a myriad of global logistics service providers to interact including ocean carriers, freight forwarders, ports and terminals, customs and government authorities, inland transportation, and the manufacturers and retailers that are their ultimate clients. Unlike the niche token-based blockchain solutions available on public blockchains, TradeLens is a permissioned blockchain with a straightforward front-end UI designed to encourage participants to move from scanned copies of shipping documents to, eventually, create transformative structured datasets. In theory, all of the essential properties of blockchain mentioned in the previous section work together to provide a basis for the creation of seamless processes that occur "within blockchain using chain code to drive cross-organizational workflow." By including a freely available API model using standard representational state transfer (REST) APIs that make it easy to integrate existing ERP systems into TradeLens without any exceptional blockchain expertise, the basis for trustless interaction of critical counterparties is made simpler freeing data from the "organizational silos" it is currently trapped in (Erdly and White 2019). Glick and Glickenhouse explain more about the importance of integration and the challenges that it has previously presented in the logistics industry.

As network members are fully onboarded and begin sending their information into the platform every day, the dream of having a single consistent view of any one member's supply chain with real-time information accessibility comes closer. TradeLens is an interesting platform worth monitoring, but given its exhaustive approach to the digital transformation of an extremely fragmented global logistics industry, it will take some time before the benefits are realized, if at all (Erdly and White 2019). But the framework is there. TradeLens provides the "technical backbone" upon which the industry as a whole can "streamline and simplify the administrative hurdles of global trade" (Pedersen and Ritter 2020).

Another popular option for enterprise blockchain development is a solution provided by R3 known as Corda. Developed by a financial services firm, Corda excels at trade finance. HSBC has already experimented with the open-source software to efficiently process trade documents on a soybean shipment from Argentina to Malaysia, reducing hours and costs spent (Lucas 2018).

All of these permissioned blockchain projects seem to emphasize an ecosystem approach. While this strategy enables blockchain solutions to transcend mere track-and-trace functionality, it also makes it more difficult to drive adoption. A single use case application, perhaps driven by a digital token on a public blockchain, has the potential to more quickly attract resources and attention to the project. Through crypto-economic incentives embedded in a properly structured utility token, these public projects with a more limited scope may also better solve one of the issues that plague ecosystem approaches – how to ensure the quality of information that is contributed to the network. If ecosystem participants are required to stake something of economic value to the network in order to fully enjoy the benefits of the network, they are much less likely to contribute faulty data. After all, as can be discerned from the maxim "trash in, trash out," datasets are only as valuable as their weakest link. The consensus mechanism for these private blockchains does not require a native cryptocurrency, as network members are already pre-approved to validate network transactions. As such, there is no competition between miners or validators to produce blocks. Permissioned members patiently wait their turn to produce them. Although this structure removes one form of incentivization, building a modular ecosystem designed for basic software engineers is better for long-term adoption, provided that the ecosystem can stay relevant long enough to attract meaningful network usage. Slowly but surely, TradeLens is positioning itself as just that kind of platform ecosystem. With the addition of CMA, CGM, and MSC to the network in mid-2019, "the TradeLens consortium now encompasses almost half of the world's cargo container shipments" (Miller 2019).

Most recently, Ethereum think tank ConsenSys has announced an important partnership with EY and Microsoft, among others, to bridge these two blockchain environments. According to a 4 March 2020 press release, Baseline Protocol will "deliver secure and private business processes at low cost via the public Ethereum Mainnet" and "enable confidential and complex collaboration between enterprises without leaving any sensitive data on-chain" (ConsenSys 2020). By leveraging zero-knowledge proofs (ZKP) to share information proofs between untrusted parties without revealing the information itself, the Baseline Protocol uses the public blockchain as an always-on middleware to interoperate between legacy systems like SAP without any data leakage of critical information to competitors. Enabling this type of interoperability should increase adoption of blockchain technology as developers do not need to replace existing systems, but rather enhance them. This collaboration challenges the usefulness of the distinction between public and private blockchains.

In the long run, blockchain-based tooling for the logistics industry will need to involve all three system elements: public blockchains, private blockchains, and legacy CRM and ERP systems. User interfaces will have to be simple and intuitive to accommodate the wide variety of service providers that interact with the network. But make no mistake, blockchain integration will drive efficiencies into the outermost corners of the global supply chain.

Conclusion

From Trusted Actors to Trustless Networks

World trade has accompanied the rise of global prosperity. Yet, while other sectors of the economy inexorably proceed at the exponential growth rates spurred on by Moore's Law, the logistics industry is stubbornly resistant to technological change. It has always been this way. The London docks that were installed to accommodate the East India trade at the beginning of the Industrial Revolution were not meaningfully upgraded until 50 years later (Lovell 1969). The containerization of the global logistics industry took almost two decades and a protracted war before it was accepted as the norm (Levinson 2006). The global supply chain operates for so long at maximum capacity, but it is eventually forced to incorporate the advances of technology in order to meaningfully expand.

A century and a half ago, when a masthead peaked above the horizon near a port city, a slumbering harborside district stirred into action. Longshoremen rushed out of their industrial neighborhood homes to meet arriving ships and provide manual labor to unload freight, the pace of life in those working-class environs dictated by the intermittent arrivals of goods-laden boats. Smokestacks would later replace sailcloth, but the basic processes remained the same. Early twentieth century ports and their immediate surroundings all over the world became "associated with crime, noise, pollution, heavy work, and what John B. Jackson has called the stranger's path, with its attendant red-light district, cheap hotels, boarding houses, bars, and money changers" (Konvitz 1994). Unions evolved to organize a variable labor force. They injected trust into an otherwise chaotic employment situation.

When the twenty-foot equivalent unit (TEU) finally took hold in the logistics industry, those port cities went into rapid decline. The merchant docks embedded into downtown manufacturing hubs were replaced by container ports in the far suburbs, essentially huge parking lots for international logistics firms. Shanghai's Yangshan deep water port, which processes more than 10 million TEU of annual container volume, goes one step further. It is built on artificial islands in the middle of Hangzhou Bay two hours from the city center and completely removed from any other economic activity. Undoubtedly, the look and feel of international ports have dramatically changed over time, but the manually driven paper-work-intensive processes of those ports have yet to be updated to reflect those changes.

Technology does not lead to digital transformation overnight. The TCP/IP protocol stack took 25 years before it finally evolved into the World Wide Web. Bell's telephone remained a curiosity until there were sufficient households tapped into its system to jumpstart the monopolistic power of network effects. Such breakthroughs have always required seasoning. Indeed, American technology historian Nathan Rosenberg noted that "innovations in their early stages are usually exceedingly ill-adapted to the wide range of more specialized uses to which they are eventually put" (Rosenberg 1976). Blockchain technology, particularly as it relates to the global logistics industry, will require this same period of tweaking and measured adoption.

Certainly, blockchain will experience initial growing pains. But that does not indicate that it is incapable of solving some of the supply chain industry's most long-standing issues. When we put new technology into practice, "the economic benefits arise not from innovation itself, but from the entrepreneurs who eventually discover ways to put innovations to practical use – and most critically, as economists Erik Brynjolfsson and Lorin M. Hitt have pointed out, from the organizational changes through which businesses reshape themselves to take advantage of the new technology" (Levinson 2006). Just as the complexity of the global supply chain forced the conversion of the bustling, chaotic ports of the nineteenth century into the quiet, methodical bayside factories of the twenty-first century, the logistics industry will necessarily evolve due to the incorporation of blockchain technology into its management systems.

At its core, this evolution will be based upon advancing the concept of trust for all stakeholders in the global supply chain industry. With blockchain, it will no longer be a requirement of doing business to place that trust within select people and institutions. Having to pay handsomely for a parsimonious dollop of trust to sanctify a transaction will no longer be necessary. Rather, the inherent trustlessness of the systems and technology governing the complex global supply chain will be enough to proceed with trade. In this upgraded flow state of blockchain-based logistics interaction, the global supply chain will function more like a network and less like a chain. Scarce resources will be automatically diverted by the ecosystem to their most efficient use. Moreover, by reorienting this supply network along the lines of a complex adaptive system rather than one based on personal relationships and long-established practices developed under a completely different technology regime, macrolevel behaviors should both develop from and influence the microlevel interactions of the elements of this emerging system (Levin 2002). This duality, of practice continuously informing design and trustless system design simultaneously shaping best practices in real time, is characteristic of a self-optimizing system. Once blockchain unleashes that ability for the entire global supply chain to nimbly correct itself, efficiency gains will multiply, and accumulated waste in logistics processes will be flushed out.

For decades, the supply chain industry has resisted such an approach due to the disruption it would have on vested interests. However, with an ecosystem approach underpinned by blockchain technology, similar to the strategy that IBM and Maersk have taken with TradeLens, the entire global logistics industry has the opportunity to get acclimated to blockchain in a measured step-by-step manner that limits industry disruption while, at the same time, provides for efficiency expansion. In such an environment, change will not come quickly. But it will assuredly come. Get ready.

Key Takeaways

- Blockchain is beginning to make its debut in the supply chain world and will not be leaving any time soon. The blockchain supply chain market is expected to grow at an annual rate of 80.2% until 2025, when the market is expected to reach a value of $10 billion.
- The logistics industry is overly reliant on manual processes for document and information sharing. This overreliance remains due to lack of trust between the parties involved in international logistics. Blockchain is the solution to the problem of trust.
- Blockchains can offer trustless transactions wherein two counterparties can enter into a binding agreement without having to rely on a trusted third party (TTP) inasmuch as one can be bound by code rather than law.
- There have been many successful projects seeking to use blockchain in logistics both for the purpose of recording track and trace data, sharing documents and information, and executing smart contracts. Some notable projects to look out for are as follows: TradeLens, created through a collaboration between IBM and Maersk, and VeChain, partnered with a variety of big firms including PwC, Walmart China, and the French government.

References

Aelker, J., Bauernhansl, T., and Ehm, H. (2013). Managing complexity in supply chains: a discussion of current approaches on the example of the semiconductor industry. *Proc. CIRP* 7: 79–84. https://doi.org/10.1016/j.procir.2013.05.014.

Big Market Research (2019). Blockchain supply chain market to huge CAGR (80.2%) by 2025: big market research. *MarketWatch*. https://www.marketwatch.com/press-release/blockchain-supply-chain-market-to-huge-cagr-802-by-2025-big-market-research-2019-07-04 (accessed 24 March 2020).

Burniske, C. and Tatar, J. (2018). *Cryptoassets: The Innovative Investor's Guide to Bitcoin and Beyond*. New York: McGraw Hill Education.

del Castillo, M. (2020). Honeywell is now tracking $1 billion in Boeing parts on a blockchain. *Forbes*. https://www.forbes.com/sites/michaeldelcastillo/2020/03/07/honeywell-is-now-tracking-1-billion-in-boeing-parts-on-a-blockchain/ (accessed 17 March 2020).

Cavin, A. (2019). What the bitcoin job market looks like in 2019 (and beyond). *Seen Indeed*. https://www.beseen.com/blog/talent/bitcoin-job-market-2019-beyond/ (accessed 24 March 2020).

Cirani, S., Ferrari, G., Picone, M., and Veltri, L. (2019). *Internet of Things: Architectures, Protocols and Standards*, 1e. Hoboken, NJ: Wiley.

ConsenSys (2020). EY and ConsenSys announce formation of baseline protocol initiative to make ethereum mainnet safe and effective for enterprises. *ConsenSys*. https://consensys.net/blog/press-release/ey-and-consensys-announce-formation-of-baseline-protocol-initiative-to-make-ethereum-mainnet-safe-and-effective-for-enterprises/ (accessed 17 March 2020).

Erdly, M. and White, M. (2019). *Shipping in the Age of Blockchain*. https://www.youtube.com/watch?v=Xwqo_fwPEJo.

Haig, S. (2020). Boeing uses blockchain to track and sell $1 billion in aerospace parts. *Cointelegraph* (accessed 3 April 2020).

Homrighausen, P.E. (1973). One large step toward less-check: the California automated clearing house system. *Bus. Lawyer* 28: 1143–1160.

IBM Blockchain (2019). IBM blockchain: transforming trade finance – and trade (accessed 1 December 2019).

IBM Blockchain (2018). Join the power of IBM food trust.

International Data Corporation (2019). Worldwide blockchain spending forecast to reach $2.9 billion in 2019, according to new IDC spending guide. *IDC Prem. Glob. Mark. Intell. Co.* https://www.idc.com/getdoc.jsp?containerId=prUS44898819 (accessed 24 March 2020).

Konvitz, J.W. (1994). The crises of Atlantic Port cities, 1880 to 1920. *Comp. Stud. Soc. Hist.* 36: 293–318.

Leiner, B.M., Cerf, V.G., Clark, D.D. et al. (2009). A brief history of the internet. *ACM SIGCOMM Comp. Commun. Rev.* 39: 10.

Levin, S.A. (2002). Complex adaptive systems: exploring the known, the unknown and the unknowable. *Bull. Am. Math. Soc.* 40: 3–20. https://doi.org/10.1090/S0273-0979-02-00965-5.

Levinson, M. (2006). *The Box: How the Shipping Container Made the World Smaller and the World Economy Bigger*. Princeton, Oxford: Princeton UP.

Lovell, J. (1969). *Stevedores and Dockers*. UK, London: Palgrave Macmillan. https://doi.org/10.1007/978-1-349-00096-8.

Lucas, L. (2018). From farm to plate, blockchain dishes up simple food tracking. https://www.ft.com/content/225d32bc-4dfa-11e8-97e4-13afc22d86d4 (accessed 17 March 2020).

Mari, S.I., Lee, Y.H., and Memon, M.S. (2015). Complex network theory-based approach for designing resilient supply chain networks. *Int. J. Logistics Syst. Manag.* 21: 365. https://doi.org/10.1504/IJLSM.2015.069733.

Markoff, J. (2006). Entrepreneurs see a web guided by common sense. *New York Times* (accessed 1 January 2020).

Miller, R. (2019). IBM-Maersk blockchain shipping consortium expands to include other major shipping companies. *TechCrunch* (accessed 2 March 2020).

Monegro, J. (2016). *Fat protocols|union square ventures*. https://www.usv.com/writing/2016/08/fat-protocols/ (accessed 3 March 2020).

Nakamoto, S., 2008. Bitcoin: a peer-to-peer electronic cash system, 9. https://bitcoin.org/bitcoin.pdf.

Pedersen, C.L. and Ritter, T. (2020). Use this framework to predict the success of your big data project. *Harv. Bus. Rev.* https://research.cbs.dk/en/publications/use-this-framework-to-predict-the-success-of-your-big-data-projec.

Rosenberg, N. (1976). On technological expectations. *Econ. J.* 86: 523. https://doi.org/10.2307/2230797.

Scola, N. (2020). Is the coronavirus killing off cash? *Politico*. https://www.politico.com/news/magazine/2020/04/17/coronavirus-cash-economy-cashless-paper-money-business-190405 (accessed 2 May 2020).

Singh, N. (2018). 35+ web 3.0 examples of how blockchain is changing the web. *101 Blockchains*. https://101blockchains.com/web-3-0-examples/ (accessed 3 May 2020).

SWIFT (2019). FIN traffic & figures. https://www.swift.com/about-us/swift-fin-traffic-figures/monthly-figures?tl=en#topic-tabs-menu (accessed 6 March 2020).

Szabo, N. (2001). Trusted third parties are security holes. https://nakamotoinstitute.org/trusted-third-parties/ (accessed 8 March 2020).

Szabo, N. (1996). Smart contracts: building blocks for digital markets. *Extropy J. Transhumanist Thought* 16.

The Economist (2015). *The promise of the blockchain: the trust machine*. The Economist (accessed 1 June 2020).

VeChain whitepaper 2.0 (2020). https://www.vechain.org/whitepaper/ (accessed 9 March 2020).

Walker, A. (2018). Blockchain security: trends, predictions, and the future. https://learn.g2.com/trends/blockchain-security (accessed 3 May 2020).

Witko, C. (2016). How wall street became a big chunk of the U.S. economy — and when the democrats signed on. *Wash. Post* (accessed 2 July 2020).

7

Digitalization Solutions in the Competitive CEP Industry – Experiences from a Global Player in China

Scott Wang[1] and Johannes Kern[2]

[1]UPS, Louisville, Shanghai, China
[2]Tongji University, Shanghai, China

Introduction

"Digitalization" is one of the most popular words for business leaders working in small- and medium-sized companies up to large multinationals. Traditional businesses are being disrupted by digitalization through innovation, productivity enhancements, and new ways of how problems get addressed and solutions developed. This is also true for logistics, which is increasingly recognized as important differentiation factor and value contributor across industries (Christopher 1993). Digitalization has profound impacts on the courier, express, and parcel (CEP) delivery services industry. With the rise of e-commerce, the number of delivered parcels has more than doubled in the past ten years. Already 10% of retail sales worldwide are made online, leading to a total volume of 1.2 trillion EUR (International Post Corporation 2018). A particular driver for growth is China, where 31.28 billion items were delivered in 2016 (Jiao 2018). To handle the pressure on the network of CEP providers that arises from these increasing parcel volumes, combined with gradually higher more sophisticated customer preferences, digitalization and automation are considered key (Pfohl et al. 2019).

Challenges in Key Logistics Segments

Logistics spent, typically measured as logistics costs share of a company's overall revenue, varies greatly across various industry segments (Handfield et al. 2013). Generally, it is the highest in the industrial manufacturing and automotive segment at average costs of 13.1%, while best-in-class companies only spend 6.6%. The average logistics spent of the high-tech, retail, and the pharmaceuticals and healthcare segment ranges from 8.3 to 11%, while best-in-class

The Digital Transformation of Logistics: Demystifying Impacts of the Fourth Industrial Revolution,
First Edition. Edited by Mac Sullivan and Johannes Kern.
© 2021 by The Institute of Electrical and Electronics Engineers, Inc.
Published 2021 by John Wiley & Sons, Inc.

companies require just 3.7–5.7%[1]. Digitalization is playing an increasingly important role in driving operational efficiency across different segments, supporting companies to reduce their logistics costs.

The *high-tech* industry is constantly evolving with considerable profits and revenues for those that can navigate swiftly changing demands. Speed to market, end-to-end visibility, lower costs, and increased customization are critical for the high-tech industry. Especially, visibility is important, considering that the goods' value is typically rather high (UPS 2017). We see changes in the *industrial manufacturing and automotive* business from time to time, with buyer demands aligning closer to those in the retail industry. Distributors need to deliver a new level of customer experience not only to win new market shares but also to defend their position. A UPS study showed that 21% of buyers are extremely likely to shift business to suppliers who offer post-sales support, 80% of buyers are likely to switch for a more user-friendly website, and 69% of millennial buyers will shift business to a supplier with a mobile app (UPS 2017). Overall, especially, the millennial age group expects services that are historically not included in industrial sales (UPS 2017). e-Commerce is disrupting traditional *retail* businesses, particularly in China. Meanwhile, shoppers in Asia are least satisfied with their overall shopping experience, both online and in-store, compared with customers in other markets. A study on retail customers shows that 57% of online shoppers indicate in-store experience, mobile application, flexible delivery options, and consistent delivery experience as top four areas to be improved in the current online retail business (UPS 2017). *Pharmaceuticals and healthcare* mainly comprises pharmaceutical manufacturers, biotech/biopharma, and specialty pharmaceutical manufacturers. Here, distribution is an important part of the industry, while there are varying shipping characteristics of each of the segments. In this industry, proof of delivery must be archived for seven years to comply with government regulations and healthcare insurance providers' requirements. The time criticality of the sensitive, high-value products requires fast delivery, often on the same day, enhanced visibility, and highly flexible delivery (UPS 2017).

Digitalization Solutions Today

To tackle these challenges, already today large CEP companies like UPS rely on a variety of digitalization solutions that support managing their operations and help their customers to manage shipping processes (UPS 2019). These include advanced shipping tools, configurable shipping solutions, visibility and reporting solutions, and returns solutions.

Integrating shipping functionality into an order system allows shippers to manage carrier compliant labels and handle customs documentation requirements across multiple carriers and transportation modes. For instance, UPS relies on an *advanced shipping tool* that was developed by its wholly owned technology subsidiary ConnectShip. It extends multi-carrier shipping software to third party transportation solution providers and large-volume shippers. Over 20% of the world's largest companies already rely on this technology to control shipments in their supply chain (UPS 2019). Customers are looking for *configurable shipping solutions* that address their unique needs, from solving international language

---

1 Figures are based on a consultancy's internal benchmark data.

barriers to balancing network loads. As traditional shipping solutions do not provide the required flexibility, companies like UPS are partnering up with software vendors and develop highly configurable, customized solutions. *Visibility and reporting solutions* offer real-time visibility in the form of instantaneous tracking of any shipment right down to its SKU[2] level and constant updates about every handoff along the transportation way (Maguire 2018). This becomes increasingly essential for many industries. For example, in the United States, UPS already offers "Follow My Delivery" to its My Choice members. With this feature, customers can follow incoming deliveries as they make their journey in real time on a virtual map (UPS 2016). Especially, firms in the high-tech industry require logistics solutions that cover not only the shipping but also the return service. Logistics service providers therefore need to offer a wide array of return options and flexible *returns solutions* technologies that create an optimal return experience and meets customers' needs. For instance, AfterShip, a UPS-ready multi-carrier shipping management system, provides the ability to generate labels, calculate shipping rates, track shipments, and manage returns (UPS 2019).

Digitalization is transforming logistics operations in various ways. Large logistics providers have been seeking digitalization solutions to enhance their competitiveness, while startups are disrupting the industry with advanced technologies and business models. In especially six areas, digitalization not only is challenging traditional sources of value but also helps to overall upgrade the CEP industry (Hausmann and Ludwig 2018). Different types of *value-added services* are introduced to the market such as online tracking, e-invoicing, customs clearance, or alternative deliveries that provide real-time tracking, transparent billing, quicker customs clearance, and flexibility to the customers. By leveraging technologies, value-added services improve the customer experience. Traditional freight companies track shipments manually as they rely on partnerships with multiple carriers and truck vendors. A lot of effort is spent on manually tracking, tracing, and reporting. Furthermore, whenever exceptions occur, it becomes a challenge for a freight forwarder to precisely predict when freight will be delivered. Digitalization can support here to improve real-time data interaction by capturing and processing information more efficiently. Just as technology disrupted the travel market, online platforms can help freight forwarders to achieve larger business scale by *bundling of volumes* and to appear as a single customer rather than multiple small- and medium-sized customers. For instance, with "Fulfillment by Amazon" (FBA), customers can enjoy better freight rates in exchange for accepting Amazon's conditions. Amazon then collects freight from multiple suppliers, which leads to a high purchasing power toward carriers through economies of scale (Amazon 2019). Another area is *real-time quotation*. Traditionally, freight forwarders collect rate, routing options, and capacity from carriers and truck vendors and then summarize it for their customers. We perceive that the lead time for a quotation and the transparency of information remains a concern for a lot of customers. Startups such as Freightos or Xeneta are connecting with carriers digitally to be able to quickly provide information to their customers with less manual processes involved. Also, utilizing data as a business resource and employing *data analytics* will allow logistics companies to better predict coming volume trends. In turn, this will lower their cost through better purchase decisions,

---

2 Stock keeping unit, a specific item for sale, e.g., a product or service, and the attributes associated with it, e.g., description, manufacturer, size.

resource allocation, and volume consolidation. Finally, digitalization helps to *streamline processes*. No matter if it is a process between different departments of a company, between forwarders and carriers, or between forwarders and customers, digitalization enables more efficient interaction and quicker response. Open application programming interfaces (APIs), which specify how software components should interact, enable different companies to interface their systems together in a fast and easy way.

E-Commerce in China and Its CEP Market

Digitalization is a particularly critical topic for the Chinese CEP market, which experienced growth rates of some 50% annually in the past years mainly due to domestic and cross-border e-commerce (Jiao 2018). Cross-border e-commerce is "the process of selling goods to a consumer located in a Foreign Country by means of online channels, either directly through a proprietary website (i.e. B2C) or through a digital intermediary such as online retailers or marketplaces (i.e. B2B2C) (Giuffrida et al., 2017)" (cf. also Chapter 25). With over 50% of global e-commerce transactions coming from China, China is the world's largest e-commerce market, overtaking that of the United States, the United Kingdom, Japan, Germany, and France combined. In 2018, some 560 million digital buyers made online retail transactions that reached a volume of 1.33 trillion USD (US International Trade Administration 2019). Chinese online shoppers are also increasingly inclined to foreign products. The Nielsen online shopper trend report revealed that 67% of consumers recently made a cross-border e-commerce purchase in 2017, up from only 34% in 2015 (Bali 2018).

However, one of the most critical barriers for e-commerce to fully take off is logistics (Giuffrida et al. 2017). Even though the industry is growing faster and bigger in China, it is still facing various operational challenges such as high costs, low efficiency, lack of integrated intermodal transport networks, and low usage of advanced technologies (Zhang and Figliozzi 2010). For instance, logistics costs as a share of GDP in China were still around 16% in 2016, a significant gap to the ca. 8% in developed countries (Jiang 2018). This is often caused by inefficient management and fulfillment of logistics companies (Rahman et al. 2019). As logistics is an information-rich industry where theoretically data can be collected at every step of the cargo movement, advanced technology can play an important role in improving its efficiency.

Recognized as an important support factor for the national economy, improving logistics efficiency and reducing logistics costs have become a target of China's government. Therefore, in 2016, the executive meeting of the State Council[3] together with the National Development and Reform Commission and relevant departments deployed the "Internet + efficient logistics implementation advice" to the Ministry of Transportation, Ministry of Commerce, and Ministry of Industry and Information Technology to create a better policy environment for smart logistics in China (The Development of Intelligent Logistics Industry 2017). Smart logistics started to transform the country's industry. So far,

---

3 The State Council of the People's Republic of China (中华人民共和国国务院) is the highest organ of the State administration and an executive branch of the central government.

already more than 400 million trucks, pallets, containers, and warehouses are equipped with GPS tracking devices and other sensors for enhanced data collection. Analyzing the gathered data increases operation accuracy and helps to forecast demand during peak season such as China's largest shopping festival "Double Eleven" (Maguire 2018).

While steps in the right direction were undertaken and some digital pioneers such as Alibaba, JD.com, and Baidu can show impressive use cases, the majority of small- and medium-sized logistics companies do not possess sufficient resources to invest in the digital transformation. While their modus operandi might have been sufficient in the past, it will not be adequate for the new competitive environment.

A New Competitive Environment

Various developments and megatrends are affecting CEP providers and increasing the pressure for all market players. An *increasing number of customers* are participating in global markets. The world's population will reach some 9 billion people by 2050, mainly due to growth from emerging markets. The growing global middle class, which will be able to access the Internet and presumably also enjoy online shopping, is going to require CEP providers to deliver to remote sites across the globe (World Economic Forum 2016). *Urbanization* rates are rising, and more *megacities* are developed. In 2018, 55% of the world's population lived in urban areas. This is up from 30% in 1950 and expected to reach 68% by 2050. Especially in low- and lower-middle-income countries, the most rapid urbanization is expected between now and 2050, from 32 to 50% and from 41 to 59%, respectively. Around half of the urban dwellers reside in settlements of less than 0.5 million inhabitants, but over 13% live in one of the 33 megacities with more than 10 million inhabitants (United Nations 2019). This will provide a tough challenge for the CEP industry that has to ensure swift delivery in such gridlocked metropoles. In addition, *international relations* are changing. Several events impacted global supply chains in the past years. Airport lockdowns after the COVID-19 pandemic, the US–China trade war, Brexit, or the EU sanctions on the Russian Federation after the Crimea annexation were just some of the events that had drastic events on global trade. Similarly, for the years to come, international incidents with great impact are conceivable (Global Security Review 2019). CEP providers must be able to react flexibly and agile toward shifting demand and supply situations as well as logistical changes that arise out of such rarely foreseeable events. In recent years, also *environmental sustainability* has become a major concern for all players in the logistics industry, with environmentally driven policies and regulations increasingly affecting market dynamics. For instance, in the shipping industry, new regulations for the fuel oil used in vessels brought down the Sulphur cap from 3.50 to 0.5% (UNCTAD 2019). And also, the air cargo industry is committed to achieve CO_2-neutral growth by 2020 and to reduce 50% net emissions by 2050 (IATA 2019). Pressure to slash carbon emissions is likely leading to price increases as well as price volatility and a reduction in available supply capacity. This is paired with other *higher customer expectations*. As consumers become more used to digital services, including e-commerce via Taobao/Amazon or ridesharing apps such as Didi Chuxing/Uber, they expect to receive the same quality, visibility, and flexibility of service in other industries. That means that while traditionally it was sufficient to deliver a consignment on time, customers now expect seamless, effortless, and fast deliveries (World Economic Forum 2016). With omnichannel, the distinctions between physical and online retailing are vanishing so that high-quality customer experiences must occur within and between contact channels (Brynjolfsson et al. 2013). For CEP providers, that means that goods could travel from multiple pickup locations such as a store or a distribution center to different consignees including businesses, consumers, or parcel lockers – and reverse. Aside from these new challenges, traditional ones such as cost pressure, fluctuating demands, and the overall complexity of global value chains remain (Kersten et al. 2017). In addition, enhancing customer satisfaction – in which logistics and delivery play a significant role – has become more important than ever. According to a study by the research firm Forrester, 72% of businesses say that improving customer experience is their top priority (Pattek et al. 2016). Only by leveraging advanced technologies, companies can survive in this new competitive environment.

Digitalization Solutions of the Future

To stay ahead of the curve, Industry 4.0 technologies must be adopted by CEP providers. Internet of Things (IoT), artificial intelligence (AI), and blockchain will help to create a digital enterprise in which data collected from physical systems drives intelligent action back in the physical world (Maguire 2018; Panetta 2018). With these technologies, forecasts can become dynamic and integrated, suppliers and carriers connected, and the order fulfillment process optimized (Kern and Wolff 2019).

Internet of Things (IoT)

Barcode scanning has been used by various logistics companies for decades. With every movement, scanning the barcode generates data. With the advancement of IoT, in this context, also referred to as "telematics in transportation," datasets can be created easily along the entire supply chain. The IoT is a collection of interconnected physical devices that monitor, report, send, and exchange data. IoT devices are connected to computer systems and measure and exchange environmental factors such as location, temperature, speed, or humidity via RFID, smart devices, or mobile sensors. Leveraging IoT technology, storage condition, or product status can be effectively measured throughout the supply chain (Tjahjono et al. 2017).

IoT devices installed in cargo or containers can transmit location information with GPS to reflect the real-time location of the goods. Also, IoT devices can track the speed of cargo movement, and in combination with AI, IoT technology can predict the estimated time of arrival (ETA) of goods. This eliminates transportation resource waste, helps to better utilize warehouses, and increases visibility for consumers. In the healthcare and pharmaceutical industry, goods must be stored at a certain temperature and humidity level. Here, IoT devices can trigger an alarm if certain thresholds are breached. This improves the tracking of the quality of the goods and can eliminate spoilage. IoT can also optimize asset utilization to drive greater operational efficiency. In-vehicle telematics and vehicle-infrastructure integration have been pioneering use cases for the use of sensor data. Automotive OEMs and transportation operators have been investing significantly in connected vehicles. Monitoring of equipment and people to increase safety and security is another benefit IoT can potentially bring. By placing acoustic and visual sensors to monitor the movement of the vehicle, logistics companies can reduce traffic accidents.

Together, using IoT in logistics and supply chain management could lead to numerous benefits. It can help to reassure stakeholders that goods are located where they are claimed to be, both at rest and in motion. Possible issues with goods getting lost or delayed can be early identified. IoT allows for real-time shipment and inventory visibility and tracking. It would also facilitate supply and demand planning as stakeholders know when they can expect to receive and process goods. Quality management could be improved as raw materials, and processed goods could be easier kept in optimal conditions. In warehouses, storage and distribution of products could be made more efficient as goods could be easier located.

While these factors are individually beneficial, companies would benefit most in their operation by not just focusing on IoT at a single area such as in transportation or warehousing. As data itself will not generate any value for logistics on its own, proper data analytics will be mandatory. The key to success is the understanding of the convergence of IoT with other cutting-edge technology, such as connecting IoT with AI to not only capture the data but also make full use of it. Finally, implementing IoT in logistics and supply chain business will require strong collaboration from within, which is a change in the business mindset to realize the potential of IoT (Macaulay et al. 2015). Chapter 3 goes into more detail regarding the use of IoT in logistics.

Artificial Intelligence (AI)

AI refers to human intelligence exhibited by machines, systems that approximate, mimic, replicate, automate, and eventually improve on human thinking (Gesing et al. 2018). AI leverages advanced programs and algorithms to understand the information collected to perform tasks given and provide outputs. It is a collective portfolio of machine learning (ML), planning, and problem-solving. We position AI in the middle between input and output. Data, regardless if collected by IoT, or based on some archival information or scanning, is provided as input side to an AI application. Then, by using complex algorithms to analyze and process data, output in the form of optimized planning and solutions can be generated.

ML is an important field of AI application in the logistics industry. By analyzing substantial datasets, existing trends can be identified and patterns for the future predicted. ML could help to ensure that cargo flows in the supply chain are delivered on time, to alert identify potential risks in the supply chain, and to propose alternative solutions. Another key application of ML is the forecast demand analysis. AI can source and process data from different areas and predict the future demand of the business, which could help businesses save costs and enlarge potential sales. AI could also study existing business models and derive optimal models for warehouse operation, forwarding routings, and operating hours to ensure ideal fulfillment. Today, logistics companies spend a lot of time dealing with inefficient operations, exceptions, and seasonal volumes. AI will also enable autonomous vehicles and even unmanned aerial vehicles, also known as drones. Some large warehouses now are employing robotic vehicles to pick up and transport goods inside. With the development of AI, in the future, it might be possible to see autonomous cars handling long hauls or drones that pick up and deliver time-sensitive goods.

Overall, AI could enhance logistics in various areas. Quicker data analysis that creates insights and implications for faster and more accurate business planning by analyzing different factors such as demand, supply, and customer needs, risk management through historical pattern analysis, improves logistics operation such as warehouse operation and routing selection, overall cost saving, and potential additional sales. Accuracy of data and how algorithms are built are vital success factors for the application of AI. Programmers should work closely with logistics and supply chain domain experts to understand the business model. Then, accurate input data can be collected and suitable algorithms to support the business team generated.

With the efficient setup of AI, the logistics industry and especially retail and industrial segments could benefit. Here, AI can help, e.g., e-commerce platforms to gather customer information when customer browsing on their platform, as well as provide optimized operation models in warehouses – especially in a multiple distribution center (DC) or multiple forward stocking location (FSL) setting. Finally, it will be helpful to predict customer trends and propose potential suitable actions (Gesing et al. 2018).

Blockchain

Blockchain is a technology that appears poised to disrupt the supply chain industry, just like the IoT or AI. A blockchain is a decentralized ledger of all transactions in a network. Using blockchain technology, participants in the network can confirm transactions without the need for a trusted third party intermediary (PwC 2019). Blockchain has the potential to transform logistics and supply chain management, but as described in Chapter 6, it is not abundantly clear how fast that will happen. One supply chain perspective is the flow view, which distinguished between the physical flow, information flow (often in the form of documents), and financial flow (Mangan et al. 2008). Historically, the information in logistics is shared in a very analog fashion. For instance, the transportation bill of lading (BOL) came into use in the sixteenth century and is still being used in slightly modified form today. However, consumer demand, especially in the form of e-commerce, is now demanding a change toward more flexible, digital solutions. Global e-commerce sales transactions have reached already 3.46 trillion USD in 2019, up from 2.93 trillion USD in 2018, representing an online share of retail spend north of 16% (Young 2019). e-Commerce presents a fundamental shift in how consumers shop, requiring a structural change in the underlying supply chain and movement of goods around the world in terms of assets, products flow, delivery mode, and enabling technology. Multiple transactions and data sources to enable traceability throughout the supply chain are used today. Various shipper/consignee points of friction exist in a typical supply chain, for example, consignees want proof of product history, shipper/consignee needs a chain of custody proof to meet government regulations, or shippers require traceability capability to recall products. A typical supply chain and how physical events are mirrored in software is set out in Figure 7.1.

With blockchain technology, one single view across the supply chain of a product is created, which will provide sole truth of product ownership, immutable records easy to review

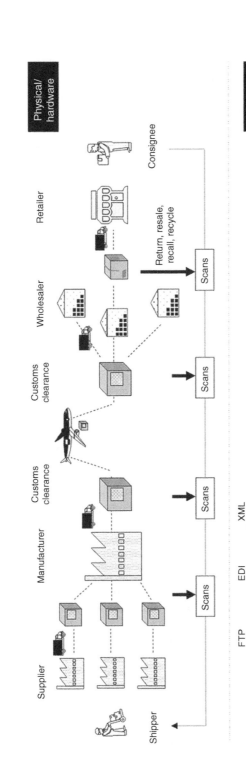

Figure 7.1 Mirroring of physical events in software.

by a consignee, and the ability to locate the product throughout the supply chain. We believe that by connecting various internal tools, blockchain could improve communication and connection of all users across the entire supply chain (Figure 7.2).

In 2018, the US Customs and Border Protection (CBP) invited members of the trade community to participate in a blockchain pilot program. The pilot was commissioned to test the potential benefits of blockchain technology in shipment processing. Objectives were to test the technology's ability to improve processing speed, reduce shipping preparation, reduce printing costs, and enable further automation. The pilot ran rather successfully, proving that communication among various protocols could be established, which allowed for smooth transactions between parties. There is a significant improvement in processing speed compared to when additional, often physical, documentation is needed. Blockchain was confirmed as a viable method of secure, authenticated documentation transfer (Rejeb et al. 2019).

To underline that such digitalization solutions are not very far away in the future and to show which benefits they might bring, a successful case study about the implementation of AI for address translation is described in the next section.

Case Study: Enhanced Address Translation Through AI

The company, where one of the authors of this chapter is working for in its China hub, is one of the largest packaging delivery companies around the globe. Partnerships with CEP providers in China are an important strategy for it as it helps the company to further grow business with less risk and the need to invest heavily in assets. By the end of 2019, the company has expanded this business model to some additional 30 cities with a very positive outcome. The partnership business model became so successful that they are now considering closing some of their own operations in underperforming cities and switching to this business model. Besides the asset-light approach, the company is also evaluating the utilization of the network of SF Express and other key domestic players to reduce their own network cost while expanding their partner's international business activities.

The IT operating and shipping system that this company provides to its customers is globally standardized. While its functionality is well proven in the United States (where most of the revenue originates from), the system causes challenges when attempted to be deployed to partners as expected in line with the local partnership strategy. The domestic market players' systems do not recognize the information that is provided in English. Therefore, the company had to look for a middleware that translates English information into Chinese before it can transmit it to its partners. It is impossible to control the entry of an import consignee address that is made overseas, and Chinese characters are pronounced and written differently. Therefore, during the software deployment, the address translation accuracy was only 30% – far from enough to be passed along to their partners. Hence, they started utilizing AI in their address translation tool. Whenever an incorrect address, for instance, road, district, or building name, is corrected by their employees or one of their partners, the system will update it for the whole database. A ML algorithm helps to improve the translation accuracy so that they now managed to achieve 80%, which is an acceptable value for the business. Also, by utilizing a cloud service, the database is shared with other

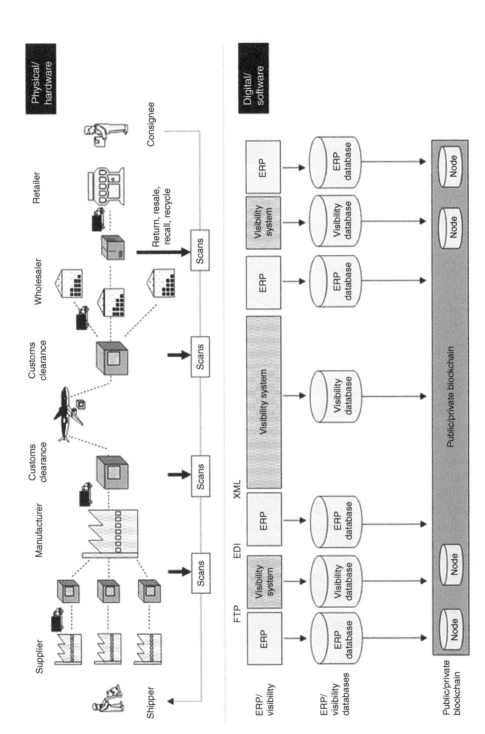

Figure 7.2 Enhancing data transparency in a typical supply chain through blockchain.

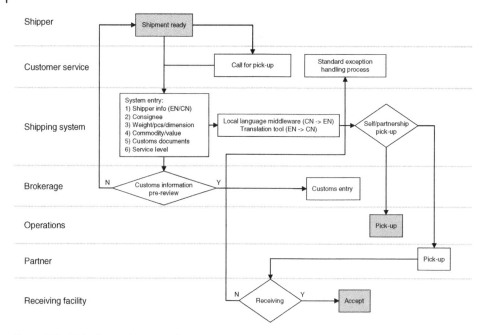

Figure 7.3 Shipping tool process flow.

cities, which helps to leverage that some road names are the same across different cities. This is a case for the company's import business into China, which is considered rather simple. Its hub is the shipper to the domestic partners, and they just need to provide the consignee address in Chinese for them to make the delivery after customs clearance is completed.

Meanwhile, the export business is more challenging as it involves more process steps. The shipper address must be given to the providers in English, together with a list of the facility addresses so that they can drop shipments into the network. The company at hand managed to develop a smartphone app that has a scanning and image uploading functionality so that its partners will be able to upload export customs documents upon pickup. The process flow above represents the steps of how the shipping tool would support the partnership business model (Figure 7.3).

The key enabler is the system itself where pre-translated customer addresses are maintained. When an order comes in, it will be first matched with the database before the translation, and then the system will determine whether pickup will be completed by the own operation team or by a partner. An order will be issued to the partner's handheld device, and as soon as the driver picks up the shipment, scanning a preprinted label will generate the partner's label. The driver will also scan customs documents and upload them to a brokerage system. A brokerage admin will then evaluate the documents before the parcel enters the company's network. An implemented exception handling process follows a standard definition where the customer service agent will handle all pickup and customs-related issues with the shipper directly.

The deployment of AI helped the company to improve the translation accuracy for both import deliveries and export shipments, while the mobile APP creates a platform between the partner's driver and the company's back-end system. As a next step, the company intends to further enhance the system connection between the shipping tool and the pricing system, so that real-time quotations can be provided to customers. This is intended to enlarge their revenue share, especially during the slack season.

Conclusion

Digitalization will lead to *enhanced productivity and speed*. By leveraging disruptive technologies, such as AI, ML, and blockchain, companies can improve tremendously in productivity, risk management, and traceability. It will also improve *visibility and accuracy*, as leveraging IoT and blockchain will allow tracking of the shipment down to each SKU. Blockchain will enable trustful visibility tracking including temperatures, locations, and the status of damage. As logistics will likely remain a labor-intensive business, by providing employees the right tools and equipment, companies who embrace digital solutions will not only improve their productivity but also improve their employee satisfaction. Lower turnover rates and easier recruitment will be a distinct advantage for any company. *Customer satisfaction* will also *improve t*hrough the usage of AI, as customers' needs can be better understood so that more proactive solutions can be provided instead of reactive actions once customer satisfaction drops. Finally, uptime benefits will derive from technologies such as IoT, supplemented by AI and human intuition. By having a clearer view of vehicle usage and performance, fleet managers will be better able to use preventive maintenance to avoid potential breakdowns (Choe et al. 2017; Kern and Wolff 2019).

CEP providers are more and more under pressure due to an increasing number of customers that are participating in global markets as well as urbanization and megacities that demand innovative delivery concepts, geopolitical changes to which CEP providers must swiftly adapt, environmental sustainability demands, and especially rising customer expectations. Together with the ever-strong cost pressure, fluctuating demands, and complexity of global value chains, further enhancing customer satisfaction will not be possible without the implementation of advanced technologies. Digitalization will not only support the improvement of customer satisfaction but also improve CEP companies' operational efficiency. Ultimately, it will enhance the overall logistics experience for all stakeholders. In China, we can see this development already today where it is happening at an accelerated speed.

Key Takeaways

- The increasing logistics requirements put pressure on various industries today. For example, the high-tech industry is struggling to achieve speed to market, end-to-end visibility while maintaining an increasingly customized product portfolio; in the industrial manufacturing and automotive business, distributors need to deliver new levels of customer experience to win new market shares and defend their positions.
- To tackle these challenges, companies like UPS rely on a variety of digitalization solutions, such as advanced shipping tools, configurable shipping solutions, visibility and reporting solutions, and returns solutions to run their operations and help customers to manage shipping processes

- However, the competitive environment is changing, with an increasing number of customers participating in global markets, a trend toward urbanization and megacities, changes in international relations, and an increasing demand for environmental sustainability
- To stay ahead of the curve, CEP providers must therefore adopt industry-advanced technologies such as Internet of Things, artificial intelligence, and blockchain. With these technologies, forecasts can become dynamic and integrated, suppliers and carriers connected, and the order fulfillment process optimized.
- An example for the successful application of such technologies is a case study of a multinational CEP provider. Using AI, they managed to boost English–Chinese translation accuracy from 30 to 80%.
- Digitalization in the CEP sector will allow to increase productivity and speed, visibility and accuracy, employee satisfaction, customer satisfaction, and uptime and flexibility

Acknowledgements

We would like to thank Prof. Dr. Tariq Samad, Mac Sullivan, and Dr. Axel Neher for their helpful comments on an earlier draft of this article. Special thanks also go to Janet Fang, without whom this chapter would not have been possible.

References

Amazon (2019). Help grow your business with fulfillment by Amazon. https://services. amazon.com/fulfillment-by-amazon/benefits.html (accessed 4 May 2020).

Bali, V. (2018). *5 things we can learn from China's e-commerce explosion*. https://www.nielsen. com/cn/en/insights/article/2018/5-things-we-can-learn-from-chinas-e-commerce-explosion/ (accessed 1 May 2020).

Brynjolfsson, E., Hu, Y.J., and Rahman, M.S. (2013). Competing in the age of omnichannel retailing. *MIT Sloan Management Review.* 54 (4): 23–29.

Choe, T., Garza, M., Rosenberger, S.A. et al. (2017). The future of freight. How new technology and new thinking can transform how goods are moved. https://www2.deloitte.com/content/dam/insights/us/articles/3556_FoM_Future-of-freight/DUP_FoM_the-future-of-freight.pdf (accessed 9 December 2019).

Christopher, M. (1993). Logistics and competitive strategy. *European Management Journal* 11 (2): 258–261. https://doi.org/10.1016/0263-2373(93)90049-N.

Gesing, B., Peterson, S.J. and Michelsen, D. (2018). Artificial intelligence in logistics. https:// www.logistics.dhl/global-en/home/insights-and-innovation/insights/artificial-intelligence. html (7 January 2020).

Giuffrida, M., Mangiaracina, R., Perego, A. et al. (2017). Cross-border B2C e-commerce to Greater China and the role of logistics: a literature review. *International Journal of Physical Distribution & Logistics Management* 47 (9): 772–795. https://doi.org/10.1108/IJPDLM-08-2016-0241.

Global Security Review (2019). Geopolitical tensions will continue to escalate over the course of the next seven years. https://globalsecurityreview.com/world-will-look-like-2025/ (accessed 8 November 2019).

Handfield, R., Straube, F., Pfohl, H.-C. et al. (2013). *Trends and Strategies in Logistics and Supply Chain Management – Embracing Global Logistics Complexity to Drive Market Advantage*. Bremen: Bundesvereinigung Logistik (BVL).

Hausmann, M. and Ludwig, B. (2018). Air-freight forwarders move forward into a digital future. https://www.mckinsey.com/industries/travel-transport-and-logistics/ our-insights/air-freight-forwarders-move-forward-into-a-digital-future (accessed 4 December 2019).

IATA (2019). Air cargo carbon footprint. https://www.iata.org/whatwedo/cargo/sustainability/ Pages/carbon-footprint.aspx (accessed 17 December 2019).

International Post Corporation (2018). 2018 Key Findings Global Postal Industry Report. https://www.ipc.be/services/market-research/market-intelligence (accessed 19 January 2020).

Jiang, X.-M. (2018). Development of China's logistics market. In: *Contemporary Logistics in China* (eds. J.-H. Xiao, S.-J. Lee, B.-L. Liu, and J. Liu), 1–24. Singapore: Springer https://doi. org/10.1007/978-981-13-0071-4_1.

Jiao, Z.-L. (2018). Development of Express Logistics in China. In: *Contemporary Logistics in China* (eds. J.-H. Xiao, S.-J. Lee, B.-L. Liu, and J. Liu), 115–135. Singapore: Springer. https:// doi.org/10.1007/978-981-13-0071-4_6.

Kern, J. and Wolff, P. (2019). The digital transformation of the automotive supply chain – an empirical analysis with evidence from Germany and China. https://www. innovationpolicyplatform.org/www.innovationpolicyplatform.org/system/files/imce/ AutomotiveSupplyChain_GermanyChina_TIPDigitalCaseStudy2019_1/index.pdf (accessed 8 November 2019).

Kersten, W., Seiter, M., Von See, B. et al. (2017). *Trends und Strategien in Logistik und Supply Chain Management – Chancen der digitalen Transformation*, 71. BVL Bundesvereinigung Logistik https://logistiktrends.bvl.de/en/system/files/t16/2017/ Trends and Strategies in Logistics and Supply Chain Management – Digital Transformation Opportunities Kersten von See Hackius Maurer.pdf.

Macaulay, J., Buckalew, L. and Chung, G. (2015). Internet of things in logistics. https:// discover.dhl.com/content/dam/dhl/downloads/interim/preview/updates/dhl-trend-report-internet-of-things-preview.pdf (accessed 14 January 2020).

Maguire, E. (2018). Logistics, supply chain and transportation 2023: change at breakneck speed. Jersey City. http://forbesinfo.forbes.com/l/801473/2019-09-23/27ns/801473/8485/ Penske_REPORT_FINAL_DIGITAL.pdf (accessed 1 December 2019).

Mangan, J., Lalwani, C., and Butcher, T. (2008). *Global Logistics and Supply Chain Management*. Wiley. https://books.google.com.hk/books?id=9bpcxQIw484C.

Panetta, K. (2018). Gartner predicts 2019 for supply chain operations. *Gartner Predicts*. https:// www.gartner.com/smarterwithgartner/gartner-predicts-2019-for-supply-chain-operations/ (accessed 26 January 2020).

Pattek, S., Fenwick, N., Matzke, P. et al. (2016). A customer-obsessed operating model demands a close partnership with your CIO. https://www.forrester.com/report/A+CustomerObsessed +Operating+Model+Demands+A+Close+Partnership+With+Your+CIO/-/E-RES122402# (accessed 9 December 2019).

Pfohl, H.-C., Wolff, P. and Kern, J. (2019) Transshipment hub automation in China's CEP sector. In: *Urban Freight Transportation Systems*, 1st edition (eds. R. Elbert, C. Friedrich, M. Boltze, and H.-C. Pfohl). Elsevier.

PwC (2019). How can blockchain power industrial manufacturing? https://www.pwc.com/cl/es/publicaciones/assets/2019/pwc-blockchain-in-manufacturing.pdf (accessed 5 December 2019).

Rahman, S., Ahsan, K., Yang, L. et al. (2019). An investigation into critical challenges for multinational third-party logistics providers operating in China. *Journal of Business Research* 103: 607–619.

Rejeb, A., Keogh, J.G., and Treiblmaier, H. (2019). Leveraging the internet of things and blockchain technology in supply chain management. *Future Internet* 11 (7): 161. https://doi.org/10.3390/fi11070161.

The Development of Intelligent Logistics Industry (2017). CeMAT ASIA. http://www.cemat-asia.com/industrynews/shownews.php?lang=en&id=3366 (accessed 23 January 2020).

Tjahjono, B., Esplugues, C., Ares, E. et al. (2017). What does Industry 4.0 mean to supply chain? *Procedia Manufacturing* 13: 1175–1182. https://doi.org/10.1016/J.PROMFG.2017.09.191.

UNCTAD (2019). Review of maritime transport 2019. https://unctad.org/en/PublicationsLibrary/rmt2019_en.pdf (accessed 17 December 2019).

United Nations (2019). *World Urbanization Prospects The 2018 Revision*. New York, NY: United Nations https://population.un.org/wup/Publications/Files/WUP2018-Report.pdf.

UPS (2016). Introducing UPS follow my delivery. https://pressroom.ups.com/assets/pdf/pressroom/infographic/UPS16-017_FollowMyDelivery_SS_p7.pdf (accessed 15 December 2019).

UPS (2017). UPS industrial buying dynamics study: buyers raise the bar for suppliers. https://www.ups.com/assets/resources/media/knowledge-center/UPS_Industrial_Buying_Dynamics_Study_2017.pdf (accessed 3 January 2020).

UPS (2019). Shipping management integrations. https://www.ups.com/us/en/help-center/technology-support/ready-program/shipping.page (accessed 3 December 2019).

US International Trade Administration (2019). China – eCommerce. https://www.export.gov/article?id=China-ecommerce (accessed 3 December 2019).

World Economic Forum (2016). Digital transformation of industries: logistics industry. http://reports.weforum.org/digital-transformation/wp-content/blogs.dir/94/mp/files/pages/files/wef-dti-logisticswhitepaper-final-january-2016.pdf (17 December 2019).

Young, J. (2019). Global ecommerce sales to reach nearly $3.46 trillion in 2019. https://www.digitalcommerce360.com/article/global-ecommerce-sales/ (accessed 11 December 2019).

Zhang, Z. and Figliozzi, M.A. (2010). A survey of China's logistics industry and the impacts of transport delays on importers and exporters. *Transport Reviews* 30 (2): 179–194. https://doi.org/10.1080/01441640902843232.

8

Understanding the Impacts of Autonomous Vehicles in Logistics

Lionel Willems

Odoo, Hong Kong, China

Introduction

Humans have thought of efficient ways of carrying goods from one location to another since the dawn of time. At each historical breakthrough, problems were solved, and new questions arose. Wheels, levers, axles, and pulleys made men discover weight relativity and laws of traction. Taming animals and building chariots dramatically reduced the physical effort needed for freight forwarding, but new challenges emerged: feeding the animals, preventing inherent fatigue, and uncertainty. As the first forms of automation, ships and railways allowed men to overcome those issues by relying on tides and combustion engines to realize their tasks. Transport by air stopped being a collective fantasy in the early twentieth century with the first aircraft, which allowed public and private enterprises to "think global" and reduce lead times drastically. Logistics is growing tremendously year over year, and yet much of the work is still reliant on human power.

With Taylor's scientific management theory, managers started to think of ways to optimize material flows and decrease labor costs on company floors since the early twentieth century. According to a 2014 DHL report on transport automation, the first concepts of autonomous guided vehicles (AGVs) for in-house logistics were introduced as early as 1940, and a decade later, both General Motors and Ford had running prototypes (DHL International GmbH 2014). They were initially designed for special missions in dangerous or practically inaccessible situations, such as high temperature and high-pressure environments (Vlahovic et al. 2016). Their usage gradually evolved to fulfill other purposes in factories and warehouses, such as realizing routine tasks, complex and precise operations on assembly lines, between job shop stations, and at loading platforms. More and more, they were seen as a way to replace labor-intensive tasks with reduced costs and uncertainty.

According to Günter Ullrich, author of the book *Automated Guided Vehicle Systems: A Primer With Practical Applications*, the era from the 1970s to the early 1990s represented an entire automation era in itself (Ullrich 2015). If the first AGVs were simply tow trucks

The Digital Transformation of Logistics: Demystifying Impacts of the Fourth Industrial Revolution,
First Edition. Edited by Mac Sullivan and Johannes Kern.
© 2021 by The Institute of Electrical and Electronics Engineers, Inc.
Published 2021 by John Wiley & Sons, Inc.

following a built-in wire on the floor rather than rails, the rapid developments in electronics and sensor technologies offered a new wave of innovation opportunities around them. Let us have a look at their computing capabilities, control functions, navigation, and energy components. AGVs in the 1970s were equipped with electronics and microprocessors, enabling quick scenario computations (e.g. block section control, preventing collisions and process blocking) and better control, using infrared and radio technologies for data transfer. Another innovation in control functions was the programmable logic controller (PLC), a major pillar of the industrial automation landscape. Experts describe the PLC as an industrially hardened computer-based unit that performs discrete or continuous control functions upon equipment, processes, motion, and batches (Romero et al. 2012). It has no display nor keyboard, yet it is a computer that is most often seen in the control panel, on the factory floor. The PLC gradually replaced panels of relays, had to be turned on and off regularly in order to maintain automation logic. The complexity of their use and programming, in addition to the unpredictability and regularity of their downtime, quickly encouraged the change to PLC. For AGV navigation, the new technological standard brought about by electromagnetics was active induction track guidance, i.e. conductors embedded in the factory floor emitting magnetic waves and inducing a current in the vehicle's coils. During this period, sensory technologies became cheaper and allowed for higher driving accuracy, positioning, and recognition within warehouse and factory grounds. Meanwhile, batteries grew more powerful, and automatic battery charging prototypes emerged.

Evolution of AGVs

After the 1990s, all AGVs contained electronic guidance and contact-free sensors and were controlled at a distance by computers. Navigation became mostly magnetic or laser-based. Data transfers increasingly transitioned from cables to wireless technologies. AGVs' computation speed increased, the technology became cheaper, more accurate, and reliable in their operations. This opened its applications to new sectors. Mostly used initially in industries related to manufacturing and warehousing of automotive, chemical and pharmaceutical products, AGVs later appeared in the food and beverage industry, hospitals, theme parks, and the world of hospitality.

At the dawn of the twenty-first century, with its introduction to the public transport sphere, automation involves new dimensions of complexity. Daily debates focus mostly on automated public transport that would be independent of human interaction and able to operate in diverse surrounding environments and climatic conditions. Such autonomous vehicles must also be capable of acting by themselves while taking all decisions and performing their control. At the core of such a system lie software and algorithms that connect, control, and make all the decisions (Shaikh 2017). If autonomous vehicles already proved their appeal in the mobility of merchandises within company floors, new questions surfaced regarding the elements that the Fourth Industrial Revolution (4IR) can provide, in terms of opportunities and challenges, to indoor and outdoor freight transport and material handling.

In order to work with a comprehensive overview of those topics, let us use Vlad Krotov's classification of 4IR elements in terms of environments and apply it to autonomous vehicles. There are three main environments (Krotov 2017).

First, let us have a look at the "technological environment," composed of hardware, software, networking, data, and industry standards. The more a vehicle is connected, the better it can perform its control and prevent all faulty scenarios. How to ensure a comprehensive control knowing that a standard car has about 30 000 parts (down to the smallest screw) (Toyota Motor Corporation 2019), supplied by many different suppliers with unique manufacturing processes? This brings up the question of hardware–software interoperability, i.e. the ability of software to recognize, act on, and control hardware independently of their supplier. There is thus a strong need for international interoperability standards for software and hardware integrations. In the field of supply chain management, an attempt to tackle this need is the GS1 EPCglobal Architecture Framework for unique identifiers of physical goods. These electronic product codes (EPC) have different formats, but all of them are to be readable by radio frequency identification (RFID) technology. From a broader perspective, standards are being built collaboratively within industry associations.

In early 2018, the Industrial Internet Consortium (IIC) and the Plattform Industrie 4.0 announced the publication of a joint paper on architecture alignment and interoperability between the two leading Industrial Internet of Things (IIoT) reference architecture models: Industrial Internet Reference Architecture (IIRA) and Reference Architecture Model for Industrie 4.0 (RAMI 4.0) (Quatromoni 2018). Once the software and hardware are integrated, questions of data management, property, and security, in addition to internet connectivity and stability (networking), remain to be answered. The software side of those questions can be answered with the capabilities of a company's enterprise resource planning (ERP) system. Recent innovations in the world of ERP made them cheaper and available for small and middle-sized companies. However, for a side of logistics that comprises vehicles, an integration between an ERP system and the physical elements comes down to a single concept: cyber-physical systems (CPS). According to Acatech, the German National Academy of Science and Engineering, CPS are "systems with embedded software (as part of devices, buildings, means of transport, transport routes, production systems, medical processes, logistic processes, coordination processes, and management processes), which directly record physical data using sensors and affect physical processes using actuators; evaluate and save recorded data, and actively or reactively interact both with the physical and digital world; are connected with one another and in global networks via digital communication facilities (wireless and/or wired, local and/or global); use globally available data and services; and have a series of dedicated, multimodal human-machine interfaces" (Acatech 2011).

Second, the "physical environment," as described by Krotov, gathers humans, objects, and even sometimes animals into various types of spatial environments. For example, different warehouse workers might come up with different ways to interact with a vehicle, as opposed to the vehicle having standard ways to consider humans in its decisions. Humans are expected to act as knowledgeable "augmented operators" with the possibility to monitor the vehicle's every action. The vehicle also needs the ability to detect other mobile objects and to scan the surrounding environment in real time via wireless sensors. These sensors will in turn collect target data and send it to a cloud via wireless networking.

Finally, the "socioeconomic broad environment" involves all stakeholders of the particular technology: customers, legislative bodies, industry associations, technology entrepreneurs, and consumer privacy groups. One might wonder what strategic attributes are needed to convince the market to use autonomous vehicles. Questions of law also hover above the responsibility of an AGV manufacturer in case of an accident and how to define a comprehensive legal framework for public infrastructure. Scalability concerns also face technology entrepreneurs on how to simplify the complexity of engineering processes and reduce the cost of their implementation. Specifically, on costs, it is crucial for stakeholders to quantify initial and recurring costs, return on investment (ROI), and business value of adopting AGVs and autonomous freight transport.

The above concepts, questions, and debates are applicable for both the automation of private passenger vehicles and commercial transport, which accounts for approximately a third of the traffic on roads (Flämig 2016). Companies have long tried to find ways to reduce the cost of freight transport, for it does not add any value to sold merchandise *per se*. Early solutions to reduce its cost were relocating factories closer to the market, ensuring full truck-loads, minimizing empty miles, outsourcing logistics, and setting up continuous drives. Today, it is interesting to have a look at the solutions provided by partial and complete automation of vehicles within and beyond factory walls, along with their opportunities and challenges in logistics. In the following subsection, those are discussed in greater detail.

Intelligent Logistics

Through the acceleration of many technologies of the Fourth Industrial Revolution, such as artificial intelligence (AI), big data, and cloud computing, logistics has entered a new era of possibilities. In July of 2017, the State Council of China announced a new program that further expanded its existing "Made in China 2025" program to have a focus on the use of AI in logistics and named it "Intelligent Logistics" (Council 2017). AGV sales in China's logistics facilities, driven by rising labor costs in large port cities, has been rising 30% a year since 2016 (Li et al. 2018). The challenges surrounding AGV adoption are surmountable but require high levels of resource investment in cutting edge technology. They will help to aid the digital transformation of the entire logistics industry (Li et al. 2018).

The Automation of Indoor Transport Systems

The use of AGVS has become popular in a majority of the industrial fields due to how flexible, reliable, safe, and productive the technology has become (Fazlollahtabar and Saidi-Mehrabad 2015). The current and prospective technologies discussed by academics and professionals show that during the last two decades, the inclusion of information and communications technology (ICT) in logistics has become increasingly vital for organizational efficiency, evidenced by the integration of ERP systems with warehouse management systems (WMS) and in the foreseeable future, their embedding in a CPS. In simpler words, if the AGV works independently, it must be connected to its environment and the merchandise to be carried over. Whether in a context of intra- or inter-company logistics, within an IIoT scenario, products will be traced in real time across the entire value chain, thanks to the collection of data via RFID technology. The data on product movements, transmitted through a wireless network to the server and stored in the ERP and a cloud, can be used by any hardware, including AGVs, at the condition that interoperability is achieved (Zhao et al. 2015).

In the warehouse, AGVs are responsible for bringing in inputs and releasing outputs via the logistic operations of offloading, deconditioning, storing, picking, consolidating, labeling, and loading. All these tasks will be subjected to the traceability and interoperability requirements mentioned above and performed in a partial or fully automated way thanks to the latest progress in AI and robotics. About the technologies that are embedded in the vehicle itself, it is important to keep in mind that vehicles are most of the time designed individually according to the specific conditions of the environment they are used in and the task assigned to them (Schulze et al. 2008). For this reason, the attributes necessary for the interoperability across the entire value chain need to be considered from the very beginning, at the design and research and development (R&D) stages for vehicles and products alike.

Opportunities for AGVs

Technologies put forth by the 4IR – AI, robotics, connectivity, interoperability, and full-on traceability – provide the AGV of tomorrow with useful features in terms of safety, energy, flexibility, and navigation capabilities.

AGVs need to avoid collision with other objects using advanced sensor and collision avoidance systems. The more connected the factory, the better the integrated GPS within the AGV will be able to recognize, in real time, nearby moving objects and their surrounding environment. To respect a safety distance with them, the vehicle will constantly broadcast its position to the network, which will then be available from any device connected to the ERP system. On the hardware side, laser scanners are currently widely used to ensure a safe distance, also of being connected to other sensor technologies such as RFID sensors and Bluetooth, providing extra layers of security and connectivity backup solutions (Ullrich 2014).

On the energy side, the vehicle runs over time, limited by the capabilities of the plant power system. A robust solution to tackle these limited capabilities lies in inductive power transfer, also called inductive coupling. It allows vehicles to transfer electrical power between two circuits (one on the vehicle, and one on the ground) through a shared magnetic field. The power system can be built on both continuous energy supply from the driving path to the AGV and an onboard battery enabling the vehicle to drive safely even in case of interruptions of the ground's inductive power supply (Runge 2006). Energy levels may be connected to the ERP system and updated in real time. Alert levels set up on the system will thus, once the conditions are met, automatically trigger the next logical step of dispatching a maintenance team or scheduling a replacement.

The modular attributes of AGVs represent an important competitive factor in terms of delivery times, flexibility, and cost of the production system. Highly modular physical properties of a vehicle will allow the CPS to run the AGV under multiple operation programs and assign it to different tasks when its main task is not necessary at a certain point in time. For example, an autonomous forklift normally responsible for offloading operations may be reassigned to picking when enough AGVs are available on the offloading site. Similarly, more robots could change programs when the value chain's bottleneck undergoes critical times and arrange their sequences to be dispatched and suggest instant throughput improvements. Similar to an autopilot system, a company will have the choice to let AI make the next logical decision or let the operations team take over in times of risk and need.

To navigate efficiently, AGVs will require intelligent routing updated in real-time to prioritize transport tasks and arrange time schedules. Path decisions need to be based on both static routing (an AGV won't perform a task already handled by another) and dynamic routing (the system will analyze the aggregated output of AGVs' use of laser technology and GPS to tackle static routing limitations and reduce the potential deadlocks and collisions). Static and dynamic routings have the potential to make the production system more flexible as a whole. In addition, this priority system can be enhanced by sensor technologies and camera systems with digital environment maps (Flämig 2016).

A structural question often appears in research papers on warehousing and manufacturing: could AGVs not be replaced by advanced conveyor belts browsing the entire factory? Conveyor belts represent a high investment in infrastructure and lack flexibility, limitations that are not critical when high and steady throughput is required (Schulze et al. 2008). Out of the four criteria of improvements presented above for AGVs in the 4IR (safety, energy, flexibility, and navigation capabilities), only those of energy and safety may seriously influence a managerial decision to implement, or not, an AGV system in a manufacturing setting. One of the main objectives of the Fourth Industrial Revolution is mass customization, which can only be achieved by a high flexibility of logistic processes within the line or job-shops organizations. Some experts introduce a flexible smart conveyor infrastructure within the Smart Factory production system as an improvement of traditional production lines (Wang et al. 2016). However, it must be taken into account that AGVs are an inherent part of this smart infrastructure and provide its overall flexibility.

AGVs such as forklifts, pallet jacks, order pickers, work robots, and other material handling vehicles, as a part of a decentralized CPS, will act according to the data received in real-time. The CPS, through the ERP system, will allow managers to monitor their status, location, repair scenarios, and analytics in real-time (Reaidy et al. 2015). Means of transport will share their position and predicted arrival time to the intelligent WMS, which will trigger the necessary preparations (docking slot, arrival, and delivery sequences and just-in-time optimizations) (Barreto et al. 2017). Each arriving load is recorded in the system with its information (e.g. what and how much) stored on the server (Szozda 2017). The WMS will then automatically allocate storage space, according to the inventory area and delivery specifics, and assign it the right type of AGV and other equipment for autonomous handling within the hub. This will provide real-time information on inventory levels, facilitating decision making in case of adjustments, and preventing out-of-stock situations. Since the CPS knows the time and amount of goods to be delivered, it always anticipates and ensures the availability of finished goods and a fortiori of raw materials through their bill of materials (BOM) (Fan et al. 2015). The two main advantages of a higher integration between AGVs and other technological elements are thus the instant responsiveness of the control system and the possibility for the vehicle to make its own decisions through AI. The other elements of the CPS will act accordingly (Fan et al. 2015).

Challenges of Adopting AGVs

For AGVs to be a pragmatic and implementable solution, their automation benefits must favorably compare with the cost, which is by no means negligible and necessitates a case study on the ROI of adopting this solution. The Fourth Industrial Revolution involves a growing dependence on technology to achieve a competitive advantage. The security of data and information is thus one of the most critical challenges to build a safe transport ecosystem (Barreto et al. 2017). The CPS needs to be under control under nearly every normal condition to avoid not only breaches and viruses but also slowdowns and breakdowns on individual AGVs or the entire CPS (Zuehlke 2010). The initial design of software, processes, and products is thus much more complex than before (Riel et al. 2017).

The concerns of data property and cooperation between a brand or manufacturer and their third-party logistics provider (3PL) require an environment where the two parties have full trust and cooperation to reach a value chain optimized for efficiency and transparency. However, even if digitization has been proven to facilitate cooperation, several experts express their concern about whether a full end-to-end integration is achievable (Hofmann and Rüsch 2017), especially when involving numbers of suppliers, co-competitors, and clients in the sharing of comprehensive raw data containing confidential information. For example, even in terms of cooperation with a vendor, having even a slight idea of its production volume will spur a logistics provider to turn the negotiation power to its advantage.

The necessity of interoperating technical and physical environments of the Fourth Industrial Revolution requires the complete integration of all CPS nodes, which will only occur when those environments can transmit and recognize the same semantic language. This implies that this semantic language is unique and widely accepted within and in-between industries.

Another challenge that is slowing the adoption of indoor AGV systems is the time-consuming and resource-draining task of measuring and identifying locations in an indoor facility to create the roadmap for the AGV to follow (Beinschob et al. 2017). Automated or semiautomated systems for setting up AGV roadmaps have been explored in hopes to reduce this barrier to entry and increase the market share of AGVs. However, in this area, human intervention is remains needed to some extent (Beinschob et al. 2017).

The Automation of Outdoor Transport Systems

The second facet of material handling comes into play after merchandise has departed from a factory or a warehouse, and before it has arrived at its destination. In this section, let us have a glance at the current state of affairs of logistics outside company grounds, i.e. outdoor transport system automation, and the important role of drones in its frame. As far as automation is concerned, outdoor transport systems share some opportunities and challenges with AGVs. However, outdoor systems are subjected to more complex legislation and are, in light of their public nature, highly risk-sensitive. Those two considerations add complexity to the logistics game. Before jumping into those concerns, let us absorb knowledge from cases where vehicles have been successfully automated, whether fully or partially. Case studies, along with their brief technical analyses, will help us understand challenges that will emerge in the fields of the logistics of outdoor transport systems.

History of the Automation of Outdoor Transport Systems

The first form of automation in public means of transport, such as in cars, trains, and airplanes, can be found in stabilization systems, e.g. through autopilot functions. Autopilot systems have been around for almost as long as aviation, with early examples dating back to 1914. In an article by the Society of Automotive Engineers (SAE), Matthew Borst explains that planes are controlled over three axes of rotation (pitch, roll, and yaw) and that the first autopilot system used gyroscopes to maintain a preset configuration of these axes in space over time (Borst 2019). Autopilot functions improved over the years to decrease human intervention while increasing the safety of air travel. Examples of such features are altitude hold, real-time path correction (or "heading select"), and the integration of navigation sources such as radio-based navigation first, then GPS. Those features are combined into what is known today as flight management systems (FMS). In the end, the capabilities of an autopilot system are determined by how integrated it is to the vehicle navigation system.

To bring us closer to present logistics, let us cast a look at two more cases (Vlahovic et al. 2016).

The MUNIN project, or "Maritime Unmanned Navigation through Intelligence in Networks," is a project that stems from European research on maritime logistics, which aims to develop an autonomous ship concept that would update today's fleet to autonomous vessels and that would be run by a combination of automated decision systems with remote control via a shore-based station. The project also focuses on environmental optimization and smart maintenance operations (TRIMIS 2020).

Driverless railway systems have been operating for years via ATO (automatic train operation) technology, but an operator was required in the front cabin for safety reasons. The first metro line to be automated was London's Victoria line, as early as 1967, and the technology has, since then, progressed to assist the driver in the control of operations. In 1981, the first fully automated and driverless mass transit rail network was introduced for the Portal Island Line in Kobe, Japan.

Operator-free automation is also being explored for air cargo. A San Francisco-based startup, Elroy (Etherington 2019), successfully launched its first 64 second flight in 2019 with the "Chaparral," a large unmanned cargo aircraft. A second prototype with flight beyond the visual line of sight is expected to happen in late 2020.

As of 2020, these examples represent only a small chunk of the myriad of innovative applications on automation that are emerging from startups and established companies across the world. The next subsection presents, from both microeconomic and macroeconomic perspectives, what opportunities the technologies of the Fourth Industrial Revolution bring about, and the current challenges preventing their broader use.

Opportunities and Challenges in Logistics

It would be difficult to assess the risks and opportunities of automation technologies on freight transport without mentioning their degree of embedded automation. Different automation levels will allow for different functions, based on the collection and analysis of traffic information, driving surface and path assessment, vehicle knowledge, and environmental conditions. According to those, the system will control longitudinal and lateral control of the vehicle, stabilization and navigation levels, steering, accelerating, braking, and decisions regarding speed. The German Federal Highway Research Institute came up with four categories to define the R&D progress for on-road vehicle automation (Gasser 2014):

1) **Assisted driving.** The driver still controls either longitudinal or lateral control. Other tasks can be automated to a certain degree. When facing certain situations, warning systems will ping the driver to perform a certain action. Automatic parking also falls under this category.

2) **Partially automated driving.** The system takes charge of longitudinal and lateral control, but the driver keeps monitoring it and should be able to take over control at any time. Assisted and partially automated driving are already common in series-production vehicles today.

3) **Highly automated driving.** This level takes into account all partially automated driving functions, but the driver is no longer required to monitor the system at all times. Changes in command must go through takeover requests.

4) **Fully automated driving.** The system always controls longitudinal and lateral control. The vehicle does not require a driver but can be monitored via a remote control center. The robot both assumes activities on the navigation level and the activities required to minimize risk and ensure safety in case of a mishap.

In the scope of logistics specifically, those degrees of automation can be explained in terms of their impact on supply chains, both on microeconomic and macroeconomic levels (Flämig 2016). From a microeconomic perspective, relying on autonomous vehicles in commercial transport could reduce both energy consumption and accident likelihood via driving reliability, while making up for a lack of driving personnel, working without interruptions and idle times, and reducing labor costs. The degree of automation within loading and unloading operations, sometimes part of the driver's responsibility, has also been rising continuously. The way AGVs within a factory will communicate with trucks will also influence the way logistics companies design them both, further shifting the human factor towards supervision and control instead of driving. This could change the way business models are built and supply chains planned from suppliers to end users. Concerning other means of transport, the introduction of drones in the aviation sector is a good example of how to rethink the distribution of transport jobs across different means of transportation. All these changes will need to be considered in terms of cost and compliance to local and sometimes international regulations. In an earlier section, the challenges of investment, cybersecurity, data property, interoperability, and implementation roadmap were presented for indoor AGVs. Those are to be considered for outdoor freight transport investment decisions as well.

From a macroeconomic perspective, autonomous transport could make up for structural flaws in current transportation systems. Since the 2008 financial crisis, in many countries, governments struggle with strict public budgeting that does not allow for revamping traffic infrastructure development. Automated driving would thus be a suitable alternative, allowing for an extended road infrastructure capacity that would benefit 3PL and other logistics companies. Indeed, fully autonomous trucks would require much less separation than with a driver and potentially sustain the same speed. Further, autonomous driving could also help companies respect environmental regulations in terms of CO_2 emissions. On the flip side, in terms of accidents, the legal question on which party will be responsible for actual mishaps remains yet to be answered. Looking ahead, the concept of "smart cities" could potentially reduce the risks of such accidents, where both passenger and commercial vehicles would be fully automated under the same citywide system. Finally, companies will likely need to manage their image with regards to job restructuring and public acceptance of implementing such a driverless system. Fears of an irreversible technological impact on jobs have burdened all industrial revolutions. This time, the speed, breadth, and depth of the transformation might create more shaking. Technological progress has two competing effects on jobs: a destruction effect that forces employees to unemployment and a capitalization effect of new occupations and businesses creation coming from the demand for new goods and services, including those related to automation. A study analyses the potential effect of technology on employment across 702 different professions and their probability to be automated (Frey and Osborne 2013). They conclude that the job destruction factor is likely to take place much faster than during the previous revolutions and towards greater polarization. High-income cognitive and creative jobs and low-income manual professions will rise, but middle-income routine jobs will decline.

Drones and Their Use in Logistics

Developments in commercial transport systems cannot be mentioned without talking about drones or small unmanned aerial vehicles (UAVs). Small UAVs might be the key to satisfy the 4IR's second important requirement: shorter time to market. Although some countries already use them for a variety of public functions (firefighting, military, police, etc.), in the context of commercial use, they are not widespread yet. Several companies have initiated pilot projects of commercial applications, including logistic companies such as DHL, UPS, and Deutsche Post, as well as e-commerce such as Amazon and Alibaba (Vlahovic et al. 2016).

The ideal scenario in logistics would see drones within the picking or shipping areas, being predesigned to be independent of human interaction and programmed to automatically arrive at their destination after receiving an input from the CPS. Drones pose major implications for diverse modes of transport (Robinson 2016). First, it is likely that transporting goods by the ocean will remain one of the main means of international transport. Some of its processes could be improved by loading and unloading drones that will make inspection and review of merchandise simpler and, thus, faster, by automatically updating the ERP while moving products on and off a cargo vessel. Second, drones deployed from a train could help dispatch their goods to reduce last-mile costs and planning efforts, even though the return scenarios of those drones still need to be optimized. Third, transport consolidation by trucks could benefit from drone support as well, by being adapted to a system where trucks would be assigned to areas where the order density is higher, and drones to areas where it is lower. This would save costs, win time, and save truckload space for trucks to further consolidate merchandise for smaller regions with high order density.

According to the World Economic Forum, using drones, even the most far-flung, hard-to-reach places can become accessible and connected (Liao 2019). UAVs will reduce the cost for companies to operate in remote areas, where using a truck is estimated to be five times costlier than operating it in urban areas. Drones also raise tremendous hopes in the public infrastructure and agriculture sectors. They are not affected by the same geographical limitations of cars, trucks, and trains and can be used in emergencies to deliver humanitarian, medical, or other critical merchandise in a matter of minutes. A very recent and encouraging example comes from the startup Zipline, which stands at the forefront of the battle against Covid-19 in Ghana in 2020 (Rustici 2020). They collect test samples from patients at rural health facilities to deliver them with a fleet of drones to medical laboratories in the country's two largest cities, and that in less than an hour. These are the first long hauls conducted by the startup since its creation in 2014.

In 2020, drones were used in Ghana at the forefront of the battle against COVID-19.

Amazon was expected to launch its Prime Air drone delivery services by the end of 2019 (Gartenberg 2019), by firstly transporting parcels across the United States that weigh less than 3 kg. However, as of May 2020, the service has yet to materialize. Amazon's early drone calling pushed other actors to speed up their R&D processes to stay competitive. According to Larry Perecko from the Science and Technology department of the US Army, the widespread use of transport drones in the logistics industry will boom within the next ten years (Williams 2016).

According to a review from Marsh, the main challenges of implementing drones in logistics are security, regulatory, and insurance concerns (Monaghan 2015). At the moment, there is a non-negligible lack of regulations surrounding the use of UAVs, in addition to discrepancies in regulations among countries. This is slowing down the worldwide adoption of drones for commercial purposes. The regulations must cover the exact area where drones are allowed to fly, and companies must improve the anti-collision technology within that restricted space.

In the United States, states and municipalities can enact individual local laws on the drone industry. This had led to extreme unevenness in drone-readiness across the 50 states, where some states like California, Nevada, and North Dakota have drone-friendly laws, and states like South Carolina, Florida, and Alabama lack clear air rights laws and avigation easement laws. Airspace lease laws allow drone users to operate in lower altitude airspace above state and local roads for commercial purposes. Without creating this drone highway, logistics companies cannot effectively fly their drones from one location to another without being the target of lawsuits. Without a national (and even supranational) commitment to accept and develop the drone industry, it will be extremely difficult for logistics companies to implement drones as anything but a side project. Local governments are currently giving out three options: supportive laws, restrictive laws, or no laws at all.

Aside from the legal standpoint, privacy concerns also feed the public debate. Situations where drones equipped with cameras would record private data aside from their main transportation objective need to be avoided. Finally, insurance products are being designed to meet the needs of drone manufacturers and operators in logistics. They cover risks of faulty deliveries, physical loss, damage to system components during normal operations, and cybersecurity.

Conclusion

The automation of transport was, is, and will remain a way for companies to optimize material flows and decrease labor costs across their supply chain. Today, the 4IR's criteria of mass customization and shorter time to market add a layer of complexity to those challenges. To tackle these challenges, two potential solutions revolve around how to maximize the current efficiency factors of automated transport outside a company's grounds, and the use of AGVs on factory floors.

4IR technology provides the AGV of tomorrow with useful improvements in terms of safety, energy, flexibility, and navigation capabilities, as well as new features of reactivity to other technological elements and the ability to take its own smart decisions through AI. Nonetheless, four main challenges need to be taken into account when considering integrating AGVs within a CPS: the cost of implementation, cybersecurity, data property, interoperability, and implementation roadmaps. It is important that the human element of logistics and supply chain management not be left out as there are substantial considerations that need to be taken into account when introducing a new form of labor to an organization (Lee and Leonard 1990).

Outdoor transport systems share some opportunities and challenges with AGVs but are to be subjected to more complex legislation and are highly risk-sensitive. As of 2020, we can find a myriad of projects working on further automation of transport by sea, on rails, by air, and on roads, with multiple degrees of automation. At the end of the day, drones represent the next transport innovation in terms of lead time reduction and market reachability. They show potential in many ways: being used in warehouse operations, compensating for other vehicles' shortcomings, and as an independent transport method. The main challenges about their implementation in actual supply chains pertain to security, regulations, and insurance concerns.

Key Takeaways

- When it comes to freight transport and material handling in logistics, two of the main challenges revolve around how to maximize the current efficiency factors of automated transport outside a company's grounds, and the use of autonomous guided vehicles (AGVs) on factory floors. Today, the Fourth Industrial Revolution's criteria of mass customization and shorter time to market add a layer of complexity to supply chains.
- The first AGV concept was elaborated in the 1940s, but it is only from the 1970s, with the rapid development of electronics, computing, and sensor technologies, that they gradually became marketable for factories. After the 1990s, their rising affordability made them attractive to nonmanufacturing sectors.

- The Fourth Industrial Revolution's mix of technologies provides the AGV of tomorrow with useful improvements in terms of safety, energy, flexibility, and navigation capabilities, as well as new features of reactivity to other technological elements and the ability to take its own smart decisions through AI.
- There are four main challenges in integrating AGVs within a CPS: the cost of implementation, cybersecurity, data property, and interoperability.
- Outdoor transport systems share some opportunities and challenges with AGVs, but they are subject to more complex legislation and are highly risk-sensitive. To realize how logistics are impacted by outdoor automated transport, both microeconomic and macroeconomic opportunities, challenges, and risks must be considered.
- After aircraft, drones represent the next transport innovation in terms of lead time reduction and market reachability. The main benefits of using drones for package deliveries are easy accessibility to hard-to-reach markets, faster time to market, and an aggregated reduction of the cost compared to trucks and trains. The main challenges of drone implementation in actual supply chains pertain to security, regulations, and insurance concerns.

References

Acatech (2011). *Cyber–Physical Systems – Driving Force for Innovation in Mobility Health Energy and Production [Position Paper]*. Munich: Acatech.

Barreto, L., Amaral, A., and Pereira, T. (2017). *Industry 4.0 implications in logistics: an overview. Procedia Manufacturing* 13: 1245–1252.

Beinschob, P., Meyer, M., Reinke, C. et al. (2017). Semi-automated map creation for fast deployment of AGV fleets in modern logistics. *Robotics and Autonomous Systems* 87: 281–295.

Borst, M. (2019). How has automation transformed aviation? *Society of Automotive Engineers (SAE International)*. https://saemobilus.sae.org/automated-connected/feature/2019/07/how-has-automation-transformed-aviation (accessed 13 March 2020).

DHL International GmbH (2014). Self-driving vehicles in logistics: a DHL perspective on implications and use cases for the logistics industry. *Delivery Tomorrow*. https://delivering-tomorrow.com/wp-content/uploads/2015/08/dhl_self_driving_vehicles.pdf (accessed 23 September 2019).

Etherington, D. (2019). Autonomous air transport startup Elroy Air completes first flights of large cargo VTOL. *TechCrunch*. https://techcrunch.com/2019/08/28/autonomous-air-transport-startup-elroy-air-completes-first-flights-of-large-cargo-vtol/ (accessed 15 March 2020).

Fan, T., Tao, F., Deng, S., and Li, S. (2015). *Impact of RFID technology on supply chain decisions with inventory inaccuracies. International Journal of Production Economics* 159: 117–125.

Fazlollahtabar, H. and Saidi-Mehrabad, M. (2015). Methodologies to optimize automated guided vehicle scheduling and routing problems: a review study. *Journal of Intelligent & Robotic Systems* 77 (3–4): 525–545.

Flämig, H. (2016). Autonomous vehicles and autonomous driving in Freight transport. *Planning and Logistics* 21071: 365–385.

Frey, C.B. and Osborne, M. (2013). *The future of employment: how susceptible are jobs to computerisation? Technological Forecasting and Social Change* 114 (100): 254–280.

Gartenberg, C. (2019). Here's Amazon's new transforming Prime Air delivery drone. *The Verge.* https://www.theverge.com/2019/6/5/18654044/amazon-prime-air-delivery-drone-new-design-safety-transforming-flight-video (accessed 16 March 2020).

Gasser, T. (2014). Vehicle automation: definitions legal aspects research needs. *UNECE workshop: towards a new transportation culture: technology innovations for safe efficient and sustainable mobility.* https://www.unece.org/fileadmin/DAM/trans/events/2014/Joint_BELGIUM-UNECE_ITS/07_ITS_Nov2014_Tom_Gasser__BASt.pdf (accessed 06 October 2019).

Hofmann, E. and Rüsch, M. (2017). *Industry 4.0 and the current status as well as future prospects on logistics. Computers in Industry* 89: 23–34.

Krotov, V. (2017). The internet of things and new business opportunities. *Business Horizons* 60 (6): 831–841.

Lee, R.J.V. and Leonard, R. (1990). Changing role of humans within an integrated automated guided vehicle system. *Computer Integrated Manufacturing Systems* 3 (2): 115–120.

Liao, J. (2019). Here's how drone delivery will change the face of global logistics. *World Economic Forum.* https://www.weforum.org/agenda/2019/01/here-s-how-drone-delivery-will-change-the-face-of-global-logistics/ (accessed 28 November 2019).

Monaghan, A. (2015). Drones: a view into the future for the logistics sector [report]? *Marsh.* https://www.marsh.com/uk/insights/research/drones-view-into-the-future-for-the-logistics-sector.html (accessed 29 November 2019).

Quatromoni, K. (2018). The Industrial Internet Consortium and Plattform Industrie 4.0 publish architecture alignment and interoperability: mapping and alignment between Industrial Internet Reference Architecture and Reference Architecture Model for Industrie 4.0. *Industrial Internet Consortium.* https://www.iiconsortium.org/press-room/02-06-18.htm (accessed 25 September 2019).

Reaidy, P.J., Gunasekaran, A., and Spalanzani, A. (2015). *Bottom-up approach based on Internet of Things for order fulfillment in a collaborative warehousing environment. International Journal of Production Economics* 159 (100): 29–40. https://doi.org/10.1016/j.ijpe.2014.02.017.

Riel, A., Kreiner, C., Macher, G., and Messnarz, R. (2017). *Integrated design for tackling safety and security challenges of smart products and digital manufacturing. CIRP Annals – Manufacturing Technology* 66: 177–180.

Robinson, A. (2016). Transport drones & autonomous vehicles: the transportation mode dance card is getting full. *Cerasis.* https://cerasis.com/transport-drones/ (accessed 28 November 2019)

Romero Segovia, V. and Theorin, A. (2012). *History of Control History of PLC and DCS.* Lunds: Lunds University. https://pdfs.semanticscholar.org/34cb/56424212448031d71588c8d74f263e31c6fd.pdf

Runge, J. (2006). Network model for the planning of plants with driverless forklift trucks. In: *Material Flow and Logistics Systems*, vol. 4 (ed. L. Schulze). Aachen: Shaker-Verlag (in German).

Rustici, C. (2020). How drone start-up Zipline is helping fight COVID-19 in Africa. *Direct Industry Magazine.* http://emag.directindustry.com/how-drone-start-up-zipline-is-helping-fight-covid-19-in-africa/ (accessed 22 May 2020).

Schulze, L., Behling, S., and Buhrs, S. (2008). *Automated guided vehicle systems: a driver for increased business performance. IMECS* II: 1–6.

Shaikh, S.A. (2017). This is the major flaw with driverless vehicles that no one is talking about. *World Economic Forum.* https://www.weforum.org/agenda/2017/08/you-haven-t-thought-of-one-of-the-biggest-problems-with-driverless-cars (accessed 25 September 2019).

Szozda, N. (2017). *Industry 4.0 and its impact on the functioning of supply chains. Scientific Journal of Logistics* 13 (4): 401–414. http://dx.doi.org/10.17270/J.LOG.2017.4.2.

Toyota Motor Corporation (2019). How many parts is each car made of? *Toyota Motor Corporation.* https://www.toyota.co.jp/en/kids/faq/d/01/04/ (accessed 03 September 2019).

Transport and Research and Innovation Monitoring and Information System (TRIMIS) (2020). The MUNIN Project. *TRIMIS.* https://trimis.ec.europa.eu/project/maritime-unmanned-navigation-through-intelligence-networks (accessed 13 March 2020).]

Ullrich, G. (2014). *Driverless Transport Systems*, 2nde. Wiesbaden: Springer Fachmedien.

Ullrich, G. (2015). *Automated Guided Vehicle Systems: A Primer with Practical Applications.* Berlin: Springer.

Vlahovic, N., Knezevic, B., and Batalic, P. (2016). Implementing delivery drones in logistics business process: case of pharmaceutical industry. *International Journal of Mechanical and Industrial Engineering* 10 (12): 4026–4031.

Wang, S., Wan, J., Li, D., and Zhang, C. (2016). *Implementing smart factory of Industrie 4.0: an outlook. International Journal of Distributed Sensor Networks* 2016 (4): 1–10.

Williams, M (2016). Move over Amazon: the US military is also developing a military drone. *IT World.* https://www.itworld.com/article/3035535/move-over-amazon-the-us-military-is-also-developing-a-delivery-drone.html (accessed 28 November 2019).

Zhao, G., Yu, H., Wang, G. et al. (2015). *Applied research of IOT and RFID technology in agricultural product traceability system. Computer and Computing Technologies in Agriculture VIII – CCTA* 2014: 506–514. https://hal.inria.fr/hal-01420266

Zuehlke, D. (2010). *SmartFactory – towards a factory-of-things. Annual Reviews in Control* 34: 129–138. https://doi.org/10.1016/j.arcontrol.2010.02.008.

9

Logistics in the Cloud-Powered Workplace

John Berry

JUSDA Supply Chain Management Corporation, Diamond Bar, CA, USA

The Cloud Revolution

We are witnessing a historic shift in how technology solutions are delivered within business environments. Cloud computing has transformed technology into a utility like water or electricity. Consumers of cloud services pay for outcomes. All the technical complexity of delivering those outcomes like servers, networking, capacity management, and security are left to the cloud provider.

The ubiquitous availability of on-demand computing has forced organizations to rethink their core competencies. In the past, the ability to execute well on software development and IT administration provided a competitive advantage. The cloud is democratizing the impact of IT. Organizations of all sizes now have instant access to highly scalable and highly available computing infrastructure. With a few hours of work, a savvy solo entrepreneur can provide a technology platform that rivals that of a Fortune 1000 competitor. Cloud-based software can be provided with nothing but a credit card and a web browser. High-quality technology services are becoming table stakes in the modern competitive business environment (Hernandez 2015).

In service-oriented industries like logistics, cloud-enabled providers differentiate their services not based on the specific technology they use, but on effective alignment of their chosen cloud services and competitive strategy. It is becoming more common that two logistics providers competing for the same business opportunity are using the same carriers to route shipments and the same cloud software to manage the data. Customer web portals, business forms, and reports offered by the two providers may be identical except for the logos. This forces logistics providers to differentiate not with what software they use, but how they use the software to deliver value to their clients.

The shift to the cloud will have a significant impact on the 5.5 million workers in US transportation and warehousing sectors (U.S. Bureau of Labor Statistics 2019). The growing use of cloud-based software is affecting the daily activities of these workers as well as the job skills employers are seeking. Although some logistics jobs are threatened by auto-mation and digitization, cloud software provides new opportunities for workers that embrace these tools. Savvy logistics workers can leverage cloud technology to shift their job focus from repetitive undifferentiated tasks to value creation.

This chapter will explore how software is used in the logistics industry and the impact cloud is having on these software tools. The principal categories of cloud-enabled value creation will be described along with an analysis of how the cloud is affecting logistics careers. Before exploring the impact of the cloud on logistics, it is essential to understand the history of these technologies and the key types of cloud computing services.

Growing Dominance of Cloud Computing

Infrastructure as a Service (IaaS)

Infrastructure-as-a-Service grew out of the practice of collocating servers in third-party data centers. Historically organizations operated servers or mainframes on premises. Maintaining high availability and security in this environment was an expensive undertaking. A simple power outage could bring an organization's systems down. It was quite common to hear stories about a janitor accidentally knocking a power cable lose at night and causing a service outage. The growth of the Internet brought further challenges. Organizations now had to run Internet circuits into their buildings to expose their web servers to the public Internet. Providing enough network bandwidth and reliability for these connections was difficult. To address these chal-lenges, collocation data centers emerged, which rented server space to organizations and offered highly reliable infrastructure and Internet connectivity. This was the first step in the progression toward public cloud infrastructure.

Another key ingredient in the creation of the cloud model was virtualization technology. In 2001 VMWare released its ESX product, which allowed pools of virtual servers to be run on a single physical machine. In data centers, virtualization made it possible to maximize hardware utilization by running tens or hundreds of virtual machines across a much smaller pool of physical servers. Clustering technology made these server pools highly resilient. If a server in the cluster failed, the virtual machine could be failed over to another node in the cluster with no downtime.

Virtualization allowed public cloud services to take off. Third-party data centers began to offer virtualized servers that were accessible over the public Internet as a subscription ser-vice. A subscriber could use a web portal to request a new virtual machine with the required CPU, memory, and storage specifications. Within a few minutes, the virtual machine would be provided on the provider's infrastructure. The subscriber was usually billed per minute of server usage. It was entirely acceptable to provide a new virtual machine, use it for a day, and decommission it, resulting in only one day of fees. This model became known as Infrastructure-as-a-Service (IaaS).

In 2006 Amazon released its Elastic Compute Cloud service (Amazon.com Inc. 2006), which became the dominant IaaS product over the next several years. Amazon was already investing heavily to build out data centers to support its eCommerce business, and Elastic Compute Cloud made it possible for customers to rent virtualized servers and storage run-ning on this infrastructure. Starting with sales of around $21 million in 2006, Amazon proceeded to double or triple revenues annually for the next six years (Morgan 2018).

As it became clear the IaaS business model was gaining traction, new providers like Microsoft, Alibaba, Google, and IBM entered the market. As of 2018, Amazon's share of the IaaS market was estimated to be 49.4%. Microsoft's market share was 12.7%, and the rest of the providers had market shares of less than 5% (Costello and Goasduff 2019). Although Amazon is still the dominant IaaS provider, Amazon's incursion into sectors like retail and logistics is becoming a larger concern to some of its IaaS customers. In 2018 and 2019, several retailers like Walmart, Albertson's, and Walgreens began shifting their workloads from Amazon to alternative cloud providers like Microsoft (Vena 2019). These cloud consumers are trying to avoid strengthening Amazon, which is becoming a severe threat. As Amazon begins to compete directly in the logistics sector (Hampstead 2019), logistics providers may need to make similar decisions about whether to use Amazon as their IaaS provider.

Platform as a Service (PaaS)

The Platform-as-a-Service (PaaS) model offers pre-configured computing services that software developers can use to host applications without the need to configure the underlying server infrastructure. Instead of starting with an empty IaaS virtual machine and installing database software, a fully hosted PaaS database service could be provided instead. In the PaaS scenario, the configuration and management of virtual machines, storage, network, operating system are entirely outsourced to the cloud provider. This allows the customer to focus purely on functional requirements.

Google brought the PaaS model to a broad market when it introduced AppEngine in 2008 (Schofield 2008). In its initial stages, AppEngine allowed developers to create software using open source tools like Python and Django and then deploy them directly to the AppEngine service. The application could be easily scaled to handle large amounts of traffic and offered high availability. All of this was possible without the involvement of an IT administrator to provision virtual machines and set up the software.

As the cloud market developed, providers began to offer a wide array of platform services. Services like queuing, messaging, IoT processing, search indexing, application monitoring, and artificial intelligence (AI) are now available as building blocks to developers. A modern software engineer can leverage these PaaS services and deploy production-quality applications without any involvement from IT admins. The capabilities provided by PaaS have enabled many enterprises to develop custom software much more quickly and inexpensively (Lawton 2008). PaaS has also supported the massive growth of the SaaS industry by making it easier for startups to build and scale web applications.

In 2017, UPS leveraged Microsoft's Azure PaaS services and Bot Framework to develop a customer service chatbot in just over three months. This chatbot allowed customers to engage with UPS on Facebook Messenger, UPS's mobile app, and virtual assistants like Amazon Alexa. Users of the service could ask questions like "Can you upgrade my SurePost packages to Ground?" or make requests like "Tell my driver to leave my package at the front door." UPS was able to accelerate their time to market by using PaaS services like storage, mapping, cognitive services, and application monitoring. After eight months of use, the bot had logged over 200 000 conversations with UPS customers (Microsoft Corporation 2017).

Software as a Service (SaaS)

In the Software-as-a-Service (SaaS) model, the software is hosted by a vendor and made accessible to end customers over the Internet on a subscription basis. According to Gartner, total sales in the SaaS market were expected to be $72 billion in 2018, and it is the largest segment of the overall cloud market (Heiser and Santoro 2019). With an overall enterprise software market of $419 billion in 2019, it is expected that spend on SaaS offerings will continue to grow at a rapid pace (Costello 2019). Initially, SaaS became popular with smaller businesses that had few or no IT employees. SaaS providers made it possible for these companies to outsource the development and operation of business systems to third parties.

SaaS is typically provided at a monthly or annual subscription. This eliminates the need to make significant capital investments for software licenses and servers at the beginning of a project. The low upfront cost is compelling to many businesses. If the software does not deliver as promised, the customer can cancel the subscription and minimize their losses. Since SaaS products are designed to reduce implementation complexity, faster projects, and lower startup costs often result in a quicker return on investment.

SaaS products are typically designed to be *multitenant* in order to scale usage. This means that one instance of the software is serving multiple users or tenants (Knorr 2006). Underlying code is the same across all users and cannot usually be customized for an individual customer. This creates economies of scale for the SaaS provider (Levinson 2007). This multitenant architecture can make SaaS implementations simpler and faster than traditional software. New usage can often be provided instantly. The costs to build custom code is eliminated. Of course, this means some business processes may need to change to conform to the SaaS design. But many organizations see the simplicity and cost savings resulting from a one size fits all SaaS as an acceptable trade-off. Additionally, SaaS products used in a particular industry tend to evolve to support best practices. SaaS implementations can often introduce positive change by forcing organizations to move away from outdated business practices.

A successful SaaS provider could have hundreds or thousands of customers running on a single code base. This tends to result in less buggy software than a highly customized piece of software built for a single organization. Given the SaaS business model is dependent on achieving economies of scale, the SaaS product must be stable and easy to use. Otherwise, as the SaaS user base grows, the cost to provide support will make the business unprofitable. The SaaS model incorporates enhancements into the subscription cost. The consistent revenue generated from subscriptions provides vendors a reliable source of investment to enhance their products. Customers benefit from continuous upgrades and enhancements. Examples of prominent logistics software providers with SaaS offerings are Oracle, JDA, Infor, WiseTech, and MercuryGate.

Enabling New Business Models

The availability of cloud services has permanently altered the entrepreneurial process. Back office activities like accounting, HR, and customer relationship management can be automated with SaaS products like Oracle Netsuite, Workday, and Salesforce. New technology-based products and services can be developed more rapidly with cloud source control, bug tracking, and continuous integration services like Microsoft Azure DevOps. Cloud-based big data platforms like Google BigQuery, Amazon Redshift, and Microsoft Azure DataBricks support descriptive and predictive analytics that can unlock the value of a startup's data. These kinds of capabilities have become important drivers for business innovation.

Cloud technologies make it possible to experiment quickly and inexpensively. In the past, if an organization wanted to perform historical analysis of sales patterns to improve forecasts, it was often necessary to build a data warehouse first. Servers and database software needed to be purchased and installed. Database engineers then spent months designing the data warehouse, building processes to load the data from external systems, and developing reports that analysts could use to interpret results. This often took one or more years to complete. In 2004, a time when on premises enterprise data warehouses were popular, it was estimated that a typical data warehouse project cost over $1 million. One half to two-thirds of all data warehouse projects were considered failures (Watson and Haley 2004).

Now with a cloud provider, powerful data platforms can be provisioned in minutes. When a new data analysis project is identified, analysts can choose from various PaaS database services, including traditional relational databases, data warehouses, graph databases, time series databases, or Hadoop-based systems. The database engine that best fits the analysis problem can be selected. If necessary, the cloud service can be scaled to handle massive data sets with quick response times. Once the analyst has found answers or solved the business problem, the cloud data platform can be decommissioned until the next analysis problem comes along. This kind of on-demand data analysis technology is allowing organizations to leverage data far more effectively than was possible with pre-cloud technologies.

The Malaysia-based airline AirAsia gained this kind of ability to scale and iterate when it moved its data platform to Google's cloud services. A minimal number of IT engineers were able to load data into Google's BigQuery data warehouse and then quickly create reports and dashboards to help facilitate digital transformation efforts. AirAsia expects to reduce operating costs by 5–10% as a result of the project (Alphabet Inc. 2019).

How Software Drives Logistics

Organizations that sell physical products have typically operated their businesses on ERP systems. These comprehensive software solutions were designed to serve large multinational organizations. This made them extremely complex and expensive to implement. Typically the highest priority of an ERP implementation was getting the financials right. Software modules like purchasing, sales, and logistics fed the financial reporting processes, and automating these across a global organization could create significant returns on investment.

Although fundamental supply chain capabilities like inventory management and order fulfillment were handled by these ERP systems, they tended to be quite basic. Their primary purpose was to support financial processes. An ERP's inventory management module might be quite good at tracking inventory valuation for the enterprise balance sheet, but it was not exceptionally good at managing the labor in the warehouse. This created a market for best-of-breed logistics applications that could integrate with the ERP and fill in the gaps. Two primary categories of logistics applications emerged: warehouse management systems (WMS) and transportation management systems (TMS).

Warehouse Management Systems (WMS)

WMS support and optimize the daily operations in a warehouse. Key capabilities include receiving goods, putting the inventory away, optimizing the location of inventory, and fulfilling sales orders by picking, packing, and shipping (Ramaa et al. 2012). These functions require integration with the ERP's product master, purchase orders, sales orders, and inventory balances. Most WMSs include support for mobile operations on radio frequency devices that can scan barcodes or process radio frequency identification (RFID) signals. The more advanced WMS products also support task management and labor management capabilities. This helps guide a forklift driver on the most efficient path through the warehouse while tracking productivity. For larger warehouse environments, it may be necessary for the WMS to interact with automated material handling systems like conveyors and sortation systems.

The use of WMS technology can offer significant benefits to logistics organizations. A survey of 131 members of the Warehouse Education Research Council (WERC) sought to confirm the business improvements offered by warehouse technology hardware like handheld computers and WMS software. The survey confirmed that investments in this technology provided benefits like reducing inventory, improving space utilization, reducing stock-outs, improving cycle time, and improving order accuracy. WMS users that were surveyed believed the technology had made their organizations more profitable and had improved customer relations (Autry et al. 2005). Another study conducted at India's largest retailer in 2012 found that a WMS implementation combined with lean improvements resulted in a 40% reduction in labor (Ramaa et al. 2012).

Transportation Management Systems (TMS)

TMS focus on shipping goods using the most cost-effective method while complying with service requirements. Given a particular shipment and its transit time requirements, the TMS searches the rates of all known carriers and transportation modes and suggests the best method to ship. This can become quite complicated when factoring in shipment consolidation, cubic volume optimization, and risks like weather, traffic, and carrier performance variability. The TMS generates shipment documentation like labels and bills of lading, which are necessary to execute the shipment. TMS products often help audit carrier freight invoices. These functions require integration with the ERP's purchase orders, sales orders, and account payables.

A TMS can reduce transportation costs in a variety of ways. By automating the process of booking and tracking shipments, labor in the shipping department can be significantly reduced. Routing guide capabilities ensure that shipping clerks select the least-cost transport mode for each shipment. Many TMS products can find opportunities for consolidating less-than-truckload shipments into full-truckload shipments, which are far more cost-effective. Carrier invoice audit and exception management help shippers catch incorrectly rated shipments, inappropriate surcharges, or service level failures, which can result in refunds.

Shortcomings of Conventional Logistics Software

Complexity

As the WMS and TMS markets grew, providers competed for customers based on the volume of features in their software. Increasing the depth of capabilities was necessary to market the software to larger customers and those with specialized requirements. However, the added complexity made it more challenging to configure and use the software. Only users with extensive product expertise were able to set up functions like warehouse layout, product masters, picking rules, or freight tariffs.

This created an additional revenue stream for the WMS and TMS providers. More complexity in the software itself drove more customers to consume professional services like consulting and training. Additionally, many software providers sold through systems integrators that helped their clients select and implement software. A vital component of a systems integrator's revenue was configuration services and ERP integration work. Given the inertia of these business realities, software providers often had limited motivation to invest in making it easier for customers to do self-serve implementations. This sometimes resulted in a lack of user documentation and training materials.

Due to these factors, advanced capabilities were costly to implement. The professional services necessary to configure, implement, and support the software often exceeded the actual license cost of the software. These projects frequently outsourced all the technical work, and few if any internal employees were able to learn the technical nuances of the software. After implementation, changes or enhancements still required support from the consultant, which added cost and delayed the time to market for the improvement.

A case study on the effects of complexity on a TMS implementation was DHL Global Forwarding's (DHL-GF) ill-fated New Forwarding Environment (NFE). In 2011, DHL-GF began a project to modernize its legacy air freight and ocean freight management system. The plan involved consolidating forwarding operations onto a single global platform, which was based on SAP TM (SAP's TMS module). By the end of 2015, the project had been canceled, resulting in a financial write-down of EUR 345 million. A principal cause of the failure appears to have been the inability to handle such a complex implementation on a TMS that was not fully optimized for global forwarding operations. DHL-GF was a large organization with many different systems and extremely complex processes that had grown to absorb customer requirements. Additionally, the global forwarding model was far more complicated than parcel or road freight. Business processes were highly variable, and there were extensive compliance requirements in each country. Rationalizing, all of that complexity and overlaying it into a generic TMS like SAP TM became impractical (Wood 2019). Since the termination of the NFE project, DHL-GF began transitioning to the SaaS-based Cargowise product from WiseTech with much more success. By the end of 2018, DHL-GF had rolled out CargoWise in over 40 countries (King 2018).

Difficult Data Integration

Studies have shown that implementing data integrations across supply chain participants can result in increased customer service levels, better product quality, and operational cost reductions (Craighead et al. 2006). However, conventional WMS and TMS were often difficult to integrate with, making it challenging to realize these benefits. Getting data in and out of these tools usually involved working with native file formats or complex XML documents. Consuming these interfaces required deep expertise with the software and its functions. The interfaces were often not well-documented. The data available in these interfaces tended to be semantically dependent on the software itself. Master data like vendor codes or trade term codes often did not follow any standards, and code dictionary lookups needed to be maintained by the integration tool. In the best-case scenario, a few members of the internal IT department were able to learn the integration mechanisms, but often the organization was completely dependent on external consultants to perform integrations. This could result in an isolated WMS or TMS that didn't interact with other key business systems effectively.

Integrations with key external trading partners like suppliers, carriers, and logistics service providers could be challenging to implement. Often electronic data interchange (EDI) software had to be implemented, which performed data translations between the shipper's systems and their trading partner's systems. Before any messages could be exchanged, it was necessary to analyze the partner's EDI specifications, build a translator map in the EDI software, and perform extensive testing. Each EDI implementation required a custom project requiring teams of expensive IT specialists to frequently communicate about connectivity, message structure, data semantics, and testing processes.

A commonly used EDI message in the logistics sector is the Advanced Shipment Notification (ASN). This message provides details about a shipment to a receiving warehouse. The most common format for this message is the ANSI X12 856 standard (American National Standards Institute 2014). Often ASNs are paired with the requirement to label associated shipment cartons or pallets with a globally unique Serial Shipping Container Code (SSCC) barcode (GS1 2019). The ASN can reference each SSCC code and specify the SKUs and quantities contained. Most WMSs can receive against ASN data. When the warehouse worker scans the SSCC, the WMS can immediately trigger the put away process. ASNs can yield significant labor savings for the receiving warehouse. However, in the past, the cost and complexity to implement EDI was a major hurdle for smaller suppliers. The initial cost of the EDI translator license combined with the consulting fees of a system integrator was far too expensive when compared to the projected profit of the project. Many deferred on EDI implementations related to logistics such as ASNs. A lack of ASNs forced the receiving warehouse to process receipts manually, which increased operational costs.

The Impact of SaaS on Logistics

A New Technology Delivery Model

Software-as-a-Service (SaaS) has grown to be a massive industry. Gartner estimates that the SaaS market has increased by 20% every year since 2015 (Moore and van der Meulen 2018). The SaaS model has allowed smaller independent software organizations to create products for specific niches. In the logistics industry, there is a growing number of SaaS products related to TMS, WMS, CRM, and customs compliance. If a business problem exists, there is probably a SaaS solution available. There are SaaS CRMs that target 3PLs, SaaS

WMSs that are optimized for small warehouse providers, and cloud-based AI services that help automate freight brokerage back-office work.

It is often the case that one or two dominant SaaS providers emerge in a particular vertical or business function. Then a "long tail" of specialty products develops to target specific niches. As employees move from one job to the next, they have an opportunity to work with various SaaS solutions in their domain. Those workers communicate with colleagues at other organizations, and a collective consciousness forms about which SaaS providers are the industry leaders.

An obvious example of a dominant SaaS provider is Salesforce. Salesforce was one of the early entrants into the cloud CRM space and now dominates market share. Due to the ubiquity of Salesforce, understanding how to use the software has become a core job skill for anyone working in the sales and marketing domain. Best practices on how to leverage Salesforce are openly available. There is a vast array of training content available to introduce beginners to the product.

SaaS Warehouse Management Systems

The Warehouse Management System market is very mature with well-established software vendors that provide deep capabilities for highly complex warehouse operations. Most of the leading WMS providers have started offering cloud delivery models. According to Gartner, 40–50% of new WMS customers are now choosing cloud delivery instead of on premises deployments (Klappich and Tunstall 2019).

Some obstacles make cloud WMS implementations difficult in larger warehouses. Unlike on premises deployments, access to the cloud service is entirely dependent on the warehouse's Internet connection. Any network downtime brings the warehouse operation to a halt. This offers quite a bit of risk, especially when the site is in an area where network infrastructure is unreliable.

On premises implementations tend to provide better software performance due to reduced network latency. When scanning hundreds of serial numbers with an RF device, an additional 60 ms of network latency could introduce a noticeable impact on task duration. Additionally, some automated material handling systems require low latency query responses from the WMS. Conveyor systems may scan a barcode to decide the correct route path for the carton. If there is a network delay between the scan and the route response from the WMS, it may be too late for the system to react. Because of these constraints, the WMS providers that target larger warehouses are still investing in non-cloud versions of their products. However, the broader WMS market continues to shift toward a SaaS-dominant model.

In addition to the WMS market leaders, new SaaS-native WMS providers are entering the market. These products tend to target smaller warehouses with lower complexity, but this target market is quite large. These providers are having success due to the lower initial costs offered by the SaaS subscription model. These products often provide prebuilt integrations with commonly used cloud ERPs and eCommerce solutions. They offer modern user experiences and focus on self-serve setup and implementation. Although these newer SaaS-native providers often lack the depth of features provided by the entrenched market leaders, these products can grow capabilities quickly. These startups tend to be nimbler than their larger competitors and are encumbered with far less technical debt. The capabilities of the SaaS native providers could eventually achieve parity with the well-established players.

SaaS Transportation Management Systems

According to Gartner's Magic Quadrant for Transportation Management Systems report, most of the established TMS providers are transitioning their products to a SaaS model to compete with new cloud-native providers that have recently entered the market. These more modern platforms often leverage emerging technologies to deliver features like mobility, real-time visibility, and analytics. Gartner estimated that SaaS would account for 65% of the TMS market by 2022 (De Muynck et al. 2019).

Many SaaS TMS vendors have successfully differentiated by offering large networks of pre-integrated carriers as well as modern web-based user experiences. As more carriers expose their services as REpresentational State Transfer (REST) application programming interfaces (APIs), it becomes possible for TMS providers to expand the number of carriers available to shippers without requiring an expensive integration project. This allows shippers to query rates, book shipments, track shipments, and process freight invoices across a large cross-section of the parcel and LTL providers. For shipping modes like full truckload that tend to operate on a spot market, many SaaS providers provide integrations with web-based load boards. A load from the TMS can be posted to multiple boards electronically, and the resulting carrier vendor quotes can be pulled back into the TMS for review. The ability to instantly transact business over the internet with hundreds of potential carriers has dramatically streamlined the modern shipping department.

Many of the SaaS TMS providers have focused on improved user experience. Their user interfaces adopt the same design principles as consumer-facing web applications. The software designers often use an intimate understanding of their customers' daily activities and frustrations to build user interfaces that solve problems with a high level of simplicity. This makes the SaaS applications easier to learn and more enjoyable to work with on a daily basis.

APIs

Since SaaS products are not installed on a local server or PC, it has become necessary to provide data integrations over the Internet. Given the multi-tenant hosting model of most SaaS products, integration methods must be consistent across all users and easily consumable on a self-service basis. The predominant data integration approach in multi-tenant cloud systems has become web-based APIs. Based on REST principles, these APIs can be accessed over the public Internet using basic HTTP actions, which are supported by many open-source tools. This makes it possible for third-party developers to move data in and out of SaaS applications easily.

A key feature of REST-based APIs is self-serve consumption. Typically SaaS products that offer APIs provide a documentation web site that explains how to work with the available interfaces and data structures. A developer that needs to create a specific type of transaction in the SaaS can simply read the API documentation, make some experimental

API calls using tools like cURL or Postman, and see the impact in the SaaS application's user interface.

An example SaaS product that offers an API with strong self-serve support is AfterShip. Shippers or logistics providers can leverage the AfterShip API to integrate parcel shipment visibility into their customer-facing services. AfterShip provides detailed documentation about how to interact with their API (AfterShip documentation 2019). After registering and obtaining an API key, shipment tracking requests can be created, and tracking result data can be retrieved. This data could be combined with sales order status to provide a user experience similar to that of Amazon's.

API-enabled SaaS tools are making integrations across trading partners in a supply chain much easier. A buyer can refer their suppliers to their SaaS product's API documentation and request integration to be completed by a particular deadline. As the SaaS market for supply chain begins to consolidate into fewer dominant players, it is becoming more likely an organization's trading partners have already developed integrations with their SaaS tool's API in the past. Many SaaS applications are now pre-integrated with the APIs of ancillary applications. For example, a cloud ERP for small-to-medium sized businesses may already have API integrations with eCommerce applications and parcel carriers for shipping. An extensive portfolio of integrations like this is becoming a key differentiator for SaaS products.

Logistics organizations that continue operating on legacy systems and conventional EDI software are at risk of falling behind SaaS-enabled competitors. It is quite common for a conventional EDI integration project to take one to two months to complete. As customers experience faster integration lead times with SaaS-enabled providers, they will begin to expect projects to be measured in days, not months. The time to implement integrations is often a differentiating factor for logistics organizations. A 3PL that can integrate a customer's sales orders into their WMS in three days will often beat out a competitor that requires two months for the same integration. Furthermore, the legacy provider will likely incur much higher implementation costs than their SaaS-enabled competitor. These costs will need to be passed through to the customer, which further limits competitiveness.

SaaS-Enabled Value Creation

These modern SaaS applications are having a substantial impact on the logistics workplace. Cloud-based warehousing and transportation software products tend to offer improved user experiences, streamlined integration options, and data analytics capabilities. Logistics workers have new tools at their disposal to solve problems and create value for customers. SaaS tools are transforming the daily work and career skillsets required in the logistics domain.

Automation

Given the explosion of SaaS products that offer APIs, a new category of process automation tooling is emerging. These tools help nontechnical workers to create workflows across cloud services by interacting with APIs. The user can graphically design flows by connecting triggers and actions. For example, the user may specify that when a new lead is created on a lead page (trigger), a new prospect should be created in the CRM (action). Then an email should be sent to an account executive, notifying them about the lead (another action). Since the process automation tool has pre-built connectors to SaaS APIs, useful workflows can be built without writing any code. When pre-built connectors do not exist yet for a particular SaaS product, it is usually still possible to create a custom connector as long as the SaaS offers a REST API.

Examples of process automation tools include Zapier, IFTTT, and Microsoft Power Automate (previously called Microsoft Flow). Zapier was an early entrant to the market and offered one of the largest portfolios of SaaS connectors. At the time of this writing, Zapier provides connections to over 1500 SaaS applications (Zapier Inc. 2019). Microsoft Power Automate provides similar capabilities but includes tight integration with Office 365, which is a dominant SaaS solution for business applications like email, word processing, spreadsheets, and collaboration (Roth 2018). Power Automate offers connectors to Microsoft Azure's PaaS capabilities, which include databases and cognitive services like computer vision and natural language processing (Microsoft Corporation 2019).

One third-party warehousing and distribution provider was able to speed its customer onboarding process 70% by moving workflows onto Microsoft's cloud-based Azure Logic Apps automation service. Before the project, a large number of human resources in the IT department were required to implement and maintain EDI workflows with customers and business partners. The company replaced 16 on premises EDI solutions with Azure Logic Apps. The new process made it possible to automate EDI workflows without the need to write code. Additionally, a mobile app was developed for truck drivers that triggered a Logic App workflow, which sent EDI messages to a big-box retailer. This reduced the time lag for data transmission and helped the 3PL meet customer compliance requirements. The 3PL estimates that this project has allowed it to reduce related costs by 50% (Microsoft Corporation 2017).

A similar type of tool is robotic process automation (RPA), which enables users to configure software-based robots that interact with applications to execute business processes (UiPath Inc. 2018). When software does not offer an API, an RPA tool can be used to interact directly with the software's user interface to automate tasks. For example, if a WMS does not offer an API for the creation of sales orders, an RPA tool could be used to create a workflow that pulls orders from the web shopping cart API, log into the WMS user interface, and type in the sales order by simulating keyboard entry and mouse clicks. Microsoft has begun to integrate RPA capabilities into its Power Automate workflow automation service (Lamanna 2019).

These tools are democratizing automation development. Before the availability of easy-to-use process automation, it was always necessary for IT Departments to implement integrations. Complex projects often required the services of third-party system integrators. This made automation projects, both expensive and time-consuming. The lead time from the identification of a problem to creating business value was often extremely long. Process automation tools now allow nontechnical business users to build highly targeted solutions themselves. This shortens the time to market and ensures the technology solution matches the business requirements accurately.

These automation tools help individual contributors in logistics organizations increase operational efficiency and improve customer service levels. Workers can streamline workflows, reduce labor, and leverage data to optimize financial results. As those closest to the business problems become empowered by technology, the ability to deliver business value is shifting from technology departments to front line workers. When organizations used custom software, every improvement project required long time frames and big budgets. Now with automation-enabled SaaS tools, business value delivery is shifting to a daily task available to all members of the organization.

Integration

Supply chains are composed of various trading partners working together to fulfill product demand. A key barrier to increasing efficiencies has always been poor information flow between supply chain participants. Each trading partner uses its own systems and processes to execute its portion of the value stream. In the past, that meant processing an order could involve isolated processing in several different on premises ERPs, WMSs, and TMSs. Data was transmitted between these applications using paper, email, Excel files, or EDI in the best-case scenario.

As more global supply chain participants move onto SaaS platforms, these data flow obstacles are becoming easier to overcome. In certain domains like international forwarding, the SaaS market has consolidated to the point that all of the top 25 global freight forwarders now use some form of technology provided by Wise Tech Global, a leading TMS provider (WiseTech Global 2019). Wise Tech Global's SaaS-based TMS provides tools that enable shipment data exchanges between trading partners. The application offers a standard XML format for common transactions like shipments, events, and invoices. Some third-party SaaS tools can now handle these formats, which makes integrations across logistics organizations more accessible than ever.

The users of these SaaS products gain significant benefits from these improved data flows across the global supply chain. Productivity is improved since logistics workers do not need to rekey data into systems like WMS and TMS. When trading partners implement high-quality integrations between their systems, data is more accurate and timely. This keeps processes streamlined and reduces the need for humans to audit and cleanse data. There are security benefits, as well. When order, shipment, or inventory data is transmitted among parties using email or Excel, little access control can be enforced. However, when the data exchange is built into the business systems, there are usually far more access controls and audit capabilities to ensure data security.

Analytics and Artificial Intelligence

Traditional software often did not include many business intelligence capabilities other than simple collections of tabular reports. IT teams needed to develop processes to load the software's data into a data warehouse for more advanced data analysis. It is far more common for modern SaaS products to include embedded analytics capabilities. Configurable reports and dashboards can help users develop actionable insights from the SaaS tool's data.

It is also possible to connect many SaaS applications to third party data analytics tools using APIs or other types of connectors. Popular tools like Tableau, Microsoft Power BI, and QlikView offer powerful data visualization capabilities and are designed to work well with cloud software and databases. Workers can use these types of tools to extract data, prepare it for analysis, create visualizations, and analyze data with methods like correlation analysis, anomaly detection, and forecasting.

Many SaaS applications are beginning to offer AI-enabled features. In 2017, Salesforce introduced a service that used AI to prioritize sales leads based on which were most likely to close. The company's goal was to make AI more accessible to its customers, most of which did not have the resources to acquire its own IT expertise (Rosenberg 2017). Many other SaaS providers have followed suit by adding new features to their solutions enabled by their data assets and AI tools.

These new data-driven SaaS features are helping front line logistics workers deliver value to their organizations. Analysis of past data can reveal business patterns that help workers plan more effectively. These *descriptive analytics* improve understanding of what happened in the past (Bertolucci 2013). For example, a warehouse supervisor may use descriptive analytics to analyze historical daily order counts to understand the impact of the financial quarter ends on labor. *Predictive analytics* uses statistical analysis and machine learning techniques on past data to make predictions (Bertolucci 2013). The previously mentioned warehouse supervisor could use a forecast function in Excel or similar tools to predict the daily order counts at the end of the next quarter. *Prescriptive analytics* recommends a course of action (Bertolucci 2013). A WMS enabled with prescriptive analytics may recommend to the warehouse supervisor how much headcount will be necessary to handle the predicted order load at the end of next quarter.

As these capabilities continue to grow and mature in the logistics SaaS market, organizations that leverage these technologies will receive significant business benefits. With AI-enabled capabilities, it will be possible for supply chain organizations to optimize inventory levels, refine pricing with competitor analysis, and predict failures of factory equipment or material handling systems (Madhavanur 2018). Logistics workers that develop expertise in these capabilities will be in a unique position to innovate and deliver significant business improvements to their organizations.

Conclusion

Over the last 15 years, cloud technology has evolved into the dominant technology delivery model. The software that powers the logistics industry is now leveraging the cloud to overcome previous weaknesses and deliver business value more effectively. This seismic shift is impacting the daily activities of workers' transportation and logistics sector. This will create new career opportunities for those that are willing to cultivate the skills employers are seeking. At the same time, new technologies like robotics, autonomous vehicles, and AI could cause significant disruption and reduce the number of jobs in the sector. Savvy logistics workers can use the opportunities provided by the cloud revolution to protect their careers from technological disruption and find rewarding work in the logistics sector.

Key Takeaways

- The three key cloud computing models are Infrastructure-as-a-Service (IaaS), Platform-as-a-Service (PaaS), and Software-as-a-Service (SaaS)
- The two primary categories of logistics software are Warehouse Management Systems (WMS) and Transportation Management Systems (TMS)
- The SaaS model is helping overcome common shortcomings of conventional WMS and TMS software.
- SaaS WMS and TMS products provide improved automation, integration, and analytics capabilities, which enable logistics organizations to create value for their customers.

References

AfterShip documentation (2019). AfterShip documentation. *AfterShip*. https://docs.aftership.com/api/4/overview (accessed 29 September 2019).

Alphabet Inc. (2019). AirAsia: turning to google cloud to refine pricing, increase revenue, and improve customer experience. *Google*. https://cloud.google.com/customers/airasia/ (accessed 29 September 2019).

Amazon.com Inc. (2006). Announcing amazon elastic compute cloud (amazon EC2) – beta. *Amazon.com, Inc.* https://aws.amazon.com/about-aws/whats-new/2006/08/24/announcing-amazon-elastic-compute-cloud-amazon-ec2---beta/ (accessed 3 August 2019).

American National Standards Institute (2014). 856 Ship Notice/Manifest. X12.org. http://www.x12.org/x12org/Subcommittees/X12J/ts/856.htm (accessed 2 October 2019).

Autry, C.W., Griffis, E.S., Goldsby, J.T., and Bobbit, L.M. (2005). Warehouse management systems: resource commitment, capabilities, and organizational performance. *Journal of Business Logistics* 26 (2): 173.

Avril, P. (2017). Big data quarterly. *Information Today, Inc., Fall.* http://www.oracle.com/us/products/database/changing-role-of-the-dba-4369434.pdf (accessed 13 February 2020).

Bertolucci, J. (2013). Big data analytics: descriptive vs. predictive vs. prescriptive. *Information Week* (31 January). https://www.informationweek.com/big-data/big-data-analytics/big-data-analytics-descriptive-vs-predictive-vs-prescriptive/d/d-id/1113279 (accessed 26 August 2019).

Costello, K. (2019). Gartner says global IT spending to grow 0.6% in 2019. *Gartner* (10 July). https://www.gartner.com/en/newsroom/press-releases/2019-10-07-gartner-says-global-it-spending-to-grow-06-in-2019 (accessed 29 September 2019).

Costello, K. and Goasduff, L. (2019). Gartner says worldwide IaaS public cloud services market grew 31.3% in 2018. *Gartner* (29 July). https://www.gartner.com/en/newsroom/press-releases/2019-07-29-gartner-says-worldwide-iaas-public-cloud-services-market-grew-31point3-percent-in-2018 (accessed 28 September 2019).

Craighead, C.W., Patterson, J.W., Roth, P.L., and Segars, A.H. (2006). Enabling the benefits of supply chain management systems: an empirical study of Electronic Data Interchange (EDI) in manufacturing. *International Journal of Production Research* 44 (1): 135–157.

Davidson, K. (2016). Employers find 'soft skills' like critical thinking in short supply. *Wall Street Journal*, 30 August. https://www.wsj.com/articles/employers-find-soft-skills-like-critical-thinking-in-short-supply-1472549400 (accessed 12 August 2019).

De Muynck, B., John, B. and Duran, S. (2019). Magic quadrant for transportation management systems. *Gartner*.

Dietrich, E. (2017). *Developer Hegemony: The Future of Labor*. DaedTech.

Florentine, S. (2015). Why soft skills are key to a successful IT career. *CIO Magazine* (2 February). https://www.cio.com/article/2878675/why-soft-skills-are-key-to-a-successful-it-career.html (accessed 11 August 2019).

General Assembly and Burning Glass Technologies (2015). Blurring lines: how business and technology skills are merging to create high opportunity hybrid jobs. *General Assembly & Burning Glass Technologies*.

GS1 (2019). Serial shipping container code (SSCC). GS1.org. https://www.gs1.org/standards/id-keys/sscc (accessed 2 October 2019).

Hampstead, J.P. (2019). Breaking: amazon's digital freight brokerage platform goes live. *FreightWaves* (26 April). https://www.freightwaves.com/news/breaking-amazons-digital-freight-brokerage-platform-goes-live (accessed 28 September 2019).

Heiser, J. and Santoro, J. (2019). Hype cycle for software as a service, 2019. *Gartner* (30 July). https://www.gartner.com/en/documents/3953760/hype-cycle-for-software-as-a-service-2019 (accessed 20 August 2019).

Hernandez, P. (2015). Verizon: enterprise cloud use is 'table stakes'. *Datamation* (9 November). https://www.datamation.com/cloud-computing/verizon-enterprise-cloud-use-is-table-stakes.html (accessed 19 February 2020).

John Langley Jr. P. A. I. C. (2019). 23rd annual third-party logistics study: the state of logistics outsourcing.

Jun, J. (2018). The lazy Dev's guide to API automation: airtable to survey monkey uploader (5 June). https://blog.getpostman.com/2018/06/05/airtable-to-surveymonkey/ (accessed 12 August 2019).

King, M. (2018). DHL digital forwarding rollout 'progressing rapidly'. *Lloyd's Loading List* (4 December). https://www.lloydsloadinglist.com/freight-directory/news/DHL-digital-forwarding-rollout-'progressing-rapidly'/73359.htm (accessed 26 September 2019).

Klappich, C. and Tunstall, S. (2019). Magic quadrant for warehouse management systems. *Gartner*.

Knorr, E. (2006). Software as a service: the next big thing. *ComputerWorld* (23 March). https://www.computerworld.com.au/article/151970/software_service_next_big_thing/ (accessed 23 August 2019).

Lamanna, C. (2019). Robotic process automation now in preview in microsoft power automate. *Microsoft Corporation* (4 November). https://flow.microsoft.com/en-us/blog/robotic-process-automation-now-in-preview-in-microsoft-power-automate/ (accessed 26 February 2020).

Lawton, G. (2008). Developing software online with platform-as-a-service technology. *IEEE Computer* 41 (6): 13–15.

Levinson, M. (2007). Software as a service (SaaS) definition and solutions. *CIO Magazine* (15 May). https://www.cio.com/article/2439006/software-as-a-service--saas--definition-and-solutions.html (accessed 20 August 2019).

Madhavanur, D. (2018). How JDA is reshaping supply chains of tomorrow with AI. *JDA* (12 July). https://blog.jda.com/how-jda-is-reshaping-supply-chains-of-tomorrow-with-ai/ (accessed 25 August 2018).

Microsoft Corporation (2017). UPS paves the way for better service with faster development and artificial intelligence. *Microsoft* (18 September). https://customers.microsoft.com/en-us/story/ups (accessed 29 September 2019).

Microsoft Corporation (2017). Logistics company transforms customer service, reduces costs by 50 percent with move to the cloud. *Microsoft Corporation* (14 December). https://customers.microsoft.com/en-us/story/legacy-supply-chain-services-partner-professional-services-azure-logic-apps (accessed 1 October 2019).

Microsoft Corporation (2019). Microsoft flow connectors. *Microsoft Corporation*. https://flow.microsoft.com/en-us/connectors/ (accessed 1 September 2019).

Moore, S. and van der Meulen, R. (2018). Gartner forecasts worldwide public cloud revenue to grow 21.4 percent in 2018. *Gartner* (12 April). https://www.gartner.com/en/newsroom/press-releases/2018-04-12-gartner-forecasts-worldwide-public-cloud-revenue-to-grow-21-percent-in-2018 (accessed 13 September 2019).

Morgan, T.P. (2018). Navigating the revenue streams and profit pools of AWS (5 February). https://www.nextplatform.com/2018/02/05/navigating-revenue-streams-profit-pools-aws/ (accessed 17 August 2019).

Oracle Warehouse Management Cloud (2018). Certified implementation specialist certification overview. *Oracle Corporation*. https://education.oracle.com/oracle-warehouse-management-cloud-2018-certified-implementation-specialist/trackp_540 (accessed 18 September 2019).

Overby, S. (2014). 5 hybrid IT roles your business needs to succeed in 2014. *CIO Magazine* (20 January). https://www.cio.com/article/2379397/5-hybrid-it-roles-your-business-needs-to-succeed-in-2014.html (accessed 2 August 2019).

Pande, P.S., Neuman, R.P., and Cavanagh, R.R. (2000). *The Six Sigma Way: How GE, Motorola, and Other Top Companies Are Honing Their Performance*, vol. 37, 5–9. McGraw-Hill.

Perkins, B. (2018). What is change management? A guide to organizational transformation. *CIO Magazine* (12 April). https://www.cio.com/article/2439314/change-management-change-management-definition-and-solutions.html (accessed 12 August 2019).

Ramaa, A., Rangaswamy, T.M., and Subramanya, K.N. (2012). Impact of warehouse management system in a supply chain. *International Journal of Computer Application* 54: 14–16.

Rosenberg, S. (2017). Inside salesforce's quest to bring artificial intelligence to everyone. *Wired Magazine* (2 August). https://www.wired.com/story/inside-salesforces-quest-to-bring-artificial-intelligence-to-everyone/ (accessed 25 August 2019).

Roth, C. (2018). Office 365 survey results are in! *Gartner* (9 January). https://blogs.gartner.com/craig-roth/2018/01/09/office-365-survey-results-are-in/ (accessed 24 August 2019).

Schofield, J. (2008). Google angles for business users with 'platform as a service'. *The Guardian* (16 April). https://www.theguardian.com/technology/2008/apr/17/google.software (accessed 20 August 2019).

Sigelman, M., Bittle, S., Markow, W., and Francis, B. (2019). *The Hybrid Job Economy: How New Skills Are Rewriting the DNA of the Job Market*. Burning Glass Technologies.

U.S. Bureau of Labor Statistics (2019). Transportation and Warehousing: NAICS 48–49. *Bureau of Labor Statistics*. https://www.bls.gov/iag/tgs/iag48-49.htm#iag48-49emp1.f.p (accessed 31 August 2019).

UiPath Inc. (2018). Robotic process automation. *UIPath*. https://www.uipath.com/rpa/robotic-process-automation (accessed 1 September 2019).

Vena, D. (2019). Amazon fear is driving retailers to microsoft's cloud. *The Motley Fool* (23 April). https://www.fool.com/investing/2019/01/28/amazon-fear-is-driving-retailers-to-microsofts-clo.aspx (accessed 28 September 2019).

Viscelli, S. (2018). Driverless? autonomous trucks and the future of the American truck. *Center for Labor Research and Education, University of California, Berkeley, and Working Partnerships USA* (September). http://driverlessreport.org (accessed 2 October 2019).

Watson, H.J. and Haley, B.J. (2004). Data warehousing: a framework and survey of current practices. *Journal of Data Warehousing* 2 (1): 10–17.

WiseTech Global (2019). WiseTech global 1H19 results (February). https://ir.wisetechglobal.com/FormBuilder/_Resource/_module/aQGWYiQKCEe9gcmOsTiUKA/WTC-1H19-Results-investor-presentation.pdf (accessed 26 August 2019).

WiseTech Global (2019). The CargoWise certified professional program – customers. https://video.wisetechglobal.com/the-cargowise-certified (accessed 3 August 2019).

Wood, R. (2019). Market insight: the inside story on DHL forwarding's IT transition nightmare; and how it made WiseTech what it is today. *The Loadstar* (3 March). https://theloadstar.com/market-insight-the-inside-story-on-dhl-forwardings-it-transition-nightmare-and-how-it-made-wisetech-what-it-is-today/ (accessed 26 September 2019).

Zapier Inc. (2019). Zapier app integrations. https://zapier.com/apps (accessed 24 August 2019).

Zsigri, C. and Reichman, A. (2016). The impact of could applications on the role of IT (July). https://www.bettercloud.com/wp-content/reports/451-research-bettercloud.pdf (accessed July 2019).

Section III

Platforms

10

Platforms Enabling Digital Transformation

Mac Sullivan

NNR Global Logistics, Dallas, TX, USA

The increased adoption of more sophisticated systems across the global supply chain is driving the need for standardization of communication and information. New digital sales channels have the power to dramatically change the labor needed and the margins possible for logistics service providers. Logistics companies, especially international freight forwarders, are increasingly under pressure as pricing moves online creating transparency in the market and opening opportunities for new competitors due to increasingly lower barriers to entry. Freight forwarding traditionally has been a very customer-intensive service industry which creates economic value through not only facilitating the timely delivery of goods, but also the arranging of other formalities, such as finding space with a carrier, creating and distributing documentation and lastly ensuring that the goods get through customs (Shang and Lu 2012). When matching algorithms can be used to identify carrier vendors, system-to-system connectivity can automatically distribute documents, and handheld devices at a manufacturing facility can be directly linked with a destination country customs system, freight forwarders are having to rethink their value proposition.

Traditionally, freight forwarding has very much been a relationship-based business-to-business (B2B) service industry with little sway being held by online forums, market reviews, and digital sales channels. This status quo has been disrupted by the growing popularity of digital platforms that allow businesses to manage their logistics services in a single online portal where they can get quotes, book space, and manage their shipments. In a broader sense, platforms are digital environments where access to a product or service can be offered with near-zero marginal cost (McAfee and Brynjolfsson 2018). Platforms represent a byproduct of The Fourth Industrial Revolution that is a culmination of many different technologies. As seen in other industries, a shift to a platform model will have a ripple effect throughout the logistics industry as it requires a new skill set to set up and manage, a change in customer behavior, and will open the door for innovation and competition.

In the opening chapter of this part of the book, Ruben Huber, the founder of *Ocean X*, a digital freight forwarding network, discusses the journey of digitalization and the effects

The Digital Transformation of Logistics: Demystifying Impacts of the Fourth Industrial Revolution, First Edition. Edited by Mac Sullivan and Johannes Kern.

that it will have on the labor force as well as the opportunities it will open up in terms of new business models. He highlights the importance of transparency as it will increase specialization and accelerate market segmentation. His chapter explores the concept of how digitalization will decrease the need to parse and disseminate information into different forms as it is automated.

One area where platforms and the underlying connectivity technology are lacking is in terms of gaining trust by an older demographic and within certain supply chain areas where inappropriate behaviors have been found previously. Matias Aranguiz, Dr. Xu Duoqi, Bill Tran, and Andrea Margheri, highlight the opportunity to solve these trust issues by using Blockchain technologies. Although it might not be the immediate panacea that some were thinking, Blockchain technology still has a strong group of followers and some large multinationals are getting onboard. Smart contracts are one of the most exciting applications of Blockchain and this chapter looks at areas of logistics and international trade where they could be applied.

As the technologies of The Fourth Industrial Revolution grow in popularity and more use cases become available, companies have begun to create platforms that decrease disintermediation and disrupt traditional logistics and supply chain models. Andre Wheeler highlights that governments, like China, are also starting to leverage technology to create operational efficiency and open up new revenue opportunities. Companies will need the support of governments to bring standardization to reality and vice-versa. Without the support of a common approach to technology integration, it may remain unfeasible to create a completely connected digital ecosystem.

A commonly overlooked component of the international supply chain is the activity that happens at marine ports throughout the world. Marine terminal operating systems are integral pieces of a shipment as ships and containers are continuously interacting in these hubs. The ability to operate these ports efficiently has a direct effect on the host countries trade volume. In the chapter on this topic, Ira Breskin teams up with Ayush Pandey to discuss the importance of these ports having an operating system that can connect with and augment the wealth of opportunities that the technologies of 4IR are providing.

Platforms cannot be discussed without looking at them through the lens of e-commerce and the effect that it is having on consumer behavior leading to a demand for new hyperaccelerated logistics efficiency. In Chapter 15, Simon de Raadt examines how some European governments and companies are teaming up to innovatively capture information further upstream than has been previously thought possible. De Raadt gives us a case study to show how the role of government should not be underestimated as regulation changes can cause huge disruptions with a seemingly small administrative change.

Cory Margand and Sam Heuck explore how platforms are revolutionizing international freight procurement and disrupting the status quo of traditional rate management systems. According to Margand's experience building the Simpliship platform, the emergence of Software as a Service (SaaS) offerings that help to index carrier contracts are opening up new communication methods between freight forwarders and their carrier partners. In this chapter we see the challenges and lessons learned in regards to creating a platform that can successfully disrupt the international freight landscape by integrating numerous sources of data and automating decision-making processes for all stakeholders.

As consumers are increasingly purchasing more perishable goods on online platforms, Alex von Stempel offers us insight into the challenges and opportunities within the cold chain industry as companies respond to the high levels of visibility and quality demands. In Chapter 17, von Stempel takes a look at how logistics companies are leveraging 4IR technologies, like Blockchain and IoT, to solve the complex and changing dynamics in the transportation of perishable goods.

References

McAfee, A. and Brynjolfsson, E. (2018). *Machine, Platform, Crowd: Harnessing our Digital Future*. W.W. Norton.

Shang, K. and Lu, C. (2012). Customer relationship management and firm performance: an Empiracal study of freight forwarder seervices. *Journal of Marine Science and Technology* 20 (1): 64–72.

11

The Digital Transformation of Freight Forwarders

Key Trends in the Future

Ruben Huber

OceanX, Laufenburg, Switzerland

Introduction

The Specter of the Digital Transformation

As digitalization is reshaping the logistics industry, traditional business models are questioned, and new ones enabled. This will also bring various opportunities for non-vessel operating common carriers (NVOCCs), ocean carriers that transport goods under their own House Bill of Lading without actually operating ocean vessels (FMC 2019). However, lots of actors are intimidated by this transformation and the changes it might lead to. At a recent conference, I discussed the topic with the long-standing operations director of a Spanish NVOCC. "When I started in this industry, digitalization meant using a Telex. If somebody from Asia wanted to send you a message, he had to ring the bell at the one machine in your office first. 'Are you there?' Then one of your colleagues would have to find you and you had to go there and reply 'Yes, I am here'. Only then you could slowly type messages to each other and arrange the shipment. Today, I am 24/7 available for everybody around the world thanks to this [cellphone]. It is scary." Also, a McKinsey study that analyzed the impact on jobs due to advances in robotics, artificial intelligence (AI), and machine learning came to the chilling conclusion that transportation and warehousing are among the sectors with the highest potential for automation (Manyika et al. 2017). But while there remains a large amount of fearmongering within the industry and beyond concerning digitalization, the subject should nevertheless be approached with an open mind for change, the willingness to adapt, and thus as the great opportunity it presents.

Actors in the Shipment Cycle

Figure 11.1 illustrates the shipment cycle, parties, and traditional roles, elaborating the distinction between business-to-business (B2B) freight and consumer-oriented last-mile logistics. The basis for all international transportation is in fact a trade of goods between a seller and a buyer.

The Digital Transformation of Logistics: Demystifying Impacts of the Fourth Industrial Revolution,
First Edition. Edited by Mac Sullivan and Johannes Kern.
© 2021 by The Institute of Electrical and Electronics Engineers, Inc.
Published 2021 by John Wiley & Sons, Inc.

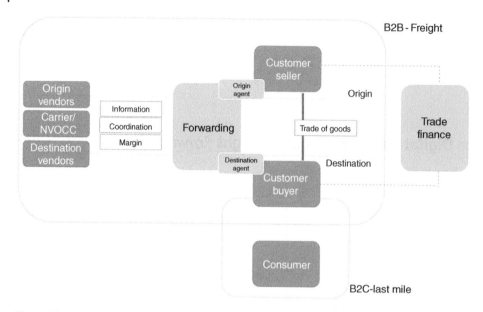

Figure 11.1 Shipment parties. *Source:* Courtesy of OceanX Network.

Key Trends That Will Shape the Future

In this ecosystem, four key trends that were triggered by digitalization are going to shape the future of international transportation:

- Technology – new tools as an opportunity and enabler.
- Specialization – focus on strengths in the face of transparency.
- Omnichannel – the selling and buying through a multitude of channels.
- Virtualization – internationalization, global expansion, and interaction.

Technology

Rather than discussing potential new technologies for transportation, like aerial unmanned vehicles, otherwise known as drones, or the Hyperloop,[1] the focus of this chapter will stay in the mature field of technology available today within maritime and containerized transportation. It is important to keep focusing on digitalization effects on this existing transport mode with a perspective on the potential future (Bamburry 2015; Ross 2015).

---

1 Hyperloop, as described conceptually by Elon Musk of SpaceX and Tesla in an August 2013 white paper, is a new, very high speed, intercity transportation mode. It consists of two (or more) very long tubes, elevated on pylons, which have been partially evacuated of their air, creating a partial vacuum. Linear induction motors propel small passenger or freight capsules riding on low-friction air bearings at very high speed within the tubes. The partial vacuum dramatically lowers aerodynamic drag resistance, which is ordinarily a key limitation on vehicle speed. (. . .) Proposals include 28 passenger pods that also carry up to three vehicles and freight pods that can hold a standard 40 ft long shipping container (Taylor et al., 2016).

Fundamentally, technology is a tool. Be it a new means of communication, the cloud, blockchain, AI, or big data analysis, they are all are essentially just tools. One may use the example of a hammer, which is lastly a simple piece of technology that enabled new ways of building houses after its invention. Beneficial technologies tend to become dominant designs and thus are lastly available to everyone; therefore we consider technological lead, rather a short-term advantage. And even employing the same tool might lead to very different results.

Technology in transportation, first and foremost, has been communication technology. The history of the bill of lading, the maritime transport document, dates back to 1397 when the main means of communication were personal messengers (McLaughlin 1926). Logistics communication evolved from using the telegram, telephone, telex, and telefax to using email and instant messaging. Yet, the document, which often constitutes a negotiable title for a cargo, remained largely an original paper document verified with stamps and on which value is bestowed upon. Customs documents, required to import and export goods, as well as related processes, have also been largely paper based and, in many countries, still are. While many of these processes have proved effective, they are lacking efficiency. Given that much of the information is communicated repeatedly between different parties, the general rule of thumb that a paper-based process is a less efficient process can be applied. Digitalization in the context of logistics means how whole domains are restructured around digital communication and media infrastructures transmitting information digitally (Brennen and Kreiss 2014) – as opposed to digitization, which simply means transferring analog information into a digital one, such as scanning a document and storing it in a digital form.

The traditional document flow required local offices to send and receive documents, making it a requirement to be local, operate local offices, hire local staff, and conduct local business at least at origin and destination. This also triggered traditions and habits. While the introduction of the telephone would theoretically have allowed being located outside of a larger city, many customers would require a local area code for their "trusted" partners. Although hard to imagine in today's world of live chat and instant communication, "geographical footprint" and office locations were 10 years ago a valid criterion for vendor choice and to some extent still are today.

Many tasks of white-collar employees in international transportation relate to communication of information: creating it, receiving it, capturing it, manipulating it, forwarding it, and taking action based on it (Manyika et al. 2017). Technology that enables automated capturing of information, accelerates its flow, enables sharing, etc. has a vital influence on the shipment handling process. In this area, B2B international shipping and freight forwarding is still far behind the standardization and digitalization of the courier and last-mile services; this is partly thanks to a direct pull from the tech-enabled consumer (PwC 2019).

A further direct impact of technology is around visibility, which has become a customer requirement (Kersten et al. 2017). Internet of Things (IoT), sensor technology, and advanced communication devices are increasingly enabling "intelligent" transportation units. As you can read in Chapter 3 on IoT, it allows to leapfrog over historic information chains and provides direct visibility for the customer, regarding their cargo whereabouts and condition, enabling faster decision making. IoT will gradually allow having "artificial intelligence" accompany cargo at every step of the way. Supercharged by this intelligence, the once analog steel box will in the long term enable new business models. Taking bike-sharing business models, such as China's Mobike, as an inspiration, where a GPS sensor combined with the tech-enabled user created a multibillion-dollar industry, containerization might finally see a true revolution, as part of the fourth industrial one (Schwab 2017).

Specialization

Transportation and logistics traditionally know many intermediary business models, where services are being resold through intermediation. Digitalization and new digital channels enable a new kind of transparency, thus making simple reselling of services, or information arbitrage, very hard (Kumar and Seppi 1994; Parker et al. 2016). The role of the traditional forwarder, which offered all services based on subcontractors, is shifting in two potential directions, (i) competence-based specialization and (ii) capability-based specialization. Figure 11.2 illustrates the possible development of the traditional "one-stop-shop forwarding sector" that will, through digital transparency and the ability of software with AI to take over coordination and information flow between many different parties, see a shift toward specialization.

Competency-based specialization will be more similar to consulting, alike a fourth-party logistics (4PL) role, possibly including digital elements and visibility services (Win 2008). Capability-based specialization will be based on specific abilities a logistics supplier has, of which many are asset based. Many of them used to have a role between the carrier and the forwarder, be it as NVOCCs operating consolidation services, specialized warehouse, and service operators, such as cool chain, fashion, or dangerous goods logistics, or actual carriers of any transport mode, such as truck or intermodal operators. Cargo does require physical movement and transparency through digitalization, as well as easier integration with customers through digital interfaces, yielding an opportunity for those parties moving up the value chain and selling directly to their customers, while the arbitrageur without the actual capability will not be able to survive (Abrahamsson 2016).

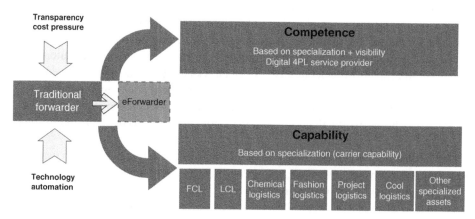

Figure 11.2 Forwarder's business model shift toward specialization. *Source:* Courtesy of OceanX Network.

Digitalization drives transparency and simplifies the coordination of different suppliers directly, in turn exposing simple information arbitrage, making it a nonbusiness model, as it is lacking any value-add. Following the efficient market hypothesis, in a more perfect and transparent market, it should be impossible that anyone sells a product cheaper than the one producing it, and digitalization will no doubt lead us closer to that (Fama 1970; Lock 2018). This effect is lastly the reason why digitalization has moved slowly in the logistics industry so far, as many business models are still based on information arbitrage and have no impetus to change. Indeed, some stakeholders in the logistics industry are fighting innovation to protect their revenue streams (Sandberg and Hemilä 2018).

Omnichannel

The sales channels for logistics services, which used to be mainly personal and either in the country of origin or destination, are now changing and increasingly becoming more diverse (Brynjolfsson et al. 2013). Historically, there has been a separation of responsibility for transport, often geographically split, with one agent, being an office of the logistics company or a third-party partner handling the origin side of the transport and the contact with the shipper, while the other agent, being the logistics company's own office or what could also be a partner, is managing the transport on the destination end with the consignee. The point of sales thereby is either at origin or the destination side. The organization in each location is hence built on the traditional service pillars of a logistics organization encompassing sales, customer service, operations or vendor management, and administration that includes finance and accounting.

This model that has been working for ages now faces challenges in the light of digitalization. The transparency provided by the Internet and online sales lifts the traditional separation of markets, where the one organization sells in their market and another one in the other market (Jain et al. 2017). In the digital age, an organization that is providing transportation services to or from a specific area will have to aim at selling its services to any customer that requires these services, independent of customer location. Sales will occur through a multitude of channels. Modern communication tools and changing customer behaviors support this development, partners will inevitably compete with each other in the online space, and therefore the traditional separation of markets will come under pressure.

Traditionally, the key sales channel for logistics services was local and personal; in the future, logistics services will be sold via a multitude of sales channels and brands – ranging from traditional, personal, and local channels over different own online sales channels with potentially multiple brands for different target customers to online marketplaces and digital intermediaries (Toczauer 2019).

This future change of sales channels is illustrated in Figure 11.3, where an environment with many channels, as opposed to one of few that have historically dominated the industry, will become common. The development is not unlike what started in consumer retail already earlier, and that is now finding its way into B2B services (McKinsey 2016).

This key trend, named as the "omnichannel future of shipping," is not only challenging previous organization structures across logistics organizations, particularly in sales and procurement, but also opening up significant opportunities for global expansion.

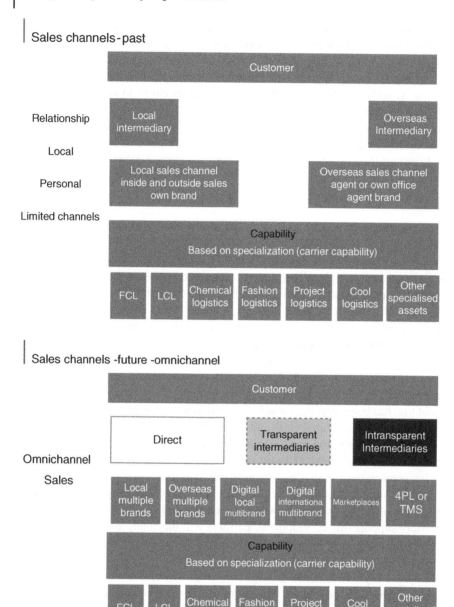

Sales channels-past

Sales channels -future -omnichannel

Figure 11.3 Shift toward omnichannel sales. *Source:* Courtesy of OceanX Network.

Virtualization

Similar to other industries, such as banking and insurance, new technologies allow questioning old practices, as services have the possibility of "virtualization" (Liao et al. 1999). Services that traditionally had a strong local component, be it through physical document flows, relationship sales, preferred communication means, or business custom, have the opportunity to be conducted in new ways in the future. The first examples for this become visible through business models of digital freight forwarders and marketplaces that have virtual presence rather than physical in many markets. While their progress and adoption might still move gradually, they are showing the way for the future of the industry (Seeßle 2019). This is also challenging the traditional ways of overseas cooperation, be it between own branches or international partners. The tech-enabled consumer, which is driving innovations in retail and the last mile, is also an employee. The younger generation is pushing the gig economy, questioning the traditional place, means, ways, and measurements of work, at a time when roles and tasks are changing.

The offshoring of noncustomer-facing operational tasks such as any noncustomer-facing elements of the process that have not been automated yet has been gradually centralized in low-cost countries, to facilitate cost reductions for logistics companies (Tate et al. 2009). Some examples for this are the service centers of shipping lines, such as Maersk Line in China and CMA CGM in India, or those of larger forwarders and logistics companies in India and the Philippines. There is also an increasing migration of sales and service channels toward online shippers (Shippers turn to digital booking platforms 2019) – be it self-quotations, online bookings, shipment visibility, or the fact that phone and personal contact has been increasingly replaced by email and message communication, done either directly or through digital marketplaces.

Reshaping Logistics Service

Those four elements, the disappearance of the local focus (or globalization of markets), automation, centralization, and offshoring of operations, going online of sales and service, will continue to significantly reshape this service industry. In particular, these changes are questioning the traditional organization model with local branches or local agents. If we further consider the technological advancements (e.g. transition of systems and software in the cloud) and the potential of process and workflow automation already possible today (or foreseeable on the horizon), we need to rethink the model of a future logistics organization (Kern and Wolff 2019).

Building on the above considerations, it may be anticipated that the physical location of a logistics organization will matter less in the future as a growing part of the service will be experienced online or "virtually" by the customer, rather than local and in person. Trust relationships between firms and customers will increasingly be shaped by omnichannel influences and experiences, like social media, ratings on public platforms, convenience, and responsiveness, rather than the personal meeting, and thus by brand value more than by personal relationships.

The office, traditionally the only visible part of the non-asset-based firm, is losing its importance.

Already for some time, the use of a beautiful office for the physical customer interactions that happen there serves as the basis for interesting discussions. It must be noted though that the importance in developed countries is mainly decreasing compared with developing regions, where checking out the "hardware" of a service firm still is somehow important for confidence building (Parasuraman et al. 1991), an argument however that can be satisfied with logistics assets even better, particularly considering that in reality, most customer meetings today happen at the customers' office instead of the one of logistics firms.

Let us further consider the high degree of automation to be expected in the years to come for our industry, and what will remain are three key tasks in the white-collar field:

1) Sales, but more in some sort of "customer consulting role."
2) Exception management, i.e. solving problems that go beyond the standard.
3) Local vendor management.

If we combine these elements and take a page from the playbook of modern technology firms that build effective, decentralized, global organizations, one could foresee a global logistics organization that does provide global services and coverage, but without any of the traditional structures. Activity-based work (ABW) would enable new, decentralized employment models. For example, if an Uber driver today can log in at the start of his workday, receiving jobs until he logs off again, why could a customer service employee engaged in exception management not have a similar way of conducting his working day from any location?

Today the workflow of a shipment follows a clear pattern. If one would imagine it more simplified and with higher automation, the pattern would become even clearer. Tasks could be allocated, escalated, and completed, without the need to be at one physical location (Sendler 2017). Regarding sales, companies such as OOCL have already tried the use of virtual sales personnel in inland offices, and some companies are already using shared desk spaces for them. The best place for a sales asset remains in front of the customer rather than in the office. If we are looking at retail especially those firms focusing on online sales, they are already there, handling a large area with multiple markets, e.g. Europe, via one central distribution center and billing from one central finance center and managing the website, sales, and customer support from one central or a multitude of decentralized locations.

Collaboration Is Key

The trends outlined above will change the shape of the logistics industry and traditional roles of those staff within those companies. The freight forwarder as the "architect of transport" will have to reinvent himself (FIATA 2020). With technology enabling the seamless combination and coordination of service providers, information arbitrage as a means to an income will disappear gradually. The traditional split of handling international shipments between origin and destination will increasingly blur. Transparency will drive specialization and a focus on the individual strength of every player in the supply chain. This opens vast opportunities for collaboration between specialists, following the old rule of focusing on what you are great at and outsourcing everything else (Deepen 2007). More can be seen on the subject of automation and outsourcing in Chapter 5 of this volume on The Role of Robotic Process Automation (RPA) in Logistics.

Companies that often used to be in the second row, between the actual carriers and customer-focused solution providers, will be able to move closer toward the customer, as the role of coordinating multiple providers might no longer be adding value in itself and become subject to automation, as has been seen in the growth of advanced customer-centric transportation management systems (TMS) (Rouse 2015). The asset-light model, a success story of the past, is losing its appeal, and the one carrying the cargo or having "carrier capabilities" moves more into the spotlight (Shariff 2019). As management developer Reinhard Sprenger recently outlined as key guidance for the digital age, "Customer, collaboration and creativity are the key strengths of human beings (Sprenger 2018)."

Given the trends above and the great shift that is going to happen in logistics, collaboration is the key element to success – collaboration between specialists, jointly building creative answers to the challenges but also opportunities that these developments do provide. While non-value-adding intermediaries will be squeezed out of the value chain, those providing services and adding value will become more visible, providing not only opportunities for margin and market share expansion but also asset-based differentiation. Virtualization and digital connectivity between members provide the opportunity to deliver services in new ways and through a multitude of brands, based on new ways of cooperation between different firms. That being said, still, the key value proposition of transportation to the customer remains the movement of goods from A to B, in a certain time, at a certain cost, without any damage to the cargo (Rutner and Langley Jr 2000). Visibility during transport is becoming a standard requirement that is part of this value proposition. The actual physics of how and through whom the service is provided becomes of less importance.

Better communication and seamless digital combination of services from different providers will enable much closer integration and thus cooperation of firms. TMS service providers or digital 4PL will allow those earlier described, second-row "carriers," to directly provide services to customers. By combining their specialized assets and capabilities with those of others, through flexible networks, selling, and servicing through each other's brands, their scope will enlarge. Where previously only integrators were able to provide one-stop-shop solutions, those networked firms will be able to leverage through their assets (Vatne 2018). Successful networks will stretch far beyond historic, bilateral facilitation but integrate leading technology firms, bringing those mastering application and those mastering technology together, weaving an "intelligent fabric of interactions."

Where technological advancements used to be limited to leaders in the market, based on controlling know-how and capital, the virtualization of services and Software-as-a-Service (SaaS) concepts provides new opportunities (Johnson 2019; Safari et al. 2015). Networks will have the ability to combine a group of peers, bringing a wider range of ideas together and as a specialized crowd can become a source for funding and a nurturing ground for implementation. Local firms that were at a disadvantage compared to global market leaders in terms of scope and innovation potential could be more agile and benefit from each other's different focus and expertise. To learn more about the role of SaaS providers, cf. Chapter 16.

These are just some of the considerations behind the creation of OceanX (http://oceanx.network). It essentially aims to build a community of logistics firms in the sea freight sector that differentiate themselves through leading capabilities and competence in specialized areas – a peer group of exciting companies and individuals interested in exploring new ways to work on multilateral solutions for some of these challenges and build on their strength.

Technology-enabled transparency throughout the industry will require new ideas for differentiation of services. OceanX aims to establish standards for business excellence in certain specialty fields with our members. In a time when customer experience and convenience have become key differentiators in the end consumer world and partly last-mile logistics, we try to jointly conceive ways of defining it for B2B shipping and logistics services, supporting our members to reach new customers in different ways and leveraging each other's strengths to serve them.

Digitalization Impact for White-Collar Workers

The impact on white-collar workers will be diverse. Certainly, the focus of the past 30 years in terms of efficiency will be shifting, as we will be able to increasingly delegate efficiency to AI and automate it. Any tasks related to the rehandling of information, obtaining it, taking action based on it, and forwarding it, will also increasingly disappear, eliminating jobs and functions related (OECD 2019). However, the overall advancement of digitalization will also trigger a level playing field. With all providers essentially having the same tools at their disposal, further innovation in terms of technology will become less of a competitive factor, and employing it in the most effective ways, in combination with other success factors, will be crucial (Carr 2003).

In terms of transportation, one of those key factors will remain – people, particularly referring to consultative sales and exception management (Vestergaard 2019). Employing digital tools, transportation assets, own or of partners, and combining them effectively into creative solutions will remain the key differentiator for the industry. Therefore, while roles, tasks, and tools will change, the essential mission of efficiently moving goods will remain.

New organization models will be triggered by these developments and provide many benefits for white-collar workers. There remains a lack of logistics and supply chain personnel today across many markets in the world. Especially as trade continues to boom, attracting talent has become one key industry challenge (McKinnon et al. 2017). A younger generation that is increasingly looking for purpose and not interested in conducting those "efficiency tasks" of the past will be attracted by companies that adapt fast to the future and provide a flexible environment that is purpose driven, letting go of legacy structures.

White-collar workers that are not physically interacting with cargo and that have various communication means today to cooperate with colleagues, vendors, and customers will not be required to conduct their work at a specific location, which requires to waste time in "soul-crushing traffic" or other forms of commute (America is wasting billions sitting in soul-crushing traffic 2013). Innovative employers will increasingly use digital means to manage their workforce and, instead of operating offices in specific locations, will have a network of employees in virtual locations across their market, not only improving flexibility for their workers but also thereby moving closer to customers, vendors, and partners.

Virtualization of the workplace will enable employees to have more flexible systems of employment, too (Beblavý et al. 2012). Using our Uber driver as an example again, who may choose to offer his services via different brands and sales channels, we might see the same happening for competent workers in the logistics industry, offering, for example, exception management and customer-related services to a range of providers. Companies that intend to internationalize their services will no longer be required to establish branch firms overseas and open offices abroad; instead they will be able to source the local talent required directly, subcontracting specific roles and tasks, to provide their services overseas.

While these possibilities will be largely beneficial for employees of today, they will be a challenge for organizations to manage and to transition to, balancing the needs for security of the older workforce with the desire for flexibility of the younger ones that will no doubt be increasingly enabled. Not every employee will need to become an IT engineer; however, the level of technological understanding and the ability to use the tools available will be necessary. As in many other industries, continued education of the workforce, room for experimentation, and nurturing an innovative culture will become crucial success factors for firms and essential to attract talent (McKinsey 2017). Given the increasing importance of the supply chain and getting products to customers, logistics experience might even be essential for future CEOs, as Nader Mikhail of Elementum recently suggested (Mikhail 2018). Therefore, the outlook for white-collar workers in the industry seems bullish.

The Regulatory Environment

A key threat to global service firms will remain the regulatory environment. Similar to the free movement of goods being essential for global trade, the free movement of services will be essential for the globalization of the service economy. With physical location becoming less of a factor in a virtual world, not only taxation but also environment and living quality will be decision makers for individuals and logistics providers when picking the location from where they intend to provide their services.

The virtualization of services is already producing losers, for example, in the advertisement and marketing industry, the European Union tried to take on Google and Facebook that produces for the European market but largely from overseas locations. Contrasting to this, logistics firms today do operate local offices and employ local employees that pay local taxes and contribute to their local economy. Therefore, regulatory actions concerning virtual services are no doubt to be expected; however, it is to be hoped that those actions will embrace the opportunities, rather than foster a new wave of protectionism.

Conclusion

At the end of the current wave of digitalization, where technology has been a differentiator and key competitive factor, technology will come full circle and become less important. The playing field will have been leveled and logistics companies will offer many similar features. This can be already seen in the automobile industry today, where different companies and brands compete with essentially the same tools, and differentiation is achieved through means largely outside of technology.

Visibility and information will certainly become a norm in the future, a vast range of data will be available, and actionable insights from this data will become of more importance than providing the data itself. Firms will need to deliver creative solutions based on that data and creativity, essentially bringing us back to key human strength over AI. There will no doubt be significant changes across our industry with a vital impact on the role of white-collar labor. However, it will also yield significant opportunities, as tasks will become more interesting when routine tasks get automated and organizational structures more flexible. Local and specialist expertise will continue to matter and be a key differentiator of services, supported by assets and aligned across flexible networks. The shift away from an efficiency focus toward creativity is not unlike the developments seen in the blue-collar field earlier and a natural progression.

Increasing automation of efficiency-related tasks through the application of AI will in the long-term also decrease the total number of white-collar jobs – from purely a perspective of the current scope of services. However, exploring more creative ways for differentiation in this future world, it can also be envisaged that logistics will take a wider role within global trade as a whole, integrating upward and downward, thereby expanding its scope.

In the end, one key talent of the logistics industry has always been the ability to adapt to exceptional circumstances and changing requirements, an essential ability of employees in our industry that will be even more important during the times of change that we are going to experience. Therefore, despite fearmongering and almost apocalyptic-sounding studies of consulting firms, it is prudent to remain positive on what is to come. Limiting factors in the future will not be technological possibilities but instead the "human 1.0." Jessica Tyler, managing director of culture and transformation of American Airlines Cargo, summarizes it with a smile. "The beautiful thing is that this industry is full of passionate, experienced, talented professionals. It's on us to achieve the vision and to do the work to shift mindsets and work together differently to get it done. Adapt or die, folks (Branson 2018)."

Key Takeaways

- Four key trends that were triggered by digitalization are going to shape the future of international transportation:
    - Technology – new tools as an opportunity and enabler.
    - Specialization – focus on strengths in the face of transparency.
    - Omnichannel – selling and buying through a multitude of channels.
    - Virtualization – internationalization, global expansion, and interaction.
- Transparency will drive specialization and a focus on the individual strength of every player in the supply chain. This opens vast opportunities for collaboration between specialists.
- Better communication and seamless digital combination of services from different providers will enable much closer integration and thus the cooperation of firms.
- The impact on workers will be diverse. Any tasks related to the rehandling of information, obtaining it, taking action based on it, and forwarding it, will also increasingly disappear due to digitalization, eliminating jobs and functions related.
- A key threat for a golden age of global service firms will remain the regulatory environment.

References

Abrahamsson, M. (2016). *Digitalization as a Driver for Logistics Service Development.* Linköping https://www.chalmers.se/en/centres/lead/urbanfreightplatform/past-vref-conferences/vrefconf16/program2016/Documents/Digitalization as a driver for logistics service development - Presentation.pdf (accessed 14 December 2019).

America Is Wasting Billions Sitting in Soul-crushing Traffic (2013). *NBC News.* https://www.nbcnews.com/business/economy/america-wasting-billions-sitting-soul-crushing-traffic-flna1B8251555 (accessed 14 December 2019).

Bamburry, D. (2015). Drones: designed for product delivery. *Design Management Review* 26 (1): 40–48.

Beblavý, M., Maselli, I., and Martellucci, E. (2012). Workplace innovation and technological change. *CEPS Special Reports.* https://papers.ssrn.com/sol3/papers.cfm?abstract_id=2147619 (accessed 14 December 2019).

Branson, J.R. (2018). *Adapt or Die.* https://www.aacargo.com/about/digital-disruption-air-cargo.html (accessed 14 December 2019).

Brennen, S. and Kreiss, D. (2014). Digitalization and digitization, culture digitally. http://culturedigitally.org/2014/09/digitalization-and-digitization (accessed 14 December 2019).

Brynjolfsson, E., Hu, Y.J., and Rahman, M.S. (2013). Competing in the age of omnichannel retailing. *MIT Sloan Management Review.*

Carr, N.G. (2003). IT doesn't matter. *Harvard Business Review* 81 (5): 41.

Deepen, J.M. (2007). *Logistics Outsourcing Relationships.* Heidelberg: Physica-Verlag.

Fama, E.F. (1970). Efficient capital markets: a review of theory and empirical work. *The Journal of Finance* 25 (2): 383. https://doi.org/10.2307/2325486.

FIATA (2020). Who is FIATA. https://fiata.com/about-fiata.html (accessed 14 December 2019).

FMC (2019) *Ocean Transportation Intermediaries.* https://www.fmc.gov/resources-services/ocean-transportation-intermediaries (accessed 5 April 2020).

Jain, S., Chaudhary, P.K., and Patchala, S. (2017). Technology-led disruptive freight marketplaces – the future of logistics industry. https://www.tcs.com/content/dam/tcs/pdf/Industries/travel-and-hospitality/Building-Disruptive-Freight-Marketplaces.pdf (accessed 14 December 2019).

Johnson, E. (2019). *SaaS helping to bring TMS to the masses. The Journal of Commerce* https://www.joc.com/international-logistics/logistics-technology/saas-helping-bring-tms-masses_20190412.html (accessed 14 December 2019).

Kern, J. and Wolff, P. (2019). The digital transformation of the automotive supply Chain–an empirical analysis with evidence from Germany and China. https://www.innovationpolicyplatform.org/www.innovationpolicyplatform.org/system/files/imce/AutomotiveSupplyChain_GermanyChina_TIPDigitalCaseStudy2019_1/index.pdf (accessed 14 December 2019).

Kersten, W., Seiter, M., Von See, B. et al. (2017). Trends und Strategien in Logistik und supply chain management - Chancen der digitalen transformation. *BVL Bundesvereinigung Logistik*: 71. https://logistiktrends.bvl.de/en/system/files/t16/2017/Trends and Strategies in Logistics and Supply Chain Management – Digital Transformation Opportunities Kersten von See Hackius Maurer.pdf (accessed 14 December 2019).

Kumar, P. and Seppi, D.J. (1994). Information and index arbitrage. *Journal of Business* 67: 481–509.

Liao, S., Shao, Y.P., Wang, H. et al. (1999). The adoption of virtual banking: an empirical study. *International Journal of Information Management* 19 (1): 63–74.

Lock, M. (2018). The marketplace model: a proven model for success & why the shipping industry should adopt it, cerasis. https://cerasis.com/marketplace-model (accessed 14 December 2019).

Manyika, J., Chui, M., Miremadi, M. et al. (2017). MGI-A-future-that-works-executive-summary. 1–28.

McKinnon, A., Flothmann, C., Hoberg, K. et al. (2017). *Logistics Competencies, Skills, and Training: A Global Overview*. Washington, DC: World Bank Group http://documents. worldbank.org/curated/en/551141502878541373/Logistics-competencies-skills-and-training-a-global-overview (accessed 14 December 2019).

McKinsey (2016). *Digital in Industry: From Buzzword to Value Creation*, 1–9. McKinsey Digital.

McKinsey (2017). Jobs lost, jobs gained: workforce transitions in a time of automation. https://www.mckinsey.com/~/media/mckinsey/featured insights/future of organizations/what the future of work will mean for jobs skills and wages/mgi jobs lost-jobs gained_report_december 2017.ashx (accessed 14 December 2019).

McLaughlin, C.B. (1926). The evolution of the ocean bill of lading. *The Yale Law Journal* 35 (5): 548–570.

Mikhail, N. (2018). Tomorrow's CEOs will come from an unlikely place: the supply chain, fortune. http://fortune.com/2018/12/11/ceo-supply-chain (accessed 14 December 2019).

OECD (2019). *OECD employment outlook 2019. OECD (OECD Employment Outlook)* https://doi.org/10.1787/9ee00155-en.

Parasuraman, A., Berry, L.L., and Zeithaml, V.A. (1991). Refinement and reassessment of the SERVQUAL scale. *Journal of Retailing* 67 (4): 420.

Parker, G.G., Van Alstyne, M.W., and Choudary, S.P. (2016). *Platform Revolution: How Networked Markets Are Transforming the Economy - And How to Make Them Work for You*, 1e. W. W. Norton & Company.

PwC (2019). Five forces transforming transport & logistics PwC CEE transport & logistics trend book 2019. www.pwc.pl (accessed 14 December 2019).

Ross, P.E. (2015). Hyperloop: no pressure. *IEEE Spectrum* 53 (1): 51–54.

Rouse, M. (2015). Transportation management system (TMS), TechTarget. https://searcherp.techtarget.com/definition/transportation-management-system-TMS (accessed 14 December 2019).

Rutner, S.M. and Langley, C.J. Jr. (2000). Logistics value: definition, process and measurement. *The International Journal of Logistics Management* 11 (2): 73–82.

Safari, F., Safari, N., and Hasanzadeh, A. (2015). The adoption of software-as-a-service (SaaS): ranking the determinants. *Journal of Enterprise Information Management* 28 (3): 400–422. https://doi.org/10.1108/JEIM-02-2014-0017.

Sandberg, E. and Hemilä, J. (2018). Digitalization in industrial logistics and supply chains - the contemporary situation in Sweden and Finland digitalization. In: *Proceedings of the 23rd International Symposium on Logistics* (eds. K. Pawar, A. Potter, C. Chan and N. Pujawan), 222–229. Bali: Centre for Concurrent Enterprise, Nottingham University Business School.

Schwab, K. (2017). *The Fourth Industrial Revolution*. Currency. ISBN: 9781524758868.

Seeßle, P. (2019). Nachhaltige Beschaffung auf digitalen Plattformen am Beispiel logistischer Dienstleistungen–Analyse der Möglichkeiten für Einkäufer und Auswirkungen anhand der drei Dimensionen der Nachhaltigkeit. In: *Nachhaltiges Beschaffungsmanagement* (eds. W. Wellbrock and D. Ludin), 371–387. Wiesbaden: Springer.

Sendler, U. (2017). Introduction. In: *The Internet of Things: Industrie 4.0 Unleashed* (ed. U. Sendler), 1–270. https://doi.org/10.1007/978-3-662-54904-9.

Shariff, F. (2019). Paradigm shift: welcome to the new freight forwarding, Coloadx. https://www.coloadx.com/blog/paradigm-shift-welcome-change-freight-forwarding-customer (accessed 14 December 2019).

Shippers Turn to Digital Booking Platforms (2019). *Air Cargo News*. https://www.aircargonews.net/freight-forwarder/shippers-turn-to-digital-booking-platforms (accessed 14 December 2019).

Sprenger, R.K. (2018). *Radikal digital: Weil der Mensch den Unterschied macht - 111 Führungsrezepte*. Deutsche Verlags-Anstalt.

Tate, W.L., Ellram, L.M., Bals, L. et al. (2009). Offshore outsourcing of services: an evolutionary perspective. *International Journal of Production Economics* 120 (2): 512–524. https://doi.org/10.1016/j.ijpe.2009.04.005.

Taylor, C.L., Hyde, D.J. et al. (2016). *Hyperloop Commercial Feasibility Analysis: High Level Overview*. John A. Volpe National Transportation Systems Center (US).

Toczauer, C. (2019). Panalpina, DHL GF join growing list of digital freight sales platforms. https://aircargoworld.com/allposts/panalpina-dhl-gf-join-growing-list-of-digital-freight-sales-platforms (accessed 14 December 2019).

Vatne, E. (2018). *The Networked Firm in a Global World*. Routledge https://doi.org/10.4324/9781315182957.

Vestergaard, T. (2019). Kuehne + Nagel on new digital setup: technology does not replace people, ShippingWatch. https://shippingwatch.com/secure/article11578805.ece (accessed 14 December 2019).

Win, A. (2008). The value a 4PL provider can contribute to an organisation. *International Journal of Physical Distribution and Logistics Management* 38 (9): 674–684.

12

International Trade Revolution with Smart Contracts

Matías Aránguiz[1], Andrea Margheri[2], Duoqi Xu[3], and Bill Tran[4]

[1] *Law, Science and Technology, Catholic University of Chile, Santiago, Chile*
[2] *Cyber Security Researcher and Consultant Specialised in Digital Identities and Blockchain Technology, Milan, Italy*
[3] *School of Law, Fudan University, Shanghai, China*
[4] *Supply Chain Management and Information Systems at Texas Christian University, Fort Worth, TX, USA*

The Blockchain Revolution

Blockchain is a novel technology that first appeared as the infrastructure underlying the Bitcoin public distributed ledger (Nakamoto 2008). Blockchains are in essence distributed databases where data is stored into chained structures, individually named *blocks*. The use of cryptographic primitives, in particular hashing functions, and distributed consensus algorithms ensures immutable storage and decentralized control of data. Practically, blockchains prevent by design tampering with stored data or centralized control on how data is changed or shared.

Blockchains provide an infrastructure where computing programs, named *smart contracts*, can be stored and executed. As smart contracts are directly stored on the blockchain, they enjoy the benefits of *code immutability*, i.e. the logic of the programs cannot be changed, and *tamper-proof execution*, i.e. once smart contracts are executed, the results cannot be changed. The term *smart contract* was coined by Nick Szabo (Szabo 1997a) to represent a computer implementation and enforcement means for a legal contract, while the Ethereum distributed ledger was the first smart contract implementation on blockchains (Wood 2017). The introduction of smart contracts significantly increased the flexibility and expressiveness of blockchains, which are now programmable and potentially general purpose, likewise any other programming language.

Overall, the key novelty of blockchain is the commoditization of *decentralized control on data and computation*: the integration of tamper-proof data storage and distributed consensus-based program execution ensures avoid by design the presence of any centralized *trusted third party* (TTP).

Blockchain has certain distinguishing features such as the integrity of the data stored, and the execution carried out is preserved and can be easily checked via the used cryptographic primitives. This is called *accountability* and ensures, together with the use of digital signatures, *non-repudiability* (i.e. the impossibility to refuse that something happened) for all blockchain operations.

Data and smart contract operations are replicated on all the nodes, which ensure that all parties involved in a transaction are *timely informed* about the current transaction status and can validate the *authenticity* of it. Notably, this *transparency* and openness do not prevent blockchain to enforce, according to the needs, data visibility and sharing controls.

The distributed consensus at the basis of smart contract executions permits updating the status of transactions across multiple parties in real time. The *decentralization* eliminates the need for any centralized control for multiparty transactions and allows (partially) distrustful parties to collaborate: it prevents single parties from abusing their authorities on the others.

Blockchains can be deployed in either a public fashion, also known as *permissionless*, or a private one, also known as *permissioned*. Bitcoin and Ethereum are two prominent examples of permissionless blockchains where parties can freely join the ledger without any form of authentication and immediately start contributing to the creation of new blocks. This activity, known as *mining*, is based on computationally expensive cryptographic operations, the notorious *Proof of Work*, which allow newly created blocks to maintain their integrity and immutability tolerating a high number of malicious parties, the so-called 51% attack. In contrast, permissioned blockchains require parties to be authenticated to join the ledger. The distributed consensus is obtained via algorithms based on message exchanges (rather than cryptographic functions) that ensure performance and tolerance to attacks comparable with modern distributed systems.[1] More importantly, permissioned blockchains offer the means to enforce fine-grained control on how data flows among parties and how smart contracts can execute.

Multiple permissioned blockchain platforms are available such as Hyperledger Fabric,[2] R3 Corda,[3] and Ethereum-related code bases as Parity and Quorum. Specifically, Parity[4] is an established permissioned implementation of the Ethereum technology, similar to Quorum[5] that also integrates by design additional privacy and consensus features. (Hyperledger) Fabric is a modular general-purpose blockchain framework that was initially released by IBM and is now under the Linux Foundation. Corda is the result of the R3 banking consortium, and, differently from the others, it comes with a more financially driven focus. To the time of writing, Fabric and Corda can offer a higher maturity level, more advanced security features, and seamless integration with all the major cloud

---

1 Traditionally, distributed systems can tolerate up to one third of malicious parties active at the same time in the system. Although much lower than the 50% tolerance of permissionless systems, this guarantees performance in the order of thousands of transactions per second compared with tens of transactions per seconds of Ethereum. Notice also that significant trade-off between attack tolerance and performance can be obtained according to the threat model chosen for a given system.

2 https://www.hyperledger.org/projects/fabric

3 https://www.r3.com/platform/

4 :www.parity.io

5 www.goquorum.com

providers. Significant efforts and improvements are targeting Quorum and other Hyperledger blockchain platforms (e.g. Hyperledger Sawtooth)[6] and may lead to new highly mature systems soon.

Smart Contracts

Smart Legal Contracts

The concept of "smart contract" is defined as "a computerized transaction protocol that executes the term of a contract" by Nick Szabo in 1994. A smart contract is an agreement among parties that can be automatized for self-execution or using a third verification party. This automatic execution is enabled by a computer code running on "if-then" principles, which translate legal prose into an executable program. A smart contract is different from an electronic version of a traditional contract due to the implementation of two technologies: One is a *contractware* that is the "physical or digital instantiations of contract terms onto machines or other property involved in the performance of the contract." Contractware is used to digitize terms of contracts in the format of conditional statements and manage enforcement of contracts through the application of a series of rules to a set of facts, eliminating human factors during the process of performing contracts. The other one is *distributed ledger*, which is a decentralized immutable database shared among parties participating in the contracts or nodes on the blockchain platform. Under the distributed consensus protocol, these parties or nodes contribute to terms of the smart contracts and enforce contracts without the intervention of a third party.

To form a smart contract, two parties must first agree to predefined terms of the contract without ambiguity. Smart contracts can formalize the cases that enable the courts to enforce the contracts, thanks to explicitly laid-out terms that inform both parties' obligations and benefits. These terms, then, are digitized using contractware and stored on a decentralized database shared among nodes on the blockchain platform. When the predefined terms are met and approved under proof of concept by parties in the contracts, the codes of smart contracts will automatically perform themselves using rules and verifications, and the verifications are normally under the supervision of a third party. Terms on the smart contracts can be modified and updated depending on the types blockchain (e.g. if contract versioning is allowed) and if such feature has been embedded in the smart contract code. The parties can then update the contract on the blockchain directly or through a government application programming interface (API). By and large, smart contracts offer a fast, cost-effective, and secure solution for parties in several lines of codes.

Service Level Agreements

Service-level agreements (SLA) are contracts between service providers (SP) and customers, containing information about the target service level or quantity of service (QoS) requirements and obligations of both parties. In case the service level agreements are not met, SP has to compensate for the customer. However, the compensation process is not transparent and straightforward because both parties can tamper with the data to support

---

6 https://www.hyperledger.org/projects/sawtooth

their claims. Then, they have to rely on a third party such as a bank to settle the dispute and ensure the correct payment. Rather than solving this problem in a costly and bureaucratic way, SP and customers can use blockchain and smart contracts to address this issue through a tamper-proof data storage and a trustful payment system. A blockchain-based smart contract is capable of translating QoS-related SLA into conditional statements, guaranteeing a transparent SLA, and automating SLA service fee payment and compensation without TTP. Smart contracts support SP and customers to make the processes simpler, transparent, and cost efficient (Scheid et al. 2019)

Smart Contracts in International Trade

International trade has been working since the start of civilization and has been one of the centers of human development. Nevertheless, many of the steps in the cross-border trade still fall under the paper-based uses, making it difficult for parties to coordinate many parts, many processes, and many documents (Ho 2018). Every time there is a mistake or the chain system of papers and certificates stops, the effects on the delivery of goods and services get affected.

Cross-border trade also involves many risks for all parties such as credit risk, foreign exchange risk, shipping risk, country and political risk, etc. This is the reason why trusted third parties are necessary for international trade along with several tools like supply chain finance, factoring, letters of credit (L/Cs) in trade finance or trade credit insurance, marine insurance, currency insurance, etc. as insurance certificates. All of these solutions are to enhance trust and guarantee payment for the parties, especially the sellers, under all circumstances.

Among the risks mentioned above, shipping risk is one of the most likely threats in international trade, which brings up the demand for supply chain visibility during shipment. This is a problem of information distribution or the matter of what, when, and how information about the goods is distributed to the parties. Another problem associated with information distribution is information asymmetry among parties. Some parties have access to a lot of information while some have little access, resulting in unequal power in the marketplace. Blockchain-based smart contracts offer an optimal solution to tackle these issues in cross-border trade.

Paperless Trade

One of the biggest obstacles in international trade is a variety of complex, paper-intensive processes involved. For instance, in 2014, when Maersk, a shipping company, tracked the physical process of a refrigerated container filled with roses and avocados from Kenya to the Netherlands, they found that "around 30 actors and more than 100 people were involved throughout the journey, with the number of interactions exceeding 200." This is an example to demonstrate how tedious the processes involved in international trade are. It is not only the problem of costly administration and intensive coordination among parties, but this manually processed paperwork may also be the source of errors, losses, and fraud (Ganne 2018).

To address these issues, smart contracts are potentially used to reduce paperwork and enhance processes, moving international trade closer to paperless trade. Smart contracts offer (the encoded representation of) electronic documents that are electronic equivalent records of physical documents and bear electronic signatures authorized by government entities or certificate authorities. Rather than digital contracts, smart contracts are also allowing terms contingent on decentralized consensus that is immutable and (self-)executing. Instead of relying on trusted third parties, parties involved in international trade reach consensus among nodes on the blockchain under the proof-of-concept model. Therefore, smart contracts can enhance contractibility and facilitate property, service, information, or money exchange in an automated, cost-efficient, and paperless way.

Trade Finance

Trade finance is one of the most complex and tedious processes involved in international trade. This results from many different forms of financing, including L/C, factoring, forfeiting, open account, etc. Payment can be in cash or exchange for goods. Traditionally, to process a transaction, many third parties have to get involved, and many costly burdensome paper-based procedures have to take place. Moreover, existing trade finance is risky due to a lack of trust and coordination among parties or the detection and prevention of fraud or authentication of parties. All of these drawbacks motivate many banks, financial institutions, and fintech startups to utilize blockchain and smart contracts in digitizing and streamlining the labor-intensive processes of trade finance. Below is the analysis of how blockchain-based smart contracts transform two important documents: bill of lading (B/L) and letter of credit.

Bill of Lading

The B/L is defined by the Uniform Commercial Code on §1–201 as "a document evidencing the receipt of goods for shipment issued by a person engaged in the business of transporting or forwarding goods." The B/L is issued by the carrier, and it records the goods that have been sent by the shipper and received by the transportation company. It also specifies the destination and quantity of the goods and the conditions of transportation. The receiver of the goods needs to sign the document at the moment of the arrival of the goods.

Traditional B/L is paper based, which leads to four main problems:

i) *Speed*: The physical B/L needs time to travel to the destination, and in cases with several documents or multiple destinations, there are different arrival times between the goods and the B/L, producing delays in the delivery of the product. According to a report by CargoX, each B/L conventionally travels with at least three courier services in about 10 days in total.

ii) *Document-related problems*: With a traditional B/L, exporters and importers highly likely suffer from loss, misplacement, damage, or theft.

iii) *Costly*: A B/L costs an average of $100 for the express fee. High costs are incurred if the document is lost or misplaced.

iv) *Fraud*: Printed B/L might be stolen or falsified that can serve as a base for fraud and criminal activity.

Intending to tackle these problems, the company CargoX, an independent supplier of public blockchain-based solutions for logistics, has launched a blockchain ecosystem looking to transform the global supply chain industry. As the first step toward that mission, CargoX has worked on altering the traditional paper-based B/L with an electric equivalent named CargoX Smart B/L. The Smart B/L solution is *fast* because the Smart B/L can be issued instantly and immediately transferred to the legal owner of the goods, *secure* due to no central storage and public/private cryptography, *paperless and cost saving* with digital versions of Smart B/L, and *auditable and fraud proof* through a real-time mechanism that is continuously recorded on the blockchain altogether with a time stamp.

More than an electronic version of a traditional B/L, CargoX Smart B/L is integrated into a blockchain platform and ecosystem through smart contracts. These differentiate CargoX from other attempts to digitize B/L in the past by avoiding some key problems such as the need for a centralized and trusted authority to run the system. The system is operated by participants on the blockchain. Also, the rules that govern the B/L exchange process are transparent, thanks to smart contracts. Finally, the duplication of data is impossible with the immutable nature of the blockchain platform (CargoX Business Overview and Technology Bluepaper).

Letter of Credit

An L/C is a written commitment issued by a bank on behalf of the importer that payment will be made to the exporter when the terms and conditions stated in the L/C have been met. The L/C offers a flexible solution for accommodating commercial contracts in trade finance because the transactions involving L/Cs can be tailor-fitted to particular transaction risks. L/Cs are also recognized internationally as payment, and finance instruments are governed by international rules published by the International Chamber of Commerce (ICC) in Paris.

However, L/Cs are relatively costly, complicated, and labor intensive. Furthermore, the L/Cs are not foolproof; therefore, they cannot protect the parties from fraud or bank failure. To simplify this process, Bank of America, HSBC, and the Infocomm Development Authority (IDA) of Singapore had collaborated to build a blockchain application based on a permissioned distributed ledger. They claimed that the blockchain application "shows potential to streamline the manual processing of import/export documentation by reducing errors, increase convenience for all parties through mobile interaction and make companies' working capital more predictable." Therefore, all activities in the workflow can be recorded, and L/C transactions data will be reported in real time. The paper-intensive L/C transaction is also digitized with a series of self-executing smart contracts, simplifying processes and enhancing trust for all parties within the platform (Parker 2016). Smart contracts have simplified L/C transactions in particular and streamlined financial flows among parties in general, augmenting the security, speed, transparency, and reliability of trade financing.

Trade Facilitation

In cross-border trade, procedures and controls governing the movement of goods across national borders usually include business-to-government (B2G) and government-to-government (G2G) procedures. A highly secure, decentralized, and automated nature of smart contracts can enhance the exchange of information between government agencies for more effective and efficient cooperation among parties. Instead of keeping all import–export documents in a silo, government agencies are allowed to access these records through distributed ledger technology in real time and validate transactions or issue relevant documents under proof of consensus. All activities are captured and distributed among ledgers of all authorize participants. More specifically, blockchain and smart contracts could improve the following:

i) *Certification and licensing:* The approval workflow of certificates (certificate of origin, conformity assessment certificate, etc.) is digitized and streamlined. Smart contracts can enable the parties to automatically render an import/export permit invalid upon the expiration of its validity period. The immutable nature of smart contracts helps parties prevent unauthorized duplicates, improve efficiency, and minimize the cost of certificate verification.

ii) *Release and customs clearance of goods:* Customs documents are analyzed and assessed based on predetermined requirements embedded in smart contracts via API.

iii) *Revenue collection and accuracy of trade data:* Smart contracts enable customs or intermediaries to collect duties.

iv) *Post-clearance audit:* Based on the predefined criteria, smart contracts can be used to screen documents to optimize the audit process.

v) *Compliance management and identity management:* The immutable nature of smart contracts and distributed ledger technology allows the provenance of products and verification of identities.

Smart Import Declaration

In cross-border trade, one of the customs duties is an import declaration before a shipment enters a customs territory. Import declaration is a document that includes information on consignor and consignee, date of issue, description of goods, country of origin, goods classification, invoice value, duties and taxes, incoterm, and means of transport. Customs authorities require exact details of the goods as described on the import declaration for risk assessment and customs valuation. However, the data provided by the import declaration tend to be vague or inaccurate because this document contains aggregated data from multiple documents, leading to document fraud that can cost billions of dollars annually.

Due to the complexity of cross-border shipping, exporters and importers usually hire freight forwarders as a declarant to manage mandatory paperwork and lodge import declarations with customs authorities. Freight forwarder keeps all relevant trade documents, including the purchase order, packing list, B/L, and pro forma invoice. The pro forma invoice is a bill of sale sent to buyers before the delivery of goods. This document is one of the most important documents during import declaration procedures. However, customs authorities have to cross-validate data elements of the pro forma invoice and other documents manually, meaning that they manually check the specified incoterms, quantity of goods, means of transport, etc. These documents are also manually exchanged among consignor, consignee, and customs authorities. It is highly likely that these manual processes result in the manipulation of data for financial gain or customs undervaluation of transaction value on goods.

With distributed ledger technology in smart contracts, all parties own a copy of documents on their ledger. Customs authorities can ensure the security of these documents due to immutability and enable to exchange them, especially pro forma invoice, much easier. An import declaration is also automatically generated by aggregating data from exchanged pro forma invoices and other documents. Instead of a declarant manually generating an import declaration, the import declaration is automatically generated based on the self-executing algorithms of smart contracts. That means if the predetermined terms of smart contracts that are invoice value, incoterms, etc. are met, an import declaration is produced.

Blockchain and smart contracts move the concept "Trade Single Window" of trade facilitation closer to international trade. The use of Trade Single Windows has been useful in the centralization of documents and procedures. The implementation of this kind of platform has produced at least three positive effects: reduce to half the time that government agencies spend in processing documents, reduce in one-third the overall time spent in compliance, and improve the transparency and user experience for merchants (World Economic Forum 2019). Trade Single Window systems represent a clear improvement in the improvement of the current dispersion of document and process, while blockchain and smart contracts represent an opportunity to develop the international trade process to the next steps of efficiency and security.

Trusted Payment

Trade financing is one of the riskiest processes in global trade. With the increasing volume of today's trade and the growing economy, the risks for exporters and importers are higher and higher. Conventionally, a TTP is introduced to get rid of the payment risk and the supply risk, meaning that it provides exporters with the payments based on the agreement while importers with the fulfillment of orders. In other words, the TTP vendor, usually a bank or an insurance company, acts like an intermediary supervising the trade in exchange for an amount of fee. However, the current method of trade of finance is prone to lack of trust and transparency as well as errors and frauds (Chandra Gupta et al. 2019). According to joint report between Boston Consulting Group and the International Chamber of Commerce of 2019 titled "Digital Ecosystems in Trade Finance", the biggest "pain points" for banks of current trade finance are the complexity of KYC compliance, highly manual documentation, limited digitization across different entities in transaction connectivity between parties, and high operational costs, which can be addressed with the implementation of blockchain-based smart contracts.

Digital payment across long distances or among unknown parties in cross-border trade is risky due to a lack of trust among parties involved. To tackle this problem, the Society for Worldwide Interbank Financial Telecommunication (SWIFT) provides financial institutions a secure, standardized, and reliable environment to send financial messages in a highly secure way without holding accounts for its members or performing any form of clearing or settlement. SWIFT only supports its customers to send payment orders or exchange banking transactions but not to facilitate funds transfer.

The application of blockchain-based smart contracts potentially facilitates the following:

i) *Faster and secure payment*: Blockchain-based smart contracts offer a highly secured and cost-efficient solution, cutting down on the demand for supervision from third parties. The tamper-proof record of transactions on distributed ledgers under proof-of-concept mechanism among nodes on the blockchain reduces risks of "double-spending" issues. Furthermore, under the "if-then" principle, smart contracts only allow money to be transferred (under the assumption that money transfer is executed within the blockchain ecosystems, e.g. paying with asset tokens) to the exporter only when certain requirements agreed on by all the parties in the blockchain ecosystem (or a group of them according to the design and implementation choices) are met, e.g. that importers receive goods that 80% are in good condition. Smart contracts also eliminate intermediaries in transferring assets and reducing asset exchange fees. With decentralized databases of tokenized digital assets, real-world assets can be easily exchanged across the parties with low transaction costs and high tokenized securities.

ii) *Clearance and settlement systems*: Distributed ledger technology enables transactions to be settled directly without bypassing a complicated system of intermediaries, which makes it much more efficient for all the parties. This also allows for transactions that are clear and settled immediately when a payment is made, while in current banking systems, a transaction is clear and settled days after payment.

iii) *Transparent and efficient process*: The use of smart contracts in trade finance along with distributed ledger technology brings visibility into shipments, mitigating risks for trade parties and automating processes, reducing the possibility of missed, lapsed, or mortgaged shipments (CBInsights 2018).

we.trade Case Study

we.trade is a blockchain trading solution that aims at challenging the model of banks serving as intermediaries and eliminating barriers between businesses and international trade. Twelve leading European banks have joined with IBM, including CaixaBank, Deutsche Bank, Erste Group, HSBC, KBC, Natixis, Nordea, Rabobank, Santander, Societe Generale, UBS, and Unicredit. we.trade also wants to simplify processes in international trade to support all companies of all sizes to trade more easily and securely. To accomplish these goals, we.trade develops a blockchain platform with two main features: distributed ledger technology and smart contracts. Distributed ledger technology allows all relevant parties to have access to real-time information on their ledgers. we.trade's built-in smart contracts also reduce payment or supply risks from trading parties. The fact that the contracts are encoded into the platform removes the need for any underpinning laws or enforce mechanisms.

With the collaboration with IBM cloud services, we.trade's blockchain platform not only guarantees scalability and flexibility of their services but also attracts many banks to partner with them, thanks to IBM's reputation. Since July 2018, we.trade platform has begun full operations with real transactions, and now it has reached 13 countries and partnered with 15 banks in total (IBM Case Study).

Information Distribution

Information Transmission

Cross-border trade is a complex and dynamic group of processes and interactions among many parties. Therefore, there is an increasing demand for transparency and visibility into the supply chain, especially during ocean freight. Over the last two decades, an electronic data interchange (EDI) has been established to enable an automatic electronic document exchange among trading parties, and supply chain operating networks (SCONs) have been developed to allow members across the network manage global transactions. Although these solutions stretch information flow capabilities for enterprises by being integrated into their enterprise systems, the two biggest gaps that are unfulfilled are real-time data sharing and validation of shipment status during the transport of goods. To tackle these problems, IBM and Maersk have collaborated and combined blockchain with the Internet of Things into the Tradelens platform.

Tradelens Case Study

Tradelens is a collaboration between IBM and Maersk to develop a blockchain-based shipping solution to promote more efficient and secure global trade through a platform for various parties that support data sharing and supply chain transparency. IBM blockchain technology laid the foundation for a digital supply chain platform that empowers multiple trading partners, including shippers, shipping lines, freight forwarders, port, and terminal operators, inland transportation, and customs authorities, to have real-time access to shipping data. Using built-in smart contracts, Tradelens allows all parties involved in international trade to cooperate in cross-organizational business processes and information exchanges. It is reported that Tradelens can reduce the transit time of a shipment of packaging materials to a production line in the United States by 40%. Generally, Tradelens provide all parties in its ecosystem with an opportunity to prevent delays caused by documentation errors or information lag and visibility into the shipping process. So far, more than 154 million events created by more than 90 organizations were captured on the platform with a growth rate of 1 million per day. Tradelens leader for Maersk said that "Our joint collaboration allows us to better address key feedback from ecosystem participants. We strongly believe this will maximize industry adoption" (Maersk and IBM Introduce Tradelens Blockchain Shipping Solution).

Information Symmetry

In the traditional world, the imperfect nature of information makes it difficult for trading parties to sign detailed contracts, showing that no one can check the authenticity and reality of certain variables that characterize relationships between contractors. Unclear wording, ambiguity, and the omission of key points are unavoidable with traditional contracts. The incompleteness of conventional contrast usually drives from two factors: the limited rationality of parties and the uncertainty of events. Consequently, the process will be inefficient, wrong decisions will be made, and the cost will be significant. All of these problems are caused by asymmetrical information.

Information asymmetry, then, results in the "hold-up problem" (HUP). HUP can be demonstrated by a case: when two companies, which are an exporter and an importer, build a long-term business relationship, they have to agree on a long-term contract at the moment of investment. However, they are unable to predict all possible outcomes that make the buyer completely trust the investment, leading to tons of hardships establishing a contract. The HUP causes negative efficiency during the negotiation phase for all trading parties.

A smart contract, which is a complete contract, is a perfect solution for information asymmetry. For the case of a long-term contract between an exporter and an importer, both parties need to predefine the requirements with the cost and the quality of the products first. Only when these conditions are met, smart contracts will automatically send messages or carry out operations that are captured and shared on the blockchain. The smart contracts offer a complete solution, eliminate information asymmetry, and provide greater transparency into the processes.

Smart Contract Initiatives

Government Initiatives

Chinese Central Bank
In May of 2018, President Xi Jinping called blockchain "a breakthrough technology" (Shi-Kupfer and Ohlberg 2019). China took an extreme dualistic vision on DLT. For one side, in December 2016, the State Council of China added to the quinquennial plan the promotion of blockchain technologies as part of the national policy development, and the average funding for blockchain-related projects in China was USD 25 million between 2010 and 2018 (Shi-Kupfer and Ohlberg 2019). For the other side in September of 2017, seven government regulatory agencies issued an announcement of banning the activities related to ICOs and crypto trading (Wang et al. 2019). Then the Chinese government endorsed the use of blockchain but banned the cryptocurrencies and trading platforms.

The People's Bank of China launched the "Bay Area Trade Finance Blockchain Platform" (BATFB) that until June of 2019 had processed $4.5 billion (Ledger Insights, 2019). To read more on China's technology ventures related to logistics, see Wheeler's chapter on Belt and Road Initiative.

South Korea
South Korea is planning to use blockchain technology to completely digitalize trade finance by 2021. In October of 2019, South Korean Trade Finance Minister Hong Nam-ki announced the developmental plan for digital trade. According to the plan, the entire trade finance ecosystem will be upgraded by 2021, moving all paper-based documents to a blockchain network and implementing a Digital Trade Blockchain Council whose members are trade associations, banks, and the Korea Trade Network (KTNet). Hong said that "We will build a digital trading platform that can be easily and conveniently used in all stages of export, such as contracts, customs, and logistics." He also mentioned that one of the goals of the initiative was to encourage blockchain-based foreign exchange transactions and a target of bringing together seven banks to establish the service was set to be accomplished by December 2019.

Private Initiatives

Mizuho Bank Using IBM

Mizuho Financial Group and IBM Japan have collaborated to build a blockchain-based trade financing platform to simplify processes for parties involved in international trade, including importers, exporters, shippers, insurers, and governmental authorities, supporting them to get access to shipment data in real time. They are working on conducting actual trade transactions based on Hyperledger Fabric and exchange blockchain-based L/C between Japan and overseas clients.

Marco Polo Trade Initiative

Marco Polo is the largest and fastest growing trade finance network with many world's leading financial institutions and their corporate clients. In 2017, Marco Polo launched a blockchain-based open enterprise software platform that allows banks and corporations to exchange data and assets seamlessly and securely. Each participant of the network controls its data and sets up applications and platforms relevant to its demands. This platform can be integrated with an enterprise resource planning (ERP) system, offering three main products that are receivable discounting, factoring, and payment commitment. Marco Polo targets large enterprises and focuses on growing the number of participants of the network. To gather the entire trade ecosystem and digitize supply chain finance, Marco Polo is providing a solution for trading parties to raise revenue and reduce costs, time, and risk, driving trade and working capital finance innovation.

Several initiatives are using smart contracts and blockchain to change the landscape of trade finance. These initiatives show a need to change the current system of trade finance and the opportunities that are being exploited by the public and private world. This change will be implemented with the force of different operators walking in the same direction. It is an individual work but aims for a common objective that will be the improvement of the international trade finance and standardization of good practices for the industry.

Risks and Challenges of the Implementation

Technological Challenges

Scalability

Scalability is one of the biggest obstacles for blockchain platforms and built-in smart contracts. The rapid growth of transactions has led to an increasing number of solutions to scalability issues. These solutions include adjustment of the computing parameters of public blockchains, an example being doubling the size of each block or regulating the numbers of nodes on the blockchain that can create blocks, or more structural changes as the integration with lightweight consensus methods and highly efficient distributed databases like distributed hash table. As many blockchain systems integrate with distributed Internet-connected devices, such as with Tradelens, the bandwidth can also create bottlenecks, leading to scalability challenges.

Sustainability

In the case of permissionless blockchains, energy consumption has become one of the hottest debates about blockchain. The Proof-of-Work process used to validate transactions consumes a lot of electricity. According to a report by WTO, in 2014 the Bitcoin blockchain consumed approximately as much electricity as Ireland. However, these consumptions have been highly influenced by the global outbreak of cryptocurrency trading. Although consumptions seem to have reached a peak, the main public blockchains are working toward new technical solutions that can mitigate energy consumption. An example is the so-called Proof-of-Stake consensus method that ensures the same security guarantees but with significantly less intensive computational operations.

In the case of permissioned blockchains, the main sustainability challenge arises from the proliferation of blockchain platforms that are not interoperable. As there is no clear market leader at this stage, the adoption of a single blockchain may lead to a vendor lock-in problem. In the longer-term, the established blockchain solution will be able to start focusing on system interoperability as a means for application sustainability.

Security

The distributed consensus and cryptographic methods used by blockchain create an underpinning security infrastructure upon which smart contract data and computation can be stored and executed. While permissioned blockchains share security guarantees of a typical distributed computing system, permissionless blockchains can benefit from the fact that there is a low possibility that 51% of nodes collaborate and collude for malicious financial gains.

Like any other IT infrastructure, blockchain platforms require to reach an adequate maturity level to ensure cyber-secure applications and libraries. Many of the trade finance applications we present have operated for some years and are passing their "trial by fire." Differently, an emerging security threat is advanced by new computing technology like quantum computing that will be able to reverse the cryptographic methods of blockchain easily.

Interoperability Challenges

Compatibility

An important issue in the integration of blockchains is that many blockchain platforms are being built without communication mechanisms. This issue matters specifically for international trade because a single party will interact with various ledgers such as finance, logistics, customs, and provenance.

Although there are several initiatives looking to make easier the compatibility between different platforms and protocols, it is still a risk to be solved. Interoperability is maybe one of the biggest challenges for every technical implementation in the digital world. Nevertheless, initiatives that aim to increase interoperability are being developed, and blockchain should be walking in that direction.

Public initiatives such as the one being developed by China or Korea should put a particular focus on technological neutrality. There is a real risk in national initiatives limiting compatibility and excluding systems with different standards; this kind of situation would have strong negative effects in suppressing innovation. Each national initiative should be structured under the logic of technology neutrality. In that way, current development would be included or easier to interconnect as well as the possibility to have better G2G communications.

Data Standardization

The interoperability challenge does not only happen at the level of interfaces but also the semantics. Data or information is exchanged through different blockchain platforms that have different standards. Enterprises that want to use these data across the platforms are required to clean and filter the dirty data and put them into relational databases. The challenge of developing standard datasets that cover all data used for information exchange is a huge obstacle for parties in international trade to get benefits out of blockchain. To see more on the challenges of data standardization in logistics, see Glick and Glickenhouse's chapter.

Legal Issues

Validity

The legal status of blockchain transactions and smart contracts remains uncertain. Various governments are now working on legislation to legally validate blockchain signatures, smart contracts, and financial instruments. Various proposals relevant to blockchain have also been submitted to WTO, but some concerning issues are electronic authentication, recognition of e-signatures, cybersecurity, encryption, and protection of personal information.

Privacy

There is also an intense debate over requirements or restrictions on data flows on the blockchain. People are increasingly concerned with the issues of cross-border data transfers and data privacy. Blockchain also exposes trading parties to the problem of personal data protection and new forms of identity management. In that sense, advances that can be made in digital identity are key to better safeguard the protection of personal data of parties related to international transactions.

It is important to be clear that there is no an international agreement regarding the protection of personal data, but rather different regulatory frameworks that protect it, such as European regulations (GDPR), California regulations (CCPA), or Chinese approach to data protection and data sovereignty. This lack of compatibility makes it particularly difficult for agreements to be reached regarding personal data protection standards, which is why the world should move toward a global standardization of how we understand the protection of personal data and its international transfer.

Conclusion

Technological innovations are becoming major forces that shape the world and leverage customers' demand. This makes a huge impact on all parties participating in international trade, requiring them to become better versions of themselves to survive. Given the potentials of blockchain-based smart contracts, this technology is likely to support companies to meet new customers' demands and build sharper competitive advantages.

Blockchain today presents us with an alternative and development opportunity in different industries. It is the responsibility of the private and public world to create new, more efficient, and secure ways to improve the development of nations. International trade has always been a key element in eliminating poverty, which is why its development is not only an opportunity but a moral obligation for both entrepreneurs and governments. Blockchain is now opening a door to the future, and it has allowed us to rethink institutions that long ago needed to be updated.

Key Takeaways

- Smart contract technology may be a game changer for many companies, but it can also possess many implicit risks. With all the risks and challenges of smart contract technology, it will create more problems than governments and businesses must face.
- Smart contracts seem like the most feasible solution to reduce paperwork and enhance the whole process of international finance by increasing transaction speed between financial and nonfinancial institutions and eliminating bureaucracy that will increase efficiency and prices for users.
- Governments, like China and South Korea, and Fortune 500 companies, like Walmart and IBM, will be crucial to the adoption of blockchain technology; however their efforts over the past five years show the inherent difficulties that come with adopting such a novel technology.

References

CBInsights. (2018). How Blockchain could disrupt banking. https://www.cbinsights.com/research/blockchain-disrupting-banking (accessed 6 October 2020).

Chandra Gupta, V., Agarwal, M., and Mishra, A. (2019). When trade finance meets Blockchain technology. In: *International Journal of Innovative Science and Research Technology*. www.ijisrt.com (accessed 8 March 2020).

Ganne, E. (2018). Can Blockchain revolutionize international trade? www.wto.org (accessed 27 February 2020).

Ho, H. (2018). Trade finance: an introduction to the key challenges. *Medium* (31 July). https://medium.com/hashreader/trade-finance-an-introduction-to-the-key-challenges-8f6545771b87 (accessed 6 October 2020).

Ledger Insights. (2019). China's Central Bank Blockchain trade finance platform processes $4.5 billion. https://www.ledgerinsights.com/oracle-bee-blockchain-supply-chain (accessed 6 October 2020).

Nakamoto, S. (2008) Bitcoin: a peer-to-peer electronic cash system. https://bitcoin.org/bitcoin.pdf (accessed 6 October 2020).

Parker, L. (2016). Bank of America Merrill Lynch, HSBC, and IDA Develop a Blockchain prototype solution for trade finance, brave new coin. https://bravenewcoin.com/insights/

bank-of-america-merrill-lynch-hsbc-and-ida-develop-a-blockchain-prototype-solution-for-trade-finance (accessed 6 October 2020).

Scheid, E., Bruno R., Burkhard S. (2019). *Toward a policy-based blockchain agnostic framework.* IFIP/IEEE Symposium on Integrated Network and Service Management (IM).

Shi-Kupfer, K. and Ohlberg, M. (2019). China's digital rise, challenges for Europe. Mercator Institute for China Studies, No 7, April, 55 p.

Szabo, N. (1997a). *Formalizing and Securing Relationships on Public Networks.* First Monday https://firstmonday.org/ojs/index.php/fm/article/view/548 (accessed 6 October 2020).

Wang, Y., Ren, J., Lim, C., and Lo, S.-W. (2019). A review of fast-growing Blockchain hubs in Asia. *The JBBA* 2 (2): 16.

Wood, G. (2017). *Ethereum: A Secure Decentralised Generalised Transaction Ledger*, 32. Ethereum Project Yellow Paper.

World Economic Forum (2019). Windows of opportunity: facilitating trade with Blockchain technology, 24 p. http://www3.weforum.org/docs/WEF_Windows_of_Opportunity.pdf (accessed 6 October 2020).

13

Exploring China's Digital Silk Road

Andre Wheeler

Asia Pacific Connex, Perth, Australia

Introduction

The Belt and Road Initiative (BRI) is made up of two physical parts, namely, the Silk Road Economic Belt and the 21st-Century Maritime Silk Road – the strategic pairing of ports and land-based transport with complementary infrastructure that essentially connects China with Europe, Asia, and South East Asia. It is expected to bring financial benefits to China and BRI participants while augmenting global supply chains and logistics (Thurer et al. 2019). The initiative has been framed as an economic development plan based on a vision of a shared human destiny where emerging nations can lift themselves out of poverty through collaborative connectivity to international markets (Chen 2016). This connectivity is enabled through the provision of reliable infrastructure.

There are concerns that the BRI constitutes "debt diplomacy" through the financing of infrastructure projects in underdeveloped countries (Brautigram 2020). However, China does offer financial and structural support to those poorer countries, with favorable terms and conditions. This assists those that do not meet the rigorous commercial requirements set by the World Bank and other Western-based financial institutions. Funds are secured at preferential rates, often with interest rates being below commercial bank rates. China has also demonstrated a willingness to renegotiate terms and conditions (Hillman 2018). This is particularly evident in the new COVID-19 world in which those countries currently encountering financial stress received extended payment terms and interest rates being reduced to below 2% (Toh 2020).

With the rise of globalization, coupled with the rise of containerized shipping, the World Bank and the Asian Development Bank poured $1.3 billion into port projects in developing countries during the 1970s (Levinson 2016). Initially global trade was bound by the economic model limited to ocean transport. This has changed over time as shippers realized that this economic model was not appropriate to current global trade driven by e-commerce and technology in which just in time delivery considerations take

The Digital Transformation of Logistics: Demystifying Impacts of the Fourth Industrial Revolution,
First Edition. Edited by Mac Sullivan and Johannes Kern.
© 2021 by The Institute of Electrical and Electronics Engineers, Inc.
Published 2021 by John Wiley & Sons, Inc.

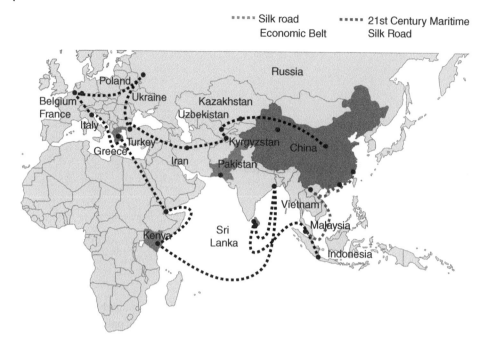

Figure 13.1 Map of One Belt One Road. *Source*: Courtesy of Silk Road Briefing. © Dezan Shira & Associates.

precedence. Logistics is now enabled by digitization to plan and manage last- and first-mile considerations.

The BRI has expanded this view and is seen as changing global trading patterns as new markets become accessible through better connectivity (Hillman 2018). Moody's Analytics Global Macroeconomic Model indicates that the level of productivity and economic growth is directly proportional to the level of these investments (Moody's Analytics 2020). In a sense, it has enabled offshoring of manufacturing industries to one inaccessible production markets. This is evident not only in countries such as Myanmar and Laos but also in more developed economies such as the Philippines and Indonesia (Figure 13.1).

Digital Integration and Supply Chain Along China's BRI

Progress within the BRI highlighted several communication gaps, creating delays and congestion. For example, Shanghai had introduced a successful port automation platform; however, this created congestion as the maritime operating system did not interface with land-based operations. Systems interface between these operations is central to the success of the maritime/land logistics pairing espoused by the BRI. Its importance was confirmed in 2015 when Xi Jinping announced the introduction of the Digital Silk Road (DSR), which was aimed to create a single platform for digital communication along the BRI. It seeks to integrate all providers along the BRI economic corridors and improve efficiency in all

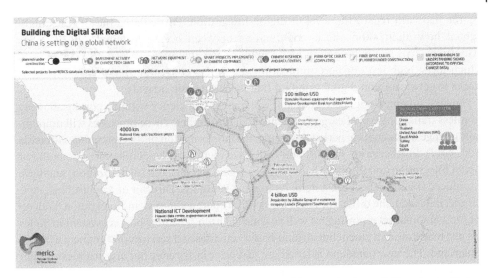

Figure 13.2 The Digital Silk Road. *Source*: Courtesy of the Mercator Institute. © Mercator Institute for China Studies (MERICS) gGmbH.

processes associated with cross-border trade. This includes efficiency in payment systems and the choice in the mode of cargo transport (Figure 13.2).

The DSR is a key driver in China's development of smart ports and smart cities within the BRI. It will allow access to data that will enable seamless movement of cargo from ocean vessels, through the port ecosystem into the hinterland. Importantly, the DSR scope has recently been upgraded to the goal of having a completely integrated transport, port, and city network by 2025. Built on Huawei's 5G and the BeiDou satellite navigation system, this poses potential trade barriers for those countries that want to trade with China but would not have an adequate platform interface with these systems.

China began emerging from its period of isolation under Mao in 1978. Essentially, under the leadership of Deng Xiaoping, the country shifted from a political movement to a focus on economic development. Initial steps included the abandonment of the collective brigade production processes to contracting collectively owned lands to the household. Special economic zones were also introduced in 1979, initially called export zones (Garnaut et al. 2018); these were regarded as the building blocks for reform and development of China. Xi Jinping extended the reform process under the OBOR, seeking to restore China's global trade preeminence under the old Silk Road that had been lost under Mao's rule. The OBOR was the vehicle by which China would achieve this. Initially, constructions of physical transport infrastructures, such as roads, bridges, and railways, were key factors to enhance global trade. Significant gaps in the infrastructure built were identified, particularly with regard to the interface planning where transport met. Improved information and communication technology (ICT) innovations offered solutions to these gaps, with the DSR established in 2015. The OBOR iteration was changed to the BRI in 2016 as it now included digital infrastructure that connects trade through improved ICT (Garnaut et al. 2018).

There is still a gap between what the service providers offer and which is being demanded by shippers in the digital world. What is clear is that BRI has identified how big the gap with regard the level of digital integration. Particularly, the BRI has highlighted how digitization affects coordination between ocean, terminal, and landside operations. This gap is becoming increasingly noticeable in the likes of Shanghai port, which has a high level of port automation yet suffers congestion problems as the port works to get a common digital standard that measures ship arrivals.

The DSR is being designed such that it will increase efficiency within a port by allowing seamless transshipment of cargo from one mode to another. These processes will dramatically change in the coming period as significant transformation processes emerge (Brotchie et al. 2017). Furthermore, the BRI and DSR have identified the development potential of digital integration and interface between oceans and rail logistics (digitalization) and have introduced the smart port ecosystem. As Ioannis Potamitis, Director of Applications, PwC Greece, indicated, the difficulties they face as a BRI port are in systems integration, budgeting, and automation reporting, as well as consolidation. The new port ecosystem includes not just maritime movement but all land-based activities and services that connect with the hinterland. The new paradigm is based not just on transparency but also seamless freight movement that incorporates last- and first-mile considerations.

The centrality of cyberspace and digitalization was highlighted in Xi Jinping's announcement at the first BRI summit in 2017. He stated that big data would be integrated into the BRI through the ongoing building of a "21st-Century Digital Silk Road."

(For the purposes of clarity, future reference to the BRI includes earlier iterations of OBOR.)

The BRI is now bringing advanced IT infrastructure to participating countries, such as broadband, e-commerce hubs, and smart cities. The literature discussing the BRI has largely understated the importance of digitally based communication in this change process (Shen 2018). In particular, very little is understood about how digital integration will impact the interface between maritime and rail logistics along China's BRI and how it will delay delivery of a key objective of the BRI, namely, optimal intermodal logistics choices.

Smart supply chain and logistics management are instrumental in an improved understanding of China's complex BRI strategy. Focused on Asia, Africa, and Europe, the planned economic corridors account for almost 64% of the entire world population and 30% of the world's GDP. This initiative aims to reconfigure China's external geopolitical arrangements to enable its own economic growth. While it is still too early to assess the impact of this ambitious initiative, the initial implementation seems to have the potential to transform the underdeveloped "Belt and Road" region into a new vibrant economic pillar.

In its initial phases, the BRI was seen to have severe limitations. While being a potential contributor opening emerging market economies, the initiative faced obstacles including a lack of a central coordination mechanism, geopolitical tensions due to disparate and beliefs, and the commercial viability of cross-border projects (Huang 2016). However, the DSR is helpful in coordination efforts and focusing projects toward holistic connectivity as it moves to a centralized data platform.

Since BeiDou's introduction in 1994, China has more satellites than that of the US GPS (Kida and Hashimoto 2019). With 70% of Chinese smartphones equipped to tap into the BeiDou's positioning services, it will play a vital role in 5G wireless communications. Integration of BeiDou and 5G is an important development in China's future IT plans (Weissberger 2019). This will deliver trade advantages to trading partners along the BRI that have participated in construction of complementary ICT networks, such as Pakistan.

Not only has optic fiber cable been laid, but there also has been increased investment in its 5G capability. At present it has 35 BeiDou satellites in play as well as 300 000 5G land base stations. The plan is to have 600 000 5G base stations and 45 BeiDou satellites by 2025. China is currently developing a 6G platform that will take the BRI further ahead in digitally based trade.

This digital infrastructure is geared toward fast tracking the use of blockchain, digital currency, and customs/border controls along the BRI. These poses both a threat and an opportunity to global trade, particularly as issues of cybersecurity start taking center stage. It is evident that there needs to be a capability for all current logistics software applications to have effective application programming interface (API) that can communicate with the centralized Chinese BRI/DSR operating systems.

Why would this BRI interface be important? In simple terms, the BRI covers an emerging and growing economic super zone. Studies show that the share of global consumption in emerging markets has risen by 50% over the past decade (Lund et al. 2019). This includes an increase of trade flows between Europe and Asia with growing container traffic within Eurasia. This sharp increase in trade volume across the globe also led to increased requirements for all logistics. It has increased the prerequisites needed for designing, planning, implementing, and managing the freight movement using ocean carriage (Dodgson 2018; Panayides 2006), as well as other forms of freight transport. With digitalization being an integral part of the BRI (Kozlowski 2018), the interplay between logistics and supply chain management is now at an important crossroad (Guo et al. 2018).

With digital connectivity being the extension of the global digital economy (Fishwick 2017), this level of connectivity inevitably needs participatory digital platforms (Seele et al. 2019). As the current container industry has been focused on maritime and port digital applications, we see a rush to implement smart port technology. However, the BRI requires the focus to be on multimodal logistics. There is increasing pressure for logistics/freight forwarding providers to adopt sustainable digital systems that increases operational efficiency within the newly defined BRI port ecosystem. This is being addressed in China with the establishment of the global shipping network that is building a digital platform that connects container operators, terminal operators, customs officials, freight forwarders, and cargo owners.

Challenges of Digitization for the Freight Industry Along the BRI

China has been the leader in adopting technology for everyday commercial use. Not only does 80% of the population make use of online shopping, but it has also largely become a cashless economy. It has taken advantage of today's fast-paced world driven by e-commerce and access to the Internet. As China introduces its digital currency and blockchain platforms, the nature of cross-border trade is undergoing dynamic changes. In particular, growth in last- and first-mile delivery considerations along the BRI as more communities have access to products has seen a move from full container load (FCL) to less than container load (LCL) shipments. Access has translated into a demand for speedier delivery and a greater focus on total supply chain cost.

China's BRI has partly met this challenge with the blockchain platform, "China-Europe E-Single Link" (Chengdu Launches Blockchain Cross-border Trade Platform to Connect China and Europe 2019) dedicated to multimodal transport systems along the BRI network.

It promotes cross-border trade by removing barriers associated with regulatory requirements as cargo passes from one jurisdiction to another. See more on the role of Blockchain in Chapter 6.

Customer experience is probably one of the most important factors in any business. Improved customer experience results from a better understanding facilitated by access to information. Beijing has an industrial park dedicated to innovation, with several innovation labs. New players and startups in the digital space are willing to take advantage of this. Bigger players have been using industrial parks and are stepping up to improve their customer experience for fear of being left behind (Biest 2019). China focuses on alleviating frustrations over shipment books, transit times, and delivery planning.

Challenges in China

There are several freight forwarders and third-party logistics providers (3PLs) in Shanghai port that have only recently overcome digital issues with port-to-port shipping (Muravev et al. 2019). They need to address the gap in the planning of cargo movement across the wharf. Exacerbating the issue is the Chinese government's policy shift with regard to subsidies, particularly about eliminating subsidies of westbound freight trains that had empty containers. Shipping empty containers by rail overcame a common problem that included having to extend wharf laydown areas as containers now must wait for pick-up. This leads to congestion due to the bottleneck of trucks and trains in an attempt to move them in and out of the port efficiently. Congestion causes delays that, in turn, extended transit times. It also undermines a key selling point of China's port pairing paradigm, i.e. faster transit time reduces supply chain costs. Freight could be moved more efficiently between large ports on a busy route by having shorter vessel turnarounds at either end (Levinson 2016).

With China consuming more of what they produce, global value chains are being reshaped by cross-border data flows, particularly new digital platforms (Lund et al. 2019). Companies then reassess their geographic locations and develop collaborative relationships within their supply chains. Blockchain within the maritime shipping sector is an example of how digitalization and digital integration are shaping this move toward a more collaborative business environment. Splash 24/7 (2019) identified six maritime groups undertaking such digital integration through blockchain (Figure 13.3).

Unfortunately, these developments are creating independent business silos that hinder data transfer across and within (Kosmala et al. 2016). The question is how we can use data at the interface between these business silos that would add value and improve port operations and efficiency. This is an emerging issue as China transitions its intermodal mix to being 30% rail based while looking to improve port operations (Hillman 2018). This is a central success measure within the framework of BRI's pairing of maritime and rail transport networks. Besides, customers increasingly demand greater transparency and visibility of trade supply chain and logistics processes. A recent Terradata (2019) survey revealed that real-time container visibility is a top challenge for 47% of shippers and freight forwarders. At the same time, overcapacity on trade routes is a top challenge for carriers (55% of respondents), next to capacity management/asset utilization (48%) and data sharing (33%). Seventy-one percent of cargo carriers are becoming fully digitalized, so the scale of these gaps paralyzes the business. Exacerbating the issue is the unequal progress being made as indicated by the survey. Sixteen percent said 0–20% of their operations are fully digitalized, 22% said 20–40% of their operations are fully digitalized, 25%

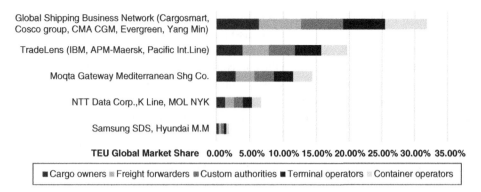

Figure 13.3 Blockchain digital integrator.

said 40–60% of their operations are fully digitalized, 24% said 60–80% of their operations are fully digitalized, and 13% said 80–100% of their operations are fully digitalized.

System Connectivity Challenges

It is quite frustrating that fragmentation leads to major inefficiencies in the logistics sector. In a study by Teradata on the State of Artificial Intelligence for the Enterprise (2019), less than one-third in the industry viewed their collaborative process as effective. On the other hand, 84% of the participants reported that although they have implemented real-time data sharing into supply chains, only 13% out of them have done so effectively. Currently, the trade technology platform Haven only offers track and trace solutions for its ocean container clients. Nonetheless, they have plans to expand it to air, rail, and intermodal clients in the future (Transmetrics 2019).

One of the major issues of sharing of information between tiers is the passing of data between disparate information systems. How can a company with a company-wide ERP system be connected to a supplier who manages their businesses on a spreadsheet? Innovations such as data collection via IoT, data analytics software, and cloud computing are making the towers of supply chain control possible. Once a dataset is designed to give supply chain partners the information, they require for shaping and identifying supply chain demand; then the data can be communicated up and down the supply chain for planning analysis (Weiner 2019).

As global trade expands, there is an increased requirement for participants within a supply chain to integrate their performance for seamless transportation systems. As trade moves to door-to-door delivery, multimodal integration, and seamless integration, access to data and information has become the core for the requirements of the new logistics paradigm. Digitization is the first step in the solution for such movement. While mid- and long-range transport (including sea transport) has proven digital platforms, these need to be integrated into a much larger transport chain to optimize door-to-door logistics requirements (Huang 2016).

Digital Silk Road as an Answer to the Digitalization Deadlock

Under the BRI, inland Belt and the maritime Road transport converge on key port infrastructure through a coordinated pairing of maritime/port/land transport routes. This convergence of transport modes has highlighted the major weakness in digital initiatives toward smart port technology. In the first instance, much of port automation/maritime digital development is taking place within the context of a transshipment facility, i.e. large container vessels arrive and offload and smaller volumes shipped elsewhere. The BRI, however, introduces a greater focus on gateway ports, in which the port connects to inland transport hubs/dry port as well as end markets. These gateway ports, in turn, have driven the changes in shipping wherein larger container vessels do not offer the once many economic advantages associated with "port-to-port" shipment, as markets move toward first- and last-mile considerations where other costs have a profound effect on final product cost.

This changed structure has led to the realization that there is a lack of seamless connectivity coupled with poor infrastructure that will challenge smart port and associated trade and logistics development. Connected networks of maritime ports and terminals with inland hubs will service major distribution and economic hubs throughout the Middle East, Central Asia, and Eastern Europe. It will boost the demand for specialized logistics, such as cold storage, custom warehouses, and on-demand transportation services.

China has introduced the coordinating element of the BRI, commonly referred to as the Digital Silk Road. Since its inception, the DSR has extended beyond simply rolling out fiber-optic cable and now includes cables, network equipment, technology, and facilitating software in the form of 5G networks. In a sense, Huawei and ZTE equipment are the new leaders in developing e-commerce trade platforms. These developments are facilitated by China's alternative tracking system, namely, BeiDou, which has upgraded the BRI into a multimodal and multidimensional supply chain and logistics ecosystem.

The benefit of the DSR is that trade with China will require an integrated and seamless technology base that will allow movement of products and containers from vessel to port to land-based transport, thus reducing the total cost of the supply chain through optimal asset utilization and lower inventory costs through efficient cargo management and visibility of the entire value chain from manufacture to end user. It can be argued that the development of independent IT silos as currently found is leading to greater congestion within ports. The DSR opens the possibility of operating ports on more than one IT operating system and platform. Integrating these will make for the smart port of the future.

The idea of a singular "end-to-end" digital platform for shipping may seem like a perfect solution, but unfortunately, it is highly unlikely to work. Even if it does, it will be bad for the industry. Experience with similar "universal" solutions in other industries suggests that stakeholders will not adopt it. As a result, this will create an environment where everyone will be categorized as winners and losers, whereas the minority will have too much power in their hands. A more effective solution is the openly available and interoperable data standards that enable communication among many digital platforms that will inevitably be built.

Last year saw many digital platform announcements that need to be observed. There is a wide list of startups looking at the problem, but some of the more outstanding announcements have come from the big 10-liner operators. This also includes IBM's Tradelens and Maersk as well as the formation of the Global Shipping Business Network (GSBN), which includes just about everyone but Maersk.

A single dominant and proprietary platform is more than likely to succeed, yet it is not desirable. This end game will be two or more incompatible and competing options, which will ultimately force everyone to make a choice. As liner operators, multimodal carriers, terminals, NVOCCs, and freight forwarders flash their colors to the mast, one side will eventually lose. However, this is not a way of losing; this could put businesses, small and large collectively, either out of business or at least put them in a position where they no longer would have control over their businesses.

A single logistics platform suppresses innovation and competition, which is one of the main disadvantages of small businesses and powerful individuals. This is also very unfortunate because the industry is already struggling when it comes to innovation, which also makes it difficult to overcome and many small players have been squeezed out of the market through acquisition and mergers. The ability to move goods profitably and safely around the world has always been the key to surviving in this business. However, survival in the twenty-first century depends entirely on one thing: data. The flow of the goods is directly dependent on its flow and future profitability lies on the ability to control and utilize this. The dominant players, in this case, are very much interested in maintaining a tight grip and keeping small innovators at arm's length.

The use and exchange of data have the power to transform many things from our industry from the operation of ships and ports to the processing of customs information. We must take full advantage of this opportunity as an industry, but we do not have to go down the route of a single expansive platform. However, there is a better way, which is the adoption of open-source data for shipping. This is where the rules of the data description that are recorded and exchanged are freely available. This also encourages competition and innovation and will keep data flowing freely, independent of the platform controlled by any one party.

Recent technological developments have empowered industries through digitization though. But they now face the challenge of the digital transformation as businesses are reconfigured beyond the realm of standalone organizations (Dodgson 2018). Badiani et al. (1988) identified that as more devices are becoming connected, new business opportunities are opened. Yang et al. (2015) highlight that this connectivity in maritime logistics is problematic due to sub-optimization together with the fragmented distribution of value beneficiaries serviced by maritime transport. As maritime transport becomes sustainable, it has become important to integrate these larger transportation networks. The emerging multimodal transport networks now offer holistic logistics solutions. These solutions are an integral part of the DSR that is built on technological advancements through the process of digitalization (Van Biljon et al. 2019).

Various measures, programs, procedures, and plans are being developed and implemented by modern shipping companies to merge their shore and seaborne transporting operations with the global supply chains. The modern shipping industry needs better operational visibility during sea voyage along with real-time access to information with a focused customer orientation. Transparency in cargo management is a key factor in ensuring the quality of service and market competition (Jerebić 2018). A common thread emerging in the debate is the need to have smart solutions not only to stay on the BRI network but also to remain competitive.

Summary

Logistics and supply chain management is undergoing a paradigm shift as the digital "Fourth Industrial Revolution" enables greater cargo visibility and transparency. There has been considerable progress in creating digital trade platforms that improve data and information transfer, in turn leading to a focus on last- and first-mile logistics delivery through multimodal transport systems. In part, this paradigm shift has been driven by China's introduction of the BRI, which is changing ports from a transshipment focus to gateway ports, moving away from the traditional port-to-port delivery.

This is particularly relevant as freight moves from FCL to LCL to cope with e-commerce retail.

For this to be successful, it requires a greater level of cargo visibility/traceability/monitoring than before. Digitization is seen as providing the solution to current information and data gaps between participants along the supply and logistics chain. To this end, China has implemented the DSR as to how this can be achieved under the BRI banner. However, what we have seen is the development of land-based versus maritime-based digital ecosystems that do not interface with each other. The creation of these data silos is leading to suboptimal utilization of space/resources and assets.

Sustainable socioeconomic development is dependent on basic services that support civil society, such as finance, energy, telecommunications, and transportation. As such, China's BRI is a global economic engagement strategy that helps deliver these key services that lift developing countries out of poverty.

As the BRI has developed, it has recognized the need for the development of ICT that connects logistics services within a complex and interdependent network. These developments are being driven by China's DSR. Unfortunately cybersecurity risks and a lack of common digital standards are creating a dual technology platform with no API or EDI interface allowing access to data and information to inform business decisions. The BRI and DSR have taken steps to address these issues within the BRI network and are creating a digital regional economy inspired by innovative thinking within the realm of the cyberspace. A global outlook is needed in the era of big data with nations shifting focus from land, capital, energy, resources, and population to competition for data (Chen 2016). In simple terms, for there to be a fully digitally integrated logistics and supply chain network, current GPS and 5G track and trace systems need to find a way to interface with the BeiDou and Huawei 5G digital framework being implemented throughout the BRI trade network.

Key Takeaways

- Digital integration has a large impact on maritime and rail logistics along China's BRI, and it could delay delivery of a key objective of the BRI.
- The shipping and logistics industry is lagging others in terms of connectivity and technological adoption; it is now under a huge amount of pressure to design, develop, and implement a sustainable digital system that should dramatically increase its operational efficiency.
- Governments, like China, are stepping in to support the building of digital platforms, like the establishment of the global shipping network, which will connect container operators, terminal operators, customs officials, freight forwarders, and cargo owners.
- Connected platforms will reduce the total cost of the supply chain through optimal asset utilization and lower inventory costs through efficient cargo management and visibility of the entire value chain from manufacture to end user.

References

Badiani, D., Bron, A., Elkington, A. et al. (1988). Use of a novel transport method for the quantification of the normal flora of the external eye. *Microbial Ecology in Health and Disease* 1 (1): 57–59.

Biest, A. (2019). 6 benefits of digitalization in the freight industry. *Cargofive*. https://cargofive.com/6-benefits-of-digitalization-in-the-freight-industry/ (accessed May 2020).

Brautigram, D. (2020). *A critical look at Chinese 'debt trap diplomacy': the rise of a meme. Journal Area Development and Policy* 5 (1): 1–16.

Brotchie, J., Hall, P., Newton, P. et al. (2017). *The Future of Urban Form: The Impact of New Technology*. Routledge.

Chen, C. (2016). Opportunities and challenges faced by the belt and road in the era of big data. In *Proceedings of International Symposium on Policing Diplomacy and the Belt & Road Initiative*, p. 142.

Chengdu Launches Blockchain Cross-border Trade Platform to Connect China and Europe (2019).ChinaBankingNewshttp://www.chinabankingnews.com/2019/10/30/chengdu-launches-blockchain-cross-border-trade-platform-to-connect-china-and-europe/ (accessed May 2020).

Dodgson, M. (2018). *Technological Collaboration in the Industry: Strategy, Policy, and Internationalization in Innovation*. Routledge.

Fishwick, O. (2017). *China in the fast lane on Digital Silk Road. China Daily*, https://www.chinadaily.com.cn/business/4thwic/2017-12/04/content_35201648.htm (accessed May 2020).

Garnaut, R., Song, L., and Fang, C. (2018). *China-40 Years of Reform and Development 1978–2018*. Australia National University Press.

Guo, H., Liu, J., Qiu, Y. et al. (2018). *The Digital Silk Belt and Road Program in support of regional sustainability. International Journal of Digital Earth* 11 (7): 657–669.

Hillman, J. (2018). *China's Belt and Road Initiative: Five Years Later. Center for Strategic & International Studies*. https://www.csis.org/analysis/chinas-belt-and-road-initiative-five-years-later-0 (accessed May 2020).

Huang, Y. (2016). Understanding China's Belt & Road initiative: motivation, framework, and assessment. *China Economic Review* 40: 314–321.

Jerebić, V. (2018). Digitalization in container shipping industry. In: *Tridesetosmi skup o prometnim sustavima s međunarodnim sudjelovanjem AUTOMATIZACIJA U PROMETU ed. Šakić, Ž. Zagreb, pp. 49–52.*

Kida, K., Hashimoto, S. (2019). China's version of GPS now has more satellites than US original. Nikkei. https://asia.nikkei.com/Business/China-tech/ (accessed May 2020). China-s-version-of-GPS-now-has-more-satellites-than-US-original.

Shen, H. (2018). Building a digital silk road? Situating the internet in China's belt and road initiative. *International Journal of Communication* 12: 19.

Kosmala, M., Wiggins, A., Swanson, A. et al. (2016). Assessing data quality in citizen science. *Frontiers in Ecology and the Environment* 14 (10): 551–560.

Kozlowski, K. (2018). BRI and its digital dimensions: twists and turns. *Journal of Science, Technology, Policy Management*. 11 (3): 311–324. doi: 10.1108/JSTPM-06-2018-0062.

Levinson, M. (2016). *The Box: How the Shipping Container Made the World Smaller and the World Economy Bigger*. Princeton University Press.

Lund, S., Manyika, J., Woetzel, J., et al. (2019). Globalization in transition: The future of trade and value chains. McKinsey Global Institute. https://www.mckinsey.com/featured-

insights/innovation-and-growth/globalization-in-transition-the-future-of-
trade-and-value-chains (accessed May 2020).

Moody's Analytics (2020). Moody's Analytics Global Macroeconomic Model. https://www.
moodysanalytics.com/microsites/model.

Muravev, D., Rakhmangulov, A., Hu, H., et al. (2019). The introduction to system dynamics
approach to operational efficiency and sustainability of dry ports main parameters.
Sustainability 11 (8), 2413. https://doi.org/10.3390/su11082413 (accessed May 2020).

Panayides, P.M. (2006). Maritime logistics and global supply chains: towards a research
agenda. *Maritime Economics & Logistics* 8 (1): 3–18.

Seele, P., Jia, C.D., and Helbing, D. (2019). *The new silk road and its potential for sustainable
development: how open digital participation could make the BRI a role model for sustainable
businesses and market. Asian Journal of Sustainable Social Responsibility* 4 (1): 1.

Splash 24/7 (2019). Blockchain roaming in the maritime industry. https://splash247.com/
blockchain-roaming-in-the-maritime-industry/ (accessed May 2020).

Teradata (2019). State of Artificial Intelligence for Enterprises. http://assets.teradata.com/
resourceCenter/downloads/ExecutiveBriefs/EB9867_State_of_Artificial_Intelligence_
for_the_Enterprises.pdf (accessed May 2020).

Thurer, M., Tomasevic, I., Stevenson, M. et al. (2019). *A systematic review of China's Belt and
Road Initiative: implications for global supply chain management. International Journal of
Production Research* (10): 1–18.

Toh, H.S. (2020). China, BRI Countries Confront Covid-19. https://www.asiasentinel.com/p/
china-bri-countries-confront-covid (accessed May 2020).

Transmetrics (2019). Logistics of the future: best supply chain visibility startups. https://
transmetrics.eu/blog/logistics-of-the-future-best-supply-chain-visibility-startups/
(accessed May 2020).

Van Biljon, W.R., Pinkham, C.C., Cloran, R.A. et al. (2019). Organizing data in a virtual
computing infrastructure. United States Patent 10282764. Available at: https://patentimages.
storage.googleapis.com/e8/f6/96/684ade1afdcecd/US10282764.pdf (accessed: May 2020)

Weiner, T. (2019). Supply chain visibilitys. *MEP supply chain*. http://www.mepsupplychain.
org/supply-chain-visibility/ (accessed May 2020).

Weissberger, A. (2019). China to complete Beidou satellite-based positioning system by June
2020- to be used with 5G. *IEEE ComSoc*. https://techblog.comsoc.org/2019/12/30/china-
to-complete-beidou-satellite-based-positioning-system-by-june-2020-to-be-used-with-5g/
(accessed May 2020)

Yang, C.S., Yeo, G.T., and Vinh, T.V. (2015). The effects of intra-and extra-organizational
integration capabilities in the container shipping freight industry. *International Journal of
Logistics Research and Applications* 18 (4): 325–341.

14

Marine Terminal Operating Systems

Connecting Ports into the Digital World

Ira Breskin[1] and Ayush Pandey[2]

[1] *SUNY Maritime College, Bronx, NY, USA*
[2] *Amazon.com Inc., Seattle, WA, USA*

Introduction

Digital transformation has been sweeping changes across all industries. Ironically for shipping, an industry that plays a key role in the global marketplace, the evolution has been progressive, but slow. This uptake has been hampered by evolving cross-border challenges, geopolitical events, worldwide economic slump, congested trade routes, operational complexities, and shrinking profit margins. Marine terminals are a key element of shipping as they provide a critical shoreside interface with arriving vessels and hence, have become a subject of significant innovation in the past decade. As such, the focus has been increasingly concentrated on efficient throughput of discharged cargo as an important factor when assessing port and terminal performance and future optimization.

For shippers, supply chains are shaped by arrival and departure times of vessels, and their choice of port is often made based on accessibility, schedule frequency and reliability, port proximity, and how quickly their cargo is discharged from both the ship and the receiving terminal. Moreover, the terminal's ability to ensure efficient cargo transfers is one central dimension of its overall function as a transport node.

In recent times, overcapacity and uncertainty in the global economic and political climate have depressed freight rates; additionally, fluctuating oil prices, vessel charter rates, and higher operating costs have added pressure on shipping lines to run a tighter ship. Rising shipper expectations amid surging demand often mean port productivity is down, and the operational costs are increasing the total landed costs of goods imported/exported. An IHS Markit report (Mooney 2019), in collaboration with eight of the world's ten largest container ship operators by volume, revealed that terminal productivity (defined as the number of container moves per vessel) at the top 30 port terminals was down by 7.5% in the year 2017. Furthermore, for 2017 and 2018, the average in-port time across all calls was 23.1 hours. The vessel's in-port time has averaged (Ducruet et al. 2014) from as high as 61 hours in some of the African ports to as low

The Digital Transformation of Logistics: Demystifying Impacts of the Fourth Industrial Revolution,
First Edition. Edited by Mac Sullivan and Johannes Kern.
© 2021 by The Institute of Electrical and Electronics Engineers, Inc.
Published 2021 by John Wiley & Sons, Inc.

as 21 hours in some of its European counterparts. Although this has been a huge improvement from the 90+ hours of in-port time about 30 years ago, to put it into perspective, a reduction of one hour per call would represent a very modest 4.3% efficiency improvement. It would equate to annual savings across these port calls of between US$320 million to US$480 million for ship operators.

In order to move more freight, and do so faster, ports with the capacity, equipment, and technologies that can support the most efficient vessel turnaround times are the need of the hour. A study (Development Asia 2020) conducted on Indonesia's Tanjung Priok port concluded that reducing cargo clearance times at the port, expanding the port, and improving its operational efficiency contribute about 1.1% to gross domestic product (GDP), thereby highlighting how an efficient marine terminal also contributes to the national economy.

Terminal Operating Systems

Improving port productivity requires a great deal of cooperation and collaboration between ship, port, and terminal operators. Marine terminals are served by laborers, supervisors, managers, liners, regulators, thus making a pragmatic case for a single integrated system called a terminal operating system (TOS). Marine terminals increasingly rely on the underlying TOS to better plan, control, and coordinate the increasingly automated movement of boxes transiting the world's shoreside facilities and beyond. A TOS has information on whether shipping lines use an electronic data interchange (EDI), or the location of the cargo (for the consignees), or when to pick up the cargo (for the truckers). An efficient TOS can also help roll-on/roll-off (RORO) terminals function more smoothly, manage mixed cargo, reduce traffic congestion, etc. It also helps terminal operators optimize operations and better strategize gate planning, equipment control, and ground stowage.

In terms of infrastructure, some terminal operators today are using outdated reporting measures, such as homegrown Excel spreadsheets, whiteboards, or in some cases, clipboards with pen and paper to record data that must be collected and processed at critical points at the gate, yard, and quay. A TOS can help streamline reporting procedures and set a standard for the same. TOS' can also be used to redesign truck gates and container yard expansions that enable ports to grow and meet evolving customer demands while maintaining efficiencies. TOS' can also be combined with process automation technologies like radio frequency identification (RFID) to help terminal operators close in on the security gaps as well as identify critical data trends in the cargo for red flags, such as point of origin. TOS' thereby generate significant operating savings by reducing delays of cargo at marine terminals, a major chokepoint in international supply chains.

Ports and Local Governments Recognizing the Value of Modern TOS

Cutting edge ports, such as the Port of Antwerp (Louppova 2018), are seeking the help of TOS to leverage the ready transfer of information about container cargo transiting their terminals to increase productivity and hence marketability. Busan New Container

Terminal (BNCT) (O'Dwyer 2019), one of the largest terminals in the world by capacity, signed an agreement with TOS provider CyberLogitech to implement OPUS TOS software across all its terminals in 2019. Port Manatee (2019), Florida's modestly sized port and the closest US deepwater seaport to the expanded Panama Canal, has adopted an integrated TOS to improve inventory levels and real time tracking of diverse cargo like liquid, dry bulk, and containers. The Ports of South Carolina (2019) recently invested $17.3 million to adopt the Navis N4 TOS as a replacement for its mature in-house system to offer a single point of entry to customers across all terminals. With the move, the South Carolina Port Authority, which owns and operates the marine terminals in South Carolina, also became the first in North America to operate all its terminals from a central location.

While the maritime industry's journey to zero carbon does not have an overnight fix, forward-thinking ports like the Port of Long Beach, California, have embraced environmental sustainability as a smart business investment. The port is currently using remote monitoring technologies in tandem with discrete TOS used by various terminal operators to improve the environmental and energy performance of terminal equipment, requiring less energy and fewer moves for equipment to transport and store containers.

An Increasingly Competitive TOS Software Market

Big container terminals have used TOS for many years to optimize their terminal traffic. While the smaller container terminals, which cannot afford the expenditure required for complex and multiple hardware and software configurations, are following at a much slower pace, some terminal operators also prefer to develop in-house solutions or use other forms of organization for the terminal operation. Marine Traffic Data (2020) currently lists about 20000 ports worldwide. Usually, ports consist of several different terminals, all acting independently, while adopting a specific organizational structure. This includes the decision-making process for implementing a software solution such as a TOS. Thus, the market for TOS is big and involves more than 20000 potential customers for special software. The Fraunhofer Center for Maritime Logistics and Services, CML, acknowledged close to 40 TOS software providers in the year 2016. The table below introduces a selection of popular products and service providers.

Company name and website	TOS software offered	Website description	Ports/terminals using the TOS
Navis	N4 Terminal Operating System, Master Terminal by Navis	The Navis N4 operating system and related applications and services improve productivity by coordinating, optimizing, and automating planning and management of complex terminal operations around the world	London Gateway Port, Port of Tanjung Pelepas, Jawaharlal Nehru Port Container Terminal

(Continued)

Company name and website	TOS software offered	Website description	Ports/terminals using the TOS
CyberLogitech	OPUS	OPUS Terminal is an integrated terminal operating system (TOS) that help ports simplify operational complexities by optimizing equipment usage, yard management and vessel planning	Busan New Container Terminal (BNCT), Derince, Turkey Safiport Derince
Jade Logistics	Master Terminal (acquired by Navis)	Master Terminal, Jade Logistics' TOS manages all cargo types, making it the TOS of choice for mixed cargo terminals	Abu Dhabi Ports, Port Nelson, CentrePort Wellington
Kalmar	OneTerminal	Kalmar OneTerminal provides an integrated automation solution delivered by one core team that brings together Kalmar and Navis software systems, equipment, and services for seamless deployment	Terminals at Port of Houston
PSA	Computer Integrated Terminal Operating System (CITOS)	CITOS processes information for allocating berths to ships, planning the stowage of containers, reading container numbers, and operating trucking gates	PSA Singapore Terminals
Tideworks	Tideworks Technology	Tideworks helps marine terminals cut operating costs by 20% with the most intuitive terminal management tools on the market	South Carolina Port Authority, Port of Cork
TGI	OSCAR	OSCAR aims to unlock the power to the intelligent movement of goods through the terminal	Kribi Port, Port of San Pedro (Cote-d'Ivorie)

Navis as a Terminal Operating System

As an example of the capabilities provided by a leading TOS, we outline here some key features of the Navis offering. Similar features are provided by some other TOSs as well, and the specifics noted here should help the reader appreciate the complexity and sophistication of modern TOS.

TOS software giant Navis LLC is helping to optimize marine-based supply chains significantly. Based in Oakland, CA, Navis is the world's largest TOS vendor with about 30% market share. A Navis survey (N4 2020) on reducing operating costs using its N4 TOS offering revealed that clients saved between 50 and 60% costs associated with yard planning and operations, as well as gate operations. For example, Ports America Chesapeake, a company that operates various terminals at the Port of Baltimore, United States (Labrut 2019), recently went live with the N4 software with an aim to handle additional volume efficiently, increase velocity, safety, and ultimately maintain its status as one of the most productive port container operators on the East Coast. Incorporating the new TOS means it is the first time Ports America has utilized a single platform for all cargo handled at its terminals, and the project is a part of a $142 million investment in equipment and infrastructure initiatives to stay competitive in the region.

The latest N4 version can manage operational events and inventory, monitor work instructions, and use a community access portal to pre-advise consignees of container availability and make gate appointments. The goal – optimize the planning and movement of containers in the yard and even beyond the yard gates. This falls in line with a McKinsey report (Murnane 2017) that stated that shippers would be willing to pay more to in return for improvements in, for example, the availability of equipment, visibility into inbound cargo pick up times, transparency, and communication. Freight forwarders can also input information into the TOS, which will allow the terminal to preclear the truck and save the trucker time. "Then the terminal knows how many trucks are coming per hour, and then they can plan the yard and labor. As a terminal operator, if I can reduce rehandles by 10 percent, I am can improve my bottom line by reducing my operating costs," says (Johnson 2018) Nishant Pillai, Vice President of Sales in the Americas for the Software-as-a-Service (SaaS) technology platform 1-Stop Connections.

The N4 software can also plan and control container discharge from a ship and storage at the terminal and facilitate timely retrieval by the consignee or its agent. For example, in the case of planning the discharge of cargo from a ship, N4 can tap into a vessel's stowage plan to determine each container's relative position on the ship. The software then follows instructions to calculate an estimated time required for the assigned gantry crane to remove the specific container from the ship. It then uses digital location data (from digital maps, not GPS coordinates) to determine the optimal workflow for yard equipment to eventually move the container to the preferred storage location in the terminal. Terminal operators using Navis N4 have thus benefitted from the N4's capabilities to optimize operations, increase operational visibility, increase throughput, and consequently increase business.

Shippers and consignees have also benefitted from the software's multifunctional use. N4 defines and enforces demurrage charges that result from a shipper's failure to pick up a container from a marine terminal within the period specified in the shipping contract. In the United States, the Federal Maritime Commission (FMC) in April 2020 published new guidance addressing what some shippers claimed was unreasonable container demurrage charges. According to the new FMC guidance, importers, exporters, intermediaries, and truckers should not have to pay demurrage charges if they cannot retrieve containers because of circumstances beyond their control. Moreover, the FMC recommends that terminal operators' demurrage policy be clear and consistent and that the terminals should notify importers when their cargo is available for retrieval. The N4 TOS can thus help calculate how much free time the consignee can store a container on the terminal property before incurring a significant "late fee", once terminal operators have notified consignees about their cargo.

Navis is also working on new technology to compliment the multifunctionality and dependability of the N4 software. The company recently introduced XVELA, a maritime business network and cloud collaborative platform designed to optimize vessel stowage planning/execution and berth utilization. Ocean Network Express (Businesswire 2019a), a carrier alliance of Japan's three major container carriers K Line, MOL, and NYK, has benefitted by implementing XVELA to better utilize stowage capacity. Another stand-alone software, Navis Smart, allows customers to customize and configure operating data to address their specific needs more easily. Ultimately, it will be able to provide data for clients to participate in the blockchain electronic permission-based supply chain system. N4 module StowMan generates stowage plans for a ship, or it can interface with a carrier's specialized hardware and software that does that job. It also communicates and interfaces with railroad stowage planning systems. A business intelligence (BI) analytics software, BI portal analytics, and BI portal of operating monitoring can also be integrated into the existing N4 software.

Breaking Down the Costs of Implementing a TOS

While the software has a myriad of uses to optimize maritime port operations, it can be difficult for managers to demonstrate that a software investment will improve productivity while lowering costs. Although the cost of the software is low compared to that of hardware like Ship-to-Shore (STS) and yard cranes, IT managers may struggle to make the case to their management for investment in new software and systems, and rightfully so. IT managers focus on increasing investment in efficiency and reducing labor costs pushes terminal operators to check automation options; however, many terminal operators still prefer more traditional operational measures based on proven system reliability and labor flexibility. A TOS' one-time license fee per location is about $6 million to $10 million. Purchasing customer service, system upgrades, and additional customization add-on about a 20% premium annually over the original licensing fee. Therefore, the total cost of ownership, including personnel, over the system's six-to-eight-year life, is five to eight times the original licensing fee.

By now, it is evident that while the hardware costs are modest, the TOS software, in itself, is modular. Customers pay for the basic product and are charged extra for bells and whistles. Navis, for example, introduced Compass in 2019, a visual workflow management module that digitizes the workflow and standardizes how terminals work to improve the quality of the planning and tracking process to improve the terminal's efficiency. Ultimately, Compass coordinates the hand off of freight to modal carriers – specifically trucks, barges, and on-dock railroads – serving the terminal's customers. Yilport, a major terminal operator with headquarters in Turkey, has installed this software at two terminals and is testing it at four others. After a successful one-month pilot (Businesswire 2019b), Yilport concluded that Compass aided in making relevant information available to all terminal actors in real time, which helped to reduce miscommunication and prevent claims across locations. TGI, another TOS vendor targeting small and medium-sized ports, also offers the add-on TGIBOX – a TOS hardware and software package providing precise Differential Global Positioning System (DGPS) geo-localization (enhanced version of the Global Positioning System [GPS]) of container handling equipment and other terminal equipment – adding another layer of transparency. A similar add-on, the Navis Smart Suite, encompassing a hands-on mobile software, optimization module, etc., has a license fee that ranges from $50 000 to $100 000.

What to Consider When Selecting a TOS?

While the operational benefits of a TOS have been valued for decades, adopting one into the supply chain comes at a considerable cost, often making it unfeasible. With small and medium terminals catching up to the big players, TOS software providers face an increasingly difficult task of tailoring the TOS to meet the specific needs of each terminal and ensuring that the TOS extrapolates and assimilates the "correct data" from the input fed by terminal employees. It is possible that the TOS is well designed to meet the needs of a terminal, but the software is fed with unreliable or incorrect data provided by the terminal operators. Therefore, to reap the benefits of such an operating system, understanding the factors contributing to the performance of the TOS, and its subsequent positive impact on terminal productivity is paramount. The infrastructure of the software that terminal operators use to perform operational activities, the vast array of data that is fed into the system and its access, and the software's future development are the foundations driving a reliable TOS.

Because replacing an existing TOS is a difficult and expensive option, the TOS performance must be judged on factors like its stability, scalability, vendor support, ease of integration and customization, and security. Since the costs involved, both operational and financial, can be significant, the end results of the TOS have to be envisaged before choosing on the right TOS.

Also, a significantly outdated TOS that a terminal may be using may have been cutting edge when it was designed, but it likely has not evolved to accommodate the increasing complexity of modern terminal operations. Therefore, a TOS should be future proof and must be able to answer some of the maritime industry's modern challenges like automation initiatives, alliance consolidations, and handling the unprecedented surge of containers being discharged at each port by today's ultra-large container vessels. The preferred TOS should possess robust data access with interpretable data that can be used to make strategic decisions. Whether its data pertaining to truck turnaround time, moves-per-hour, or availability of reefer plugs, the data provided by the chosen TOS should be used to derive patterns through analysis and drive operational process changes consequently. Moreover, keeping in touch with the positive trend of automation in general, it is important to test the TOS system and fine-tune its functions before the go-live to avoid unnecessary risks.

As previously stated in the chapter, a TOS software relies on the RFID or optical character recognition (OCR) and on the integrity of planning data that is held in the software to accurately identify containers at the terminal. Any loss, corruption, or malicious modification of this data could prevent containers from being located within the terminal or could result in containers being shipped to the wrong locations. This risk is not merely theoretical; criminal gangs can use RFID-blocking technology to mask smuggling operations, thus bringing into question the security of the terminals. Therefore, a strong network that hosts the TOS data should also be given attention to, for this protects the confidentiality and security of the terminals.

Importance of Visibility at the Terminal

TOS also are designed to mitigate potential supply chain snafus that often crop up at marine terminals. Consignees frequently do not know about in transit delays until the box is received by the terminal, after being discharged from the ship. In this case, terminal operators are often criticized, since they are a very visible stakeholder in the container shipping activity. However, terminals inherently do not have control over the various transport modes that bring cargo and containers to them. Other lingering problems that have prevented the terminals from ensuring timely cargo delivery include vessel bunching, yard congestion, and delays in container handoffs in case of intermodal networks. Note that inbound overseas cargo typically arrives late as much as 25% of the time at American marine terminals, further exacerbating the delays. Environmental factors also add to the issue, and hence, vessel schedule integrity in the major East–West trades can be anywhere between a dismal 34 and 70%. All these terminal-related snafus can further multiply the original in-transit delay. After receipt, the terminal stores the box until the consignee's agent arrives to pick it up. Before being released from the terminal, cargo must clear customs, often incurring yet another delay. All these factors indicate that poor scheduling reliability at the terminals makes it increasingly difficult for cargo owners to schedule inbound pickups from the ports. This only adds on to the agitation of the cargo owners, since now they cannot plan their truck dispatches and use warehouse personnel efficiently.

Digital Communication is Key

To ease out the frustration of cargo owners, N4 software can notify the consignee when cargo is ready for pick up, increasing the efficiency of container drayage. Thus, the N4 also reduces the costs for cargo dwell time demurrage that was covered earlier in this chapter. A related opportunity for a TOS is to link in operators of the chassis pools (that provides equipment for North American drayage drivers to deliver inbound cargo to consignees). Linking the chassis pool helps ensure that deliveries are completed on time, and thus avoid unnecessary demurrage charges.

A TOS can also reduce in-transit-related cargo expenses. The consignee generally assumes the cost of financing goods in transit; that obligation ultimately is determined by the negotiated INCO or International Commercial Terms. Excess material handling costs add to the total landed costs of the cargo. Handling containers of cargo requires extraordinary use of expensive machinery, often operated by highly paid unionized workers. Also, added moves inside the terminal are a prime cause of cargo loss and damage. The software also seeks to minimize unnecessary movement of cargo within a terminal. It also automatically incorporates container-specific handling instructions, such as alerting for required holds or permissions.

The Future of Terminal Operating Systems

Modern problems require modern solutions. The move toward Hub and Spoke operations, and mega-ships bringing in more containers are causing bottlenecks in ports and challenging the existing shoreside infrastructure to maintain productivity. To put it into context, container ships have grown by almost 1500% in carrying capacity: from 1530 TEU (20-feet equivalent units) in 1968 to 24 000 TEU in 2020. This development puts the stress on the quay, where containers are unloaded and loaded withing limited space and equipment and also scheduled tightly with quick turnarounds and small timeslots. How a rapid turnover of the containers is considered a competitive advantage for ports has already been discussed in the chapter's introduction. In the struggle to remain competitive in this fast-paced landscape, terminals are increasingly turning toward automation with one aim: to reduce the cost per handled container at the terminal. "Automation has allowed terminals to improve their operational efficiencies and ability to manage larger vessels. This allows terminals to keep up with the rest of the world with minimum dependency on human resources. This has made such a big breakthrough in our industry because terminals can now enable container handling equipment to automatically adapt to dynamic situations which optimize performance and output," mentions Harry Nguyen, CEO of Real Time Business Solutions (Bensalhia 2019).

Gone are the days when users drive a car with their eyes fixed to the rearview mirror. Industry 4.0 has created a compelling case to integrate the various systems in a port, collect, and analyze the high volumes of data from these integrated systems to forecast results for the future. In the case of the TOS, the industry trend has gravitated towards a single system solution that can meet the majority of the needs of a terminal. The rapid development of information and communication (ICT) technologies such as the Internet of Things (IoT), Machine Learning, and 5G networks offer huge opportunities for ports to optimize their intraterminal operations. A survey (Manenti and Chen 2018) carried out by the SCM World (a Gartner Community) further solidifies this case: the most important disruptive technologies enabling digital supply chains in the future are big data analysis, IoT, cloud computing, machine learning, advanced robotics, 3D printing, drones/self-guided vehicles, sharing economy and blockchains. Dashboards and BI modules capitalizing on big data, integration of 3D visualization of terminals into the TOS, the capability to handle multiple types of cargo, and the automation of terminal operations are some of the ways the shipping industry is keeping up with the surge in technological advancements.

As the world continuously shifts to the digital landscape due to consumerism and globalization, cloud technology has also had a massive impact on the container terminal industry. For example, Octopi, a TOS product offered by the Florida-based Cetus Labs catering to smaller mixed port and container terminals (less than 500 000 TEU a year), has its system linked through Amazon Web Services. Cetus Labs aims to make every feature of Octopi available in the cloud, which thus gives flexibility, speed of implementation and integration, and cuts the costs associated with installation.

TOS vendors are moving more and more towards the development of service-oriented architecture (SOA), using which they can provide a collection of loosely coupled services on a cloud platform as a SaaS. Traditionally, organizations were required to maintain on-premise TOS'. Adapting to a new software was a costly undertaking in regard to implementation and user adoption. With the advent of SaaS TOS solutions, organizations can now pay a subscription fee to use a specific software, and the software vendor is in charge of managing and hosting the software itself – rather than requiring the organization to install, host, and manage the software themselves. This means that SaaS modeled TOS' also offers more flexible add-on modules to the "core" product, thus enabling operators to gradually grow the system in accordance with their business needs and complexity. Moving to usage-based charging thus ensures that terminals have more flexible OpEx options for financing their investment on systems as against hefty CapEx investments.

Technologies such as blockchain are also playing a part, and more so, ensuring the integrity of the data being captured and providing trust that the information is true and accurate. For example, Jade Logistics' TOS, Master Terminal, is already integrated with the CargoChain, a blockchain themed information sharing platform, to achieve transparency. Technology is thus helping ports take control of their operations, thanks to the use of a TOS, and helping them provide better customer service to port customers such as shipping lines and carriers. The three largest overseas shipping companies in the world – A.P. Moller–Maersk (Maersk), Mediterranean Shipping Co. (MSC), and CMA CGM, are currently using TradeLens (developed by Maersk in collaboration with IBM), a blockchain platform to track cargo ships and containers in near real time. Since Maersk and IBM first launched the TradeLens in 2018, more than 100 shipping and cargo firms have signed on to track vessels and shipping containers via the immutable electronic ledger. The system, based on Hyperledger blockchain, uses a shared governance system through which members hold others accountable for data input and must approve of new blocks added to the ledger. Youredi, a provider of cloud-based Integration Platform as a Service (iPaaS) solutions, has successfully integrated the TradeLens platform with its TOS.

Technology is thus helping ports take control of their operations, thanks to the use of a TOS, and helping them provide better service to port customers such as carriers, shippers, consignees, and their intermodal partners.

Conclusion

With over 90% of the world's international cargo goods being transported by the sea, container traffic is projected to be on the rise, particularly in the emerging economies of the world in the new era of mega-ships and digital transformation, ports. Their marine terminals can leverage smart technologies to reinvent themselves and contribute to "making the world smaller and world economies better." While the government and privately-operated marine terminals have been appreciating and investing in TOS, heavy initial investment, data security issues, downsizing port labor, loss on flexibility, and other challenges associated with integrating a TOS to the in-house ERP system are some reasons that TOS uptake has been slow.

While the future (Barrons 2020) of port terminals is automation, a scene where terminals are fully operated by only a handful of operators, the opportunities in terminal operations are yet to be fully realized. Soon, terminals will become a cohesive unit that optimizes the entire facility, instead of a siloed focus on yard or gate operations specifically. More terminals will try to replicate the success achieved by the likes of London Gateway Port and Port of South Carolina to optimize operations and increase customer satisfaction. As the future of terminal success relies on the ability to streamline and optimize operations and manage logistics information, TOS technology will be an essential component in helping terminals gain a competitive advantage as well as overcome and thrive in the face of challenges ahead.

Key Takeaways

- A terminal operating system (TOS) is a digital system that can integrate communication across a marine terminal – from vessel to gate and enhance operational efficiency. TOS can also help with other activities like cargo pick up alerts, managing roll-on/roll-off (RORO) terminals, manage mixed cargo, ground stowage, etc. Thus, a TOS can reduce duplicate investments in equipment, redundant workforce, and non-value-adding processes, while standardizing operations.
- A robust TOS must have a strong foundation for its software architecture and must be future proof. The TOS must promote not only digitization but also data-driven decision-making. In the end, the selection of a TOS will have profound impacts on both the tactical performance and strategic viability of the terminal, its customers, and its operator.
- The main advantages of implementing a TOS include reducing bottlenecks by using advanced technology to significantly increase TEU throughput, shortening the stay of the ship at the container terminal, increasing the accuracy of the data flow to facilitate cargo movement throughout the marine terminal, and continuously monitoring containers storage to reduce dwell.
- TOS providers are moving to the Software-as-a-Service model as opposed to licensing the basic model and then charging extra for the bells and whistles.

References

Barrons, A. (2020). Terminal operating systems: driving the future of optimization with TOS. https://www.maritimeprofessional.com/magazine/story/201305/terminal-operating-optimization-211383 (accessed 16 March 2020).

Bensalhia, J. (2019). Terminal operating innovation target. https://www.portstrategy.com/news101/technology/innovation-remains-the-target-for-terminal-operating-systems (accessed 29 March 2020).

Businesswire (2019a). Ocean network express chooses TPS Valparaíso to launch XVELA Collaborative Platform in Latin America. https://www.businesswire.com/news/home/20191120005703/en/Ocean-Network-Express-Chooses-TPS-Valpara%C3%ADso-Launch (accessed 17 March 2020).

Businesswire (2019b). YILPORT to roll out Navis' compass visual workflow management application in five terminals. https://www.businesswire.com/news/home/20190506005191/en/YILPORT-Roll-%20%20%20Navis%E2%80%99-Compass-Visual-Workflow-Management (accessed 17 March 2020).

Development Asia (2020). How an efficient port impacts the national economy, 14 April 2016. https://development.asia/policy-brief/how-efficient-port-impacts-national-economy (accessed 29 March 2020).

Ducruet, C., Itoh, H., and Merk, O. (2014). Time efficiency at World Container Ports. https://www.itf-oecd.org/sites/default/files/docs/dp201408.pdf (accessed 27 March 2020).

Johnson, E. (2018). Tech providers try to illuminate chronic marine terminal black holes. https://www.joc.com/technology/tech-providers-try-illuminate-chronic-marine-terminal-black-holes_20180610.html (accessed on 29 March 2020).

Labrut, M. (2019). Ports America introduces new terminal operating system in Baltimore. https://www.seatrade-maritime.com/ports-logistics/ports-america-introduces-new-terminal-operating-system-baltimore (accessed 16 March 2020).

Louppova, J. (2018). Solvo. TOS implemented in Antwerp terminal. https://port.today/solvo-tos-implemented-antwerp-terminal/ (accessed 16 March 2020)

Manenti, P. and Chen, X. (2018). Jumpstarting your digital roadmap. http://www.scmworld.com/wp-content/uploads/2018/04/Jumpstarting-Your-Digital-Roadmap.pdf (accessed 15 March 2020).

Marine Traffic Data (2020). http://www.marinetraffic.com (accessed 17 March 2020).

Mooney, T. (2019). Port productivity insight. https://ihsmarkit.com/Info/port-productivity.html (accessed 20 March 2020).

Murnane, J. (2017). Ports and shipping: the need for solutions that cross lines. https://www.mckinsey.com/industries/travel-transport-and-logistics/our-insights/ports-and-shipping-the-need-for-solutions-that-cross-lines (accessed 17 March 2020).

N4 (2020). Improvements by the numbers. https://www.navis.com/en/about/news-events/news/n4-improvements-by-the-numbers/# (accessed 20 March 2020).

O'Dwyer, R. (2019). Busan new container terminal upgrades its operating system. https://smartmaritimenetwork.com/2019/03/06/busan-new-container-terminal-upgrades-its-operating-system/ (accessed 11 March 2020).

Port Manatee (2019). Kinder Morgan ink terminal lease extension, *Port Manatee*. https://www.portmanatee.com/2019/09/19/2567/ (accessed 20 March 2020).

South Carolina Ports (2019). Port of Charleston deploys Tideworks' terminal operating solutions. https://www.ship-technology.com/news/south-carolina-ports-tideworks/ (accessed 16 March 2020).

15

Improving Cross-Border eCommerce Through Digitalization

The Case of Compliance in B2C Shipments

Simon de Raadt[1] and Jiao Xu[2]

[1] HyperSKU, Shenzhen, China
[2] Flexport, Amsterdam, NL

Introduction

eCommerce continues to expand in Europe. As many as 270 million Europeans say that they regularly shop online (Postnord 2018). In parallel with consumers' growing habit of shopping online, cross-border eCommerce is emerging as an important and swiftly growing trend. Cross-border eCommerce is defined as international business activities executed by trade entities belonging to different customs territories. Transactions and payments are settled through eCommerce platforms, and goods are delivered through an international logistics system (Ma et al. 2018). This provides opportunities for companies to access new markets and participate in global production networks (Aydın and Savrul 2014). It plays an indispensable role in economic globalization and is becoming a more and more important part of business models and consumers' everyday lives. This trading model can happen between two businesses (B2B), two individual consumers (C2C), or between a business and a consumer (B2C). This chapter focuses on B2C transactions.

Cross-border eCommerce allows consumers to directly purchase products and services that may be unavailable or expensive in their home countries. Despite the different languages and currencies used between a consumer and a merchant from abroad, people can purchase goods located in other countries and jurisdictions via websites and online marketplaces (Hsiao et al. 2017). The lack of availability of certain products, the convenience of effortlessly ordering 24/7 from home, and more competitive prices abroad are the most cited triggers for cross-border online shopping (Ding et al. 2017; Huang and Chang 2019).

As an analysis of Cross-Border Commerce Europe (2019), a platform that activates cross-border eCommerce in Europe, showed European Union (EU) cross-border trade (excluding travel) amounted to 95 billion EUR in 2018. 55% is generated by players from within the

The Digital Transformation of Logistics: Demystifying Impacts of the Fourth Industrial Revolution,
First Edition. Edited by Mac Sullivan and Johannes Kern.
© 2021 by The Institute of Electrical and Electronics Engineers, Inc.
Published 2021 by John Wiley & Sons, Inc.

EU and 45% by retailers from outside. Among retailers outside of the EU, 80% of such cross-border sales are generated through "marketplaces," led by Amazon with 32 billion EUR. In 2018, from the cross-border parcels shipped among 41 countries worldwide, 38% were purchased from China, with the share increasing for the last three years (International Post Corporation 2018). In 25 out of the 41 countries where the survey was conducted, China was the most popular region to buy from, and in ten of them, China ranked number two.

The popularity of China's export eCommerce can be explained by customers valuing low price products and an increasingly positive attitude towards "made in China" (Cross-border Commerce Europe 2019). China surpassed the United States as the biggest eCommerce exporting country in 2016. While its current share of export goods is already impressive, studies indicate that especially B2C cross-border eCommerce from China to Europe will rise even more (Cross-border Commerce Europe 2019). Despite these advances, various barriers, especially those related to customs clearance, are still constraining the growth of cross-border transactions (Heel et al. 2014).

Customs Clearance as a Barrier Constraining Growth

Cross-border eCommerce trade is more complicated and riskier than traditional offline trade. This risk is mostly carried by the seller and the consolidator and is largely caused by information asymmetry and deficiencies in the supply chain. Today, there is a great deal of paperwork and multiple stakeholders involved for every individual cross-border eCommerce shipment. In addition, the business model is subject to multiple trade barriers (Gomez-Herrera et al. 2014). Within the EU, regulations for international trade are not harmonized, allowing every EU member country to enact its own laws and policies. These stipulate the conditions under which goods are eligible to be exported or imported.

Items imported into a country must be cleared by customs. Typically, this is done by customs brokers who represent the importer during the customs clearance procedure. For customs brokerage companies, this task involves the preparation and submission of documents required to facilitate the import procedure. These documents can include the (i) Purchase Order from the buyer, (ii) Commercial Invoice, (iii) Bill of Lading or Airway Bill, (iv) Packing List, (v) Manifest, and (vi) Certificate of Origin. Customs inspects the documents submitted as well as the physical goods to ascertain the value declared is by the international markets and approve the assessment based on the appropriate classification. The goods will then be released out of the customs bond after a successful declaration (Management Study Guide 2019). This process is inevitable in cross-border eCommerce and can, depending on different cases, take from one up to seven days.

This often manual effort is aggravated by the EU's complex tax and duty regulations. Goods originating from a non-EU country are subject to value added taxes (VAT) and import duties when entering the EU. Depending on the value of the content of a shipment, there are some rules to consider when to pay VAT and when to pay import duties. If the value of the parcel is less than 22 EUR, no VAT and no import duties apply. Above 22 EUR,

VAT will apply and from 150 EUR, both VAT and import duties will be charged to the consumer or the seller.

Most of the B2C parcels entering the EU are handled by postal operators and imported under DDU[1] terms. This means that if the parcel has a value of over 22 EUR, both VAT and/or import duties will be charged by the postal company to the consumer, as well as an admin fee for the manual clearance service provided. As there are no requirements for small parcel shipments to alert customs in advance, the parcels are typically cleared based on information provided on a shipping label applied to the parcel. Without digitized information available, the only way to determine if the declared value of a parcel is correct is by manually checking the parcel label or opening the parcel.

Entities that participate in cross-border commerce face multiple risks, including but not limited to political risk, service supply chain risk, taxation risk, risk of data acquisition failure, and regulatory invalidation of government agencies (Ma et al. 2018). This is the case especially for customs clearance companies since the trade entities in this context of business belong to different customs territories, and the clearance operation is the key to successful cross-border delivery. Hameri and Hintsa (2009) identified customs issues, including security concerns and regulations, as one of the key drivers of change in international cross-border supply chains. Failure to comply with applicable customs legislation and to pay the duty can lead to fines and custodial sentences.

Compliance in the supply chain and documentation is an integral part of a company's import program (Cook 2011). For instance, Grainger (2016) conducted a study among customs managers in multinational companies whose function was split in logistics support, supply chain management, and regulatory compliance. Several participants highlighted that their number one priority is to ensure port and border clearance without delay. Others also mentioned a focus on reducing duty liabilities and other subsequent costs of importers. They, therefore, have to make sure that correct information is provided to the customs administration and that the correct amount of duties paid (Grainger 2016). Failure to carry out compliant operations can lead to serious consequences. If goods are incorrectly declared, a whole shipment can be confiscated by customs upon arrival. It is important to note that there are other causes of possible shipment delay related to incompliance that are not covered in this chapter, in particular legal aspects related to intellectual property rights, privacy of data, and CE marking.

As most shipments are cleared as bulk instead of as individual parcels, it means that even if 1000 individual's parcels are correct, as long as one parcel is incorrect, the whole batch will be stopped. In 2016, multiple wholesalers that imported underwear and cosmetics from China went bankrupt because of a comprehensive investigation into their customs declarations and then a failure to supply orders (Grainger 2016). Despite such stringent regulations and possibly severe consequences, parcels entering the EU are often incorrectly declared. This is so severe that the European Commission concluded that the EU is missing over 5 billion EUR in VAT and import duties from B2C cross-border eCommerce shipments every year (European Commission 2017).

---

1 Delivered Duties Unpaid (DDU): the buyer is responsible for paying import duties.

Lack of Integration as the Root Cause of Customs Misdeclaration

The lack of adherence to customs regulations typically happens for two reasons. One is the deliberate or accidental *undervaluation of shipments.* Parcels imported into the EU often show a value under 22 EUR, which allows for VAT and import duty exemption. For example, importers based outside the EU fraudulently mark expensive goods such as mobile phones or tablets as costing not more than 22 EUR. This is violating regulations according to which the price of a product should be logically and reasonably in line with the product name and description. Such an incorrect value declaration can be made by any of the involved stakeholders.

Some sellers are shipping their products directly from the supplier to an end consumer, a practice known as dropshipping where a seller forwards a consumers' order to a supplier who then fills the order directly to the consumer and is paid a predetermined price by the seller. The advantages of this business model include lower costs of holding inventory, materials handling, and obsolescence. Disadvantages are that order delivery is fragmented when a single customer order involves products from different manufacturers as well as typically longer delivery times (Khouja 2001). Another aspect is the lack of data. The supplier normally arranges the shipment on behalf of the seller. For B2C shipments, the goods should be declared on the consumer. The shipper does not know the online sales price of the seller nor does the seller know at what value the goods are declared when they are shipped out. This lack of transparency is an important cause of compliance issues. Noncompliance can also originate from a declaration of an incorrect harmonized system (HS) code.

HS codes were developed by the World Customs Organization as a multipurpose international product nomenclature that describes the type of good that is shipped. The HS code is organized logically by economic activity or component material and composed of a six-digit identification code. The declared HS code will determine how much import duty must be paid for one shipment. Deceitful shippers can declare incorrect HS codes to reduce costs. For instance, an importer could declare a cotton T-shirt as "Plastic toy" because the import duty for clothing and fashion products is approximately 12%, while that for plastics toys 0%.

The EU tried to address the issue by adjusted its regulations. According to the Amendment to the Union Customs Code Delegated Regulation (EU) 2015/2446, the existing VAT exemption for goods up to 22 EUR will be removed from July 2021 onwards (European Commission Taxation and Customs Union 2020. That also means that every single shipment will have to be declared fully, including the provision of the right product information, HS code, and proof of purchase or sales. While this change might be aimed at addressing the challenges in compliance, the principal causes are not tackled as undervaluation and misdeclaration would still be possible (some people argue that the regulatory change is rather aimed at creating a level playing field for European towards non-European eCommerce players). Acting against customs misdeclaration is challenging once a shipment arrives at the customs border. Therefore, focusing on the origin of the shipment can decisively solve the problem and lead to lasting maintenance of control, correct declarations, and even speedup of the customs clearance process.

The key is to be able to isolate individual parcels, for which information must be available already before shipment arrival – digitally and in good quality. To accomplish this,

Figure 15.1 Stakeholders in B2C supply chain.

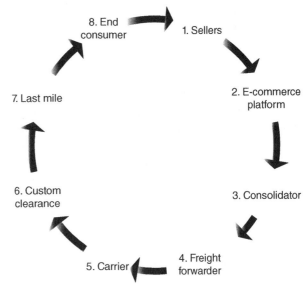

multiple players in the supply chain will have to adjust their mindset, operations, and business model to build a seamless end-to-end B2C shipping experience driven by technology. Currently, all stakeholders involved in the supply chain have conflicting earning models and little interest to align their data. The process of aligning standards for output from one company resulting in input to the other company is not aligned and is missing integration mechanism. Figure 15.1 provides an overview of the actors in the cross-border eCommerce supply chain.

Each stakeholder uses conflicting key performance indicators. The sellers (i) want to create as many sales as possible while reducing costs to maximize their profits. eCommerce platforms (ii) offer a wide range of products, ideally with visibility from purchase till delivery, at a competitive price. Consolidators (iii) collect as much volume as possible to be able to get competitive rates for both the main haul and the last mile to the customers. Freight forwarders (iv) and carriers (v) operate mostly on actual or volumetric weight. They are especially for B2C shipments very concerned about dangerous goods, substances that, when transported, are a risk to health, safety, property, or the environment. The custom clearance companies or brokers (vi) focus on an accurate matching of provided data and the actual content to be able to have compliant clearance processes. The last-mile delivery companies (vii) want as many parcels as possible to maintain their scale, without getting involved in the earlier processes of the supply chain due to the involved risks and uncertainties. Finally, the end consumers (viii) want to buy whatever item whenever and have it delivered at the selected time to the selected location. If the end consumer has a satisfying customer experience, they will buy again, recommend to others, and leave a positive review. Then the cycle repeats itself. The challenges for these stakeholders are optimizing and harmonizing the supply chain and create a benefit for the involved stakeholders.

Then the question arises: why is it that none of the stakeholders take measures to align with other stakeholders? Noncompliance is particularly an issue for shipments from China, where no one-stop solution exists beyond the express services of CEP[2] providers like DHL, UPS, FedEx, or TNT (Pfohl et al. 2019). These companies are more focused on B2B shipments and do not consider B2C shipments as their primary market segment due to its complexity and lack of profitability. Current B2C market players from China are consolidators that work on an agent model where they leverage rate cards instead of service and compliance. Multiple platforms, consolidators, freight forwarders, and last-mile service providers have jumped into this space as well. This has resulted in a race to the bottom when it comes to pricing, but not in a sustainable solving of the customs clearance problem. No one is willing to take the risk of the customs clearance as a nonowner of the goods, especially as some brokers already had to file for bankruptcy or even got jailed due to incompliant clearance processes. eCommerce platforms with access to all relevant data have not encountered any compliance challenges. For these platforms, it means that the sales prices for end consumers would go up and hence the attractiveness of the platform goes down. Besides, most platforms do not want to take any financial risk when it comes to customs clearance, particularly since many of them are stock listed and therefore have to obey special regulatory requirements. Overall, no party wants to connect their company entity to a shipment when it is unclear what the real content is or if a shipper is purposely under-declaring its value. To sum it up, there is a lack of trust and transparency in the supply chain.

In this situation, it is particularly challenging to agree on a cooperative model that secures the position of each stakeholder while, at the same time, ensures the sharing of reusable data for all. However, with increasing market pressure, this will become necessary. The blooming cross-border eCommerce market enabled consumers and provided them with multiple provider choices at the fingertip. Bargaining power is, therefore, shifting toward consumers, whose demand for service quality keeps rising. They want convenience and consistency of the overall shopping experience, accessed from one platform. For example, consumers typically go to an eCommerce website to track the logistics status for what they have purchased. From their perspective, it is not relevant that after placing an order on the platform, different stakeholders own and manage different parts of a cross-border eCommerce supply chain (Accenture 2015). Therefore, all stakeholders must connect their services to create better visibility in the supply chain, achieve customs compliance, and ultimately satisfy consumers' needs.

How the Digitalization Can Revolutionize Cross-Border ECommerce

Digitalization has the potential to significantly increase customer satisfaction through increased supply chain visibility and faster customs clearance. The relations are shown in Figure 15.2.

---

2 Courier/express/parcel.

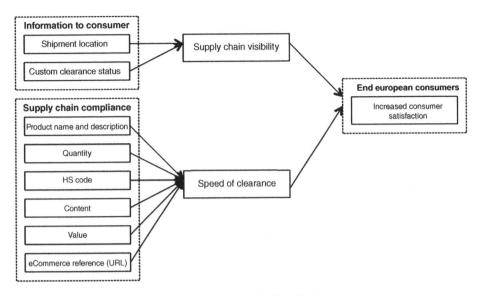

Figure 15.2 Increased customer satisfaction through digitalization.

Increased Customer Satisfaction

Hsiao et al. (2017) applied a Kansei engineering approach to cross-border eCommerce logistics service design. This method aims at the improvement of products and services by translating customers' feelings and needs into concrete product parameters and provides support for future service or product design (Schütte et al. 2004). Service quality has become a critical factor that affects how buyers evaluate sellers as well as cross-border eCommerce platforms. The Kansei approach focuses on accuracy, speed, and efficiency. This is precisely reflecting the end consumers' demands from their shopping experience and what stakeholders in the supply chain should strive to achieve. Endo et al. (2012) also concluded that consumer satisfaction in the post-purchase stage is critical. They suggested that firms need to warrant more accurate information among other distribution members as well as consumers. Since consumers cannot obtain the products immediately, customer service (e.g. on-time delivery, efficient information presented to consumers, and communications with them) becomes essential in the post-purchase stage. This is especially valid for cross-border eCommerce, where the post-purchase stage is more complex than in domestic eCommerce. Here, digitalization could improve the eCommerce customers' experience through increased visibility and faster customs clearance.

Supply Chain Visibility

Barratt and Oke (2007) define supply chain visibility as "the extent to which actors within a supply chain have access to or share the information which they consider as key or useful to their operations and which they consider will be of mutual benefit." Thus, visibility has a range of levels determined by the amount of useful information that is shared across the supply chain. Information sharing is not directly linked to improved performance, while it is through supply chain visibility enabled by high-quality information that performance can be improved.

To stay in the competition and win in today's marketplace, the stakeholders need to create transparency and visibility to present the right and adequate information not only to consumers but also to the other stakeholders to maintain the continuity of the supply chain. They also need to be compliant with customs regulations and provide accurate information to the customs authorities to deliver the parcels to the end consumers as soon as possible. DHL Supply Chain's recent research report on the evolution of eCommerce supply chains also shows that, despite understanding of the vital importance of eCommerce on consumer retention and satisfaction, most companies have failed to fully implement an eCommerce strategy due to changing consumer expectations, the pace of delivery, and limitations in existing infrastructure.

Zhou and Zhu (2010) found that information sharing can create value for the overall eCommerce market, but affects buyers and sellers very differently. Depending on the competition mode they are engaged in, it can create a win-lose situation by distorting, concealing, or disclosing information in the interest of the company. For B2B, the return on investment is higher compared to B2C shipments. Visibility in cross-border ecommerce B2C shipments has multiple owners in the supply chain waiting for platforms to make the first move to investing in supply chain visibility.

Supply chain visibility plays an important part in communications between different stakeholders. Yet it remains one of the top issues consistently mentioned in surveys of supply chain management professionals. It is crucial to note that supply chain visibility is the result of information sharing, not the other way around (Barratt and Oke 2007)

Speed of Customs Clearance

Quite often, it is the customs clearance that causes delays, which then will result in lower consumer satisfaction. Even if the supply chain has been made visible, there is no paperwork for B2C parcels available as the purchase happened online. That means there is limited reference material to determine the accuracy of the data. B2C shipments from China especially are causing more alerts than those from other countries, because of not only the big quantities but also the noncompliance of some Chinese importers.

Customs clearance companies face two challenges: first, by performing compliance in customs clearance procedure for a successful import and second, by sharing parcel status information to eliminate the incoherence in the supply chain. These companies should act as the responsible compliance examiner of data and information provided by platforms, eCommerce sellers, and consolidators. All the stakeholders in the supply chain should also try to aggregate their services and systems and take responsibility to optimize the supply chain to contribute to visibility by sharing crucial and accurate information of the shipment status. The most important information in B2C customs clearance documentation is

- Product name and description
- Quantity
- HS code

- Content
- Value
- ECommerce reference (URL)

To overcome this problem for cross-border eCommerce B2C shipments, the best proof for the online purchase is the URL or the website link to the online store. The URL shows the website of the selling product and its corresponding detailed information. Looking at this information is a method of counterchecking used by customs brokers and customs authorities to inspect if all of the information shown in the URL match what is provided on the declaration documents.

All these data are available from the online store, but simply not shared and made available to the other stakeholders. The solution lies in making this data available before the parcels leave China. By retrieving the URL of the online store and linking it to the parcel content, it is possible to identify what the content of this product should be and what it actually is. This makes it instantly much easier to check if, indeed, the data matches the actual product. To run this physical check on all the millions of parcels everyday is impossible; therefore, continuously updating the logic to determine potential frauds will be key to maintain a green flow of goods. If one or more parcels have to be taken out for inspection in China, then what remains of the shipment can still go. If you determine this upfront, there will be no unnecessary delay after arrival in the port of destination.

Technology as Enabler for a Cooperative Model

Through the use of technology, it is possible to create a cooperative model that will unlock consumer satisfaction for all types of cross-border eCommerce purchases. Customs clearance today is a challenge for cross-border eCommerce operations. This might seem surprising, considering that, for example, luggage at an airport can be brought into a country within minutes and without any data check. However, though it seems that the flow of luggage at an airport and a cross-border parcel do not differ much, a key difference is physical ownership of the luggage. If it were possible to ensure clear accounting of ownership of parcels, a faster process would be feasible. This could be achieved through digital integration, where cargo ownership could be traced back to the source or seller. With this visibility of the ownership of the goods, customs authorities would consider the data flow as reliable, put more trust in the digital checks, and thus impose fewer physical checks as well as speeding up the clearance process. This is also known as a "green flow" of goods and in this context it is not related to sustainability, but solely for the purpose of having a traffic light ranging from red (a complete black box), orange (semitransparent flow of goods), to green (a fully transparent and compliant flow of goods).

Operationally, it could be realized in the following way. Before any shipment is created, there will be a (i) compliance check. After uploading the parcel information, logic will determine the quality and accuracy of the data and the information obtained from the URL. Once the parcels have arrived in the China warehouse, the questionable parcels will be taken out and reinspected and the data will be updated if needed. Then (ii) the shipment

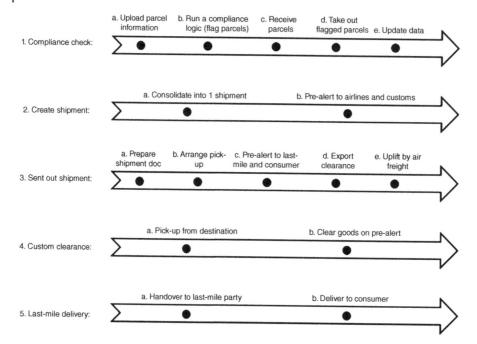

Figure 15.3 Steps needed for a green flow.

will be created, and data will be sent as a pre-alert to the airline and customs. With no red flags from the pre-alert (iii), the shipment will be ready to be sent out. The shipping documents can be created for pick up, export clearance, and uplifted by the airline. In this stage, the delivery is confirmed, and a pre-alert can be shared with the last-mile operator as well as the consumer. As data is available, the shipment will be picked up for (iv) custom clearance, and within minutes, the shipment can be released. After clearance, the B2C parcels are ready for the (v) the last-mile delivery towards the consumer (Figure 15.3).

To reach this form of transparency, data availability and quality will be key. For instance, an API could be created that is connected with all players involved in the supply chain. This would connect the party uploading the data with both the supplier and the seller. The provided information is then checked, uploaded, and shared with the other stakeholders in the supply chain. In this way, all stakeholders know that shipment is about to arrive and can plan their resources based on this knowledge. In practice, plenty of ground handlers and freight forwarders can only act once the goods have arrived, therefore delaying the flow of operations. So sharing the information would already speed up the performance after the arrival of goods.

Since the data is already provided in advance, parcels could be cleared even while they are still in the air. Regarding small parcels stored in postal bags, as long as it is known which parcels are in which postal bag, it is possible to scan each bag and put them aside if an inspection is required. With such flow, goods could be picked up from the destination port, checked, custom cleared, and handed over to the last-mile delivery service provider within just 24 hours. Ultimately, the end consumer would win. Players like Alibaba already

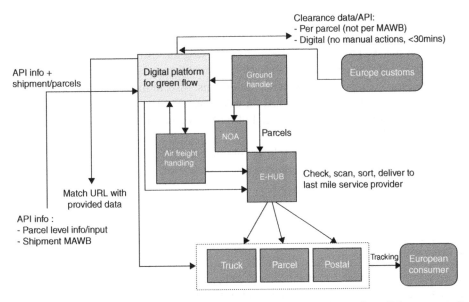

Figure 15.4 Basic technical requirements to realize a green flow *Source*: From Sales presentation, Basic technical requirements © 2020, ViaEurope.

announced that they want to achieve a global delivery of B2C parcels within 72 hours. In addition, airports such as Amsterdam's Schiphol in the Netherlands are working on paperless operations to allow the digitalization of operations in connection with surrounding logistics service providers.

Technical Requirements

To be able to share the data among the different stakeholders, all the data must be collected, stored, and made available to all involved users. This digital platform will allow a green flow. After checking the provided data with the URL, the data can be uploaded and shared with the use of an API connection. The ground handler will receive a Notice of Arrival (NOA), and the airfreight handler will be informed of the goods to be picked up. After delivering at the e-hub, the parcels can still be checked, inspected, sorted, and delivered to the last-mile operators. Each parcel is connected to the AWB until arrival in the e-hub, so each parcel automatically is updated on the latest status. Only once the goods leave the e-hub the tracking information be recorded on parcel level (Figure 15.4).

Conclusion

Data and technology are key to overcome cross-border eCommerce challenges. Current cross-border B2C logistics solutions are not sufficient to improve consumer satisfaction, increase supply chain visibility, and provide faster customs clearance operations. Through digitalization, information can be made available across the whole supply chain.

The main pain point in the supply chain for B2C shipments is the limited access to and unreliability of data for custom clearance. Typically only platforms and sellers have full information of this data as they own the data. By making the URL and shipping information available to all involved stakeholders, every stakeholder can become a market player.

The focus of this chapter has been on B2C shipments. Also, in the case of overseas warehousing, a similar solution can be created for eCommerce sellers. In case of stock movement, you can clear on parcel levels instead of bulk level and still be able to identify the quality of the data in advance.

Using technology allows a one-stop solution or one to become part of it. It creates a cooperative model that will unlock consumer satisfaction for cross-border eCommerce purchases and a green flow of goods.

Key Takeaways

- Cross-border eCommerce trade is more complicated and riskier than traditional offline trade as there is a lack of trust and transparency in the supply chain.
- A major barrier for the growth of cross-border eCommerce and headache for logistics and supply chain managers is customs compliance, which is done manually for a majority of B2C customs clearance procedures.
- All stakeholders must connect their services to create better visibility in the supply chain, achieve customs compliance, and ultimately satisfy consumers' needs.
- Digitalization will improve cross-border eCommerce by providing greater visibility in the supply chain from beginning till the final delivery as well as increasing the speed of customs declarations.
- With the use of technology, it is possible to create a cooperative model that will unlock consumer satisfaction for all types of cross-border eCommerce purchases resulting in a "green flow of goods."

References

Aydın, E. and Savrul, B.K. (2014). The relationship between globalization and eCommerce: Turkish case. *Procedia – Social and Behavioral Sciences* 150: 1267–1276. https://doi.org/10.1016/j.sbspro.2014.09.143.

Barratt, M. and Oke, A. (2007). Antecedents of supply chain visibility in retail supply chains: a resource-based theory perspective. *Journal of Operations Management* 25 (6): 1217–1233. https://doi.org/10.1016/j.jom.2007.01.003.

Bateman, T., Buhler, B., and Pharand, A. (2015). *Adding value to parcel delivery*. https://www.accenture.com/_acnmedia/accenture/conversion-assets/dotcom/documents/global/pdf/dualpub_23/accenture-adding-value-to-parcel-delivery.pdf (accessed 26 April 2020).

Cook, T.A. (2011). *Compliance in Today's Global Supply Chain*. CRC Press. ISBN: 9781420086218.

Cross-border Commerce Europe (2019). *Top 500 cross-border retail Europe.* https://www.cbcommerce.eu/news/press-release-top-500-cross-border-retail-europe/ (accessed 17 April 2020).

Ding, F., Huo, J., and Campos, J.K. (2017). The development of cross border eCommerce. *Advances in Economics, Business and Management Research (AEBMR)* 37: 370–383.

Endo, S., Yang, J., and Park, J. (2012). The investigation on dimensions of e-satisfaction for online shoes retailing. *Journal of Retailing and Consumer Services* 19 (4): 398–405. https://doi.org/10.1016/j.jretconser.2012.03.011.

European Commission (2017). *Corrigendum to Commission Staff Working Document on Significant Distortions in the Economy of the People's Republic of China for the Purposes of Trade Defence Investigations*, Brussels. http://trade.ec.europa.eu/doclib/docs/2017/december/tradoc_156474.pdf (accessed 3 May 2020).

European Commission Taxation and Customs Union (2020). Modernising VAT for cross-border e-commerce. https://ec.europa.eu/taxation_customs/business/vat/modernising-vat-cross-border-ecommerce_en# (accessed 22 December 2020)

Gomez-Herrera, E., Martens, B., and Turlea, G. (2014). The drivers and impediments for cross-border eCommerce in the EU. *Information Economics and Policy* 28: 83–96. https://doi.org/10.1016/j.infoecopol.2014.05.002.

Grainger, A. (2016). Customs management in multinational companies. *World Customs Journal* 10 (2): 17–35.

Hameri, A.P. and Hintsa, J. (2009). Assessing the drivers of change for cross-border supply chains. *International Journal of Physical Distribution & Logistics Management* 39 (9): 741–761. https://doi.org/10.1108/09600030911008184.

Heel, B., Van Lukic, V., et al. (2014). Cross-border eCommerce makes the world flatter. http://image-src.bcg.com/Images/Cross-Border_ECommerce_Makes_The_World_Flatter_Sep_2014_tcm108-82788.pdf (accessed 12 March 2020).

Hsiao, Y.-H., Chen, M.-C., and Liao, W.-C. (2017). Logistics service design for cross-border eCommerce using Kansei engineering with text-mining-based online content analysis. *Telematics and Informatics* 34 (4): 284–302. https://doi.org/10.1016/j.tele.2016.08.002.

Huang, S.L. and Chang, Y.C. (2019). Cross-border eCommerce: consumers' intention to shop on foreign websites. *Internet Research* 29 (6): 1256–1279. https://doi.org/10.1108/INTR-11-2017-0428.

International Post Corporation (2018). Cross-border eCommerce shopper survey 2018. https://www.ipc.be/services/markets-and-regulations/cross-border-shopper-survey (accessed 19 March 2020).

Khouja, M. (2001). The evaluation of drop shipping option for eCommerce retailers. *Computers & Industrial Engineering* 41 (2): 109–126. https://doi.org/10.1016/S0360-8352(01)00046-8.

Ma, S., Chai, Y., and Zhang, H. (2018). Rise of cross-border eCommerce exports in China. *China & World Economy* 26 (3): 63–87. https://doi.org/10.1111/cwe.12243.

Management Study Guide (2019). A brief on customs brokerage. https://www.managementstudyguide.com/customs-brokerage.html (accessed 19 Febraury 2020).

Pfohl, H.-C., Wolff, P., and Kern, J. (2019). Transshipment hub automation in China's CEP sector. In: *Urban Freight Transportation Systems*, 1e (eds. R. Elbert, C. Friedrich, M. Boltze, and H.-C. Pfohl). Elsevier.

Postnord (2018). ECommerce in Europe 2018: consumers' buying behavior increasingly global. https://www.postnord.com/en/media/publications/eCommerce/ (accessed 18 January 2020).

Schütte, S.T.W., Eklund, J., Axelsson, J.R. et al. (2004). Concepts, methods and tools in Kansei engineering. *Theoretical Issues in Ergonomics Science* 5 (3): 214–231. https://doi.org/10.1080/1463922021000049980.

Zhou, Z.Z. and Zhu, K.X. (2010). The effects of information transparency on suppliers, manufacturers and consumers in online markets. *Marketing Science* 29 (6): 1125–1137. https://doi.org/10.1287/mksc.1100.0585.

16

Enabling Platform Business Models for International Logistics

Sam Heuck[1] and Cory Margand[2]

[1] Atlas Sourcing Partners, Hong Kong, HK
[2] SimpliShip, Boston, MA, USA

Introduction

A platform acts as a medium of securing products and services, usually in some combination of price, service, and quality. In supply chain specifically, this has been an area that has been least impacted by the digital movement leaving buyers and sellers to work in archaic means such as spreadsheets, phone calls, and face-to-face meetings for negotiations. Software can and will forever change the way procurement is managed by providing a platform that creates the most efficient means to bring this process online and by leveraging data so that each party spends less time on unnecessary activities. Unlike platforms, marketplaces are a place where people physically met to exchange goods and services. Today, all is driven by platforms that leverage technology to match buyers and sellers more efficiently and scalable. When utilizing filters, a marketplace can ensure that only the most relevant information is provided to the user ensuring a more efficient and pleasing experience. In this chapter, we will be specifically referring to the new age of marketplaces that are utilizing software to create a digital marketplace where users can buy and sell goods and services.

Freight procurement has historically been handled in a very non-digital and very much siloed fashion. From a pricing perspective, two types of freight rates can be procured, spot, and contract rates. This includes face-to-face negotiations, phone calls, and emails with very little being driven by insightful data. This has led to the following inefficiencies:

1) Lack of understanding of the current and future market.
2) Poor profitability management from ocean carriers, freight forwarders, and NVOCCs. An non-vessel operating common carrier (NVOCC) is a company that signs contracts directly with the carriers that own the vessels, like Maersk or COSCO. An NVOCC can then sell the space they have contracted with the carriers to freight forwarders or directly to a beneficial cargo owner (BCO).

The Digital Transformation of Logistics: Demystifying Impacts of the Fourth Industrial Revolution,
First Edition. Edited by Mac Sullivan and Johannes Kern.
© 2021 by The Institute of Electrical and Electronics Engineers, Inc.
Published 2021 by John Wiley & Sons, Inc.

3) The inability of BCOs[1] to ensure that they are always optimizing price and service levels. BCOs are companies that are acting as the Importer of Record. They are responsible for the specific legal requirements for the country where they will be importing. For example, the BCO has the legal responsibility to ensure their product follows all legal requirements as well as customs requirements like a bond and documentation requirements.
4) Decreasing service levels of ocean carriers represented by slow steaming or canceled port calls.

The costs associated with traditional direct sales methods are expensive not only to acquire new customers but also to maintain them. Freight forwarding Boston station manager, Greg Hebard, explains, "Our typical sell-in process takes about 10–12 months from first engagement to securing some business. I estimate it costs us roughly $500–$1,000 weekly per account to maintain the relationship from a sales perspective and not including operational costs." (Hebard 2019). There are many reasons why this occurs, but the most concerning is the sell-in process that often takes a year or more, which is very costly from an expense standpoint but also an enormous hurdle to hitting sales goals. This also leads to poor service levels for small and medium-sized businesses (SMBs) as their freight forwarders are focused on trying to not only manage profitability but also maintain their top-level revenue targets. To give an example from the parcel shipping industry, it costs significantly less from an administrative point of view, approximately $0.50 per shipment.

"Traditional forwarders that wish to survive have no choice but to digitize. By our estimate, automating manual processes now could reduce certain back-office and operations costs by up to 40%, while digitizing significant parts of the sales process could reduce related direct costs even more. And as the new business models gain traction over the long term, digital capabilities will open up tremendous opportunities to win in the marketplace (Riedl et al. 2018)."

So, if we consider the opportunity for digital transformation from an operational and sales perspective, the opportunity is enormous and something that cannot be ignored.

Lack of a data-driven approach can lead to many downfalls for both parties involved in the procurement process. In 2015, we saw the ocean freight market collapse due to many drivers but perhaps the most important being overcapacity. Refer to Chinese containerized freight index, shown in Figure 16.1. Traditionally, the ocean carriers have tried to create unit economics by purchasing larger and larger ships, which created the perfect storm in 2015 when more capacity hit the market along with slowing global demand (Petersen 2015). For those BCOs that locked into contracts in Q1 of 2015, this created a multimillion dollar lost opportunity to capitalize on the market (Richter 2016). This was driven by two factors: their requirement to fulfill their minimum contracted requirements and the BCO's inability to efficiently contract on a short-term basis and/or procure spot rates (Richter 2016). A large BCO can leverage its forecasted annual volumes by submitting a Request for Proposal (RFP), which may contain certain requirements expected by the BCO like free time, transit

---

1 Beneficial cargo owners (BCOs) are companies that act as the importer of record. They are responsible for the specific legal requirements for the country they are importing into. For example, the BCO has the legal responsibility to ensure their product follows all legal requirements as well as customs requirements like a bond and documentation requirements.

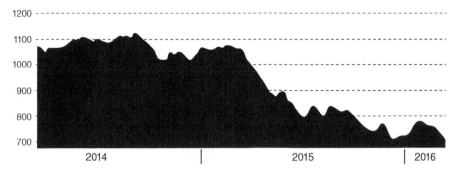

Figure 16.1 Chinese containerized freight index 2014–2016. *Source*: Richter (2016). © Wolf Street Corp.

times, and technology capabilities (Jung et al. 2018). The challenge is that this process is very similar to any other industry in which hedging is used. It can be unpredictable and leads to many inefficiencies on both sides but does make for an easier operational process.

SMBs are reliant heavily on the spot market. So, they do not necessarily have to deal with contracted rates, but they do have the challenge of efficiently procuring spot rates while sifting through offers from their provider portfolio. It is not uncommon for a logistics manager to wait days to receive rates back, and when received, they spend hours determining the best path forward on a per shipment basis. This can be due to many different reasons, but the main challenge for forwarders is that the rates are owned and maintained by the origin office and are not distributed effectively to their satellite offices. These non-value-added activities are costly both in terms of hours wasted and distract from their operational responsibilities ranging from managing on-time delivery to communicating to both internal and external stakeholders.

For a company to ensure that they are getting the best rates and service, they need to create a provider portfolio that may include carriers, forwarders, and NVOCCs, depending on their annual shipping volumes. This creates challenges post procurement as these BCOs are now managing multiple external parties who have their own platforms and reporting mechanisms. This creates inefficiency when trying to track critical milestones in the journey of an ocean container or pay invoices. In particular, invoice accuracy has been an enormous challenge for all parties involved due to the reduction in carrier back-office staffing and the fluctuating market dynamics. The CEO of Maersk, the world's largest ocean carrier, explained at the JOC's 13th annual TPM conference in Long Beach, California, that the company's invoice accuracy rate was at the 88.2 percent mark and "has been, for Maersk Line, the highest customer service issue for some time." As CEO, he said, he was distressed "to know that 12 percent of our invoices are not accurate (Morley 2016)."

Current International Logistics Technology Landscape

In logistics, SaaS offerings have not been tied to specific forwarders or carriers. However, these are very much siloed offerings based on functionality. For example, a BCO can utilize contract management or RFP software from companies like Oracle or SAP to manage their

Table 16.1 International Logistics Technology Landscape.

Contract and rate management software	Marketplaces
Catapult	SimpliShip
Tender tool	Freightos
Descartes	Cargobase
Cargosphere	Freightbro
7L	Cogoport
SimpliShip	Searates
Visibility	**Transportation management systems (TMS)**
INTTRA	GT Nexus
Ocean Insights	SAP
Crux Systems	BlueJay
Vizion	Oracle
Vessel Tracker	Kuebix

normally offline negotiations. The other option is to utilize a tender management platform specifically made for logistics. For years, the industry has been heavily reliant on electronic data interchange (EDI) for track and trace but have recently seen improvements in this area due to automatic identification system (AIS) satellite data being available at the vessel level (ORBCOMM 2019). EDI has been the primary medium of data transfer used between different supply chain systems. EDI relies heavily on information technology (IT) expertise and can be extensive in terms of mapping of terms from one system to another. This technology is also reliant on pushing data from one platform to another rather than an application programming interface (API), which provides a more standardized and quicker form of instant data transfer. There are instances where EDI, sent in batches, has sent data that is outdated by the time it is received by the end user. To understand more about the intricacies of EDI and APIs, Glick and Glickenhouse's Chapter 4 offer extensive coverage.

In Table 16.1, there are a few of the providers that are offering software solutions in the logistics industry.

The above is not an all-encompassing list of providers to date. But it does give a high-level look into the current solution providers that vary from niche focuses to transportation management systems (TMS) who utilize specialists to integrate into their software to provide an end-to-end product for their customers (Rouse 2015; Trelleborg Marine Systems 2018). As an example, many software providers and BCOs rely on INTTRA's EDI integration with the carriers to provide ocean container visibility. INTTRA integrates with many of the leading ocean carriers so that a BCO does not have to. In other words, the BCO simply does one integration with INTTRA rather than having to integrate with the many carriers they are doing business with, which could change during any time.

Recently, we have seen the emergence of two game-changing SaaS offerings, which include marketplaces and those focused on digitizing both forwarders and carriers (Jain, Chaudhary and Patchala 2017; Lock 2018). Marketplaces are changing the international freight landscape by integrating numerous sources of data and automating the decision-making process for stakeholders. Forwarders and carriers typically do not consider technology as one of their core competencies, which in turn creates a major challenge as marketplaces have to figure out how to get access to the data in a seamless and scalable fashion. The exceptions to this rule are digital forwarders who have a strong digital infrastructure that allows for better seamless data integration and customer experience from an online usability perspective. Some of the leading digital freight forwarders are Flexport, Twill icontainers, ExFreight Zeta, Freight Hub, Global Forwarding, Zencargo, and YQN Line. Technology providers like Kontainers are taking advantage of this movement to online freight procurement by creating software that enables those forwarders and carriers an online platform for their customers through a white-labeled, licensed website (Kontainers 2019).

Freight forwarders are not the only players selling online. Carriers are increasingly providing instant rates and booking functionality through their websites. Vincent Clerk, Chief Commercial Officer for AP Moller–Maersk, said, "We are now making it as easy for our customers to book a container as booking a flight ticket. Instant booking confirmation makes it faster, easier, and simpler for our customers to interact with Maersk." The importance of all of this is that the trends are clear and that technological adoption is making system-to-system integrations possible. Therefore, more data is widely available to optimize rate and service levels in real time.

An industry-wide push to overhaul backbone technologies of logistics is overdue. An example of the old way of doing things is EDI, which relies on an expensive mapping process that is done on an individual customer basis (*SAP EDI Document Standard* 2018) and relies on pushing data from one platform to another. Kayak, an example of a travel-related platform, requires a customer to fill in certain information like from and to cities that are used as initial filters to ensure the results are a fit. The instant options that surface are the result of API technology that allows the platform to call the airline carriers and hotels for instant individualized results that are specific to the filter.

An API is a way for software to communicate with each other without a platform having direct access to a database. An example of an API action would be when a user comes onto Kayak's website (www.kayak.com); they fill in their information and instantly can see flights relevant to the user. Kayak can provide this information seamlessly to the user by leveraging API technology to integrate with the airlines. When Kayak receives the user's requirements, they make an API call to the airlines, which returns the schedules, rates, and other amenities in real time. If this were EDI, the user would have to wait until the data was pushed to them by the data source. EDI also would not allow for specific, relevant, on-demand data to be available.

If Kayak, for example, was reliant on EDI technology, this would not be possible, or at the very least, the cost to serve their customers would be much more expensive. The user would wait for an EDI data push, which would then, in turn, need to be highly filtered by Kayak to ensure a match between the data provided and the filters. Today, API's are far superior from an integration perspective, and most important is the ability to share information in real time that is specific to the customer's request (Hughes 2019). As Eric Johnson, Senior Editor Technology for *The Journal of Commerce*, outlines below, "Global shippers are uniform in their desire to go from a collection of disconnected systems to platforms where they can more effectively manage, compare, and act on data. While those disconnected systems might individually function well, shippers need to manage more data on fewer systems (one if possible) to truly understand their total cost and the performance of their transportation and logistics suppliers."

APIs can enable data sharing specific to international logistics in several ways, all of which allow real-time sharing of valuable data that will save both sides of the industry to save time and money. We previously discussed some of the challenges in the industry that include procuring rates, visibility, and invoice auditing. The use of integrating APIs in this process can lead to significant savings, such as knowing in real time that a shipment has been pulled for an intensive exam by customs or the BCO's free time is about to expire, and the trucker has not pulled the container from the port. If the BCO receives an EDI update days after their free time has expired, they have incurred hundreds of dollars in added expenses, but with an API, the BCO would have immediate visibility to the issue and could resolve it the same day.

Another example is overbookings and no shows (Huynh and Walton 2008) due to the inability of BCO's and carrier's inability to share real-time data. There are many reasons why a container may not make the original vessel booked that range from production delays, quality issues, document issues, or simply the vessel has been overbooked. APIs have the potential to solve this issue by integrating into the enterprise resource planning (ERP) and sourcing platforms used by BCOs and their supplier network. ERP is the software backbone for many companies that handles everything from procuring (Purchase Order) to pay.

The software generally does a great job of linking core departments such as finance/accounting to operational departments such as purchasing and logistics. Generally speaking, they do not do a great job at specific departmental requirements, which gave rise to SaaS offerings such as TMS or warehouse management systems (WMS). In many large enterprises, production data are housed in a sourcing platform that is integrated with the BCO's ERP (Yao Yurong and He Houcun 2000). These data should be shared with their logistics partners. If an ex-factory date, the target completion date, is shared, then it becomes evident whether a booking will need to be canceled and allocated to another BCO, which will make the cutoff dates. API technology can make this all possible by integrating with these platforms and the carriers to ensure real-time data sharing.

Now that we have discussed the underlining need to upgrade the technologies utilized in international shipping, we will take a deeper dive into the impact of marketplaces and why they will make the single largest impact on the industry since the invention of the ocean container. First, let's understand that a marketplace's goal is to ensure a perfect fit between the BCO and the carrier or forwarder. This means that both customers are happy with the outcome. In other words, the BCO has optimized their freight spend and service levels while the provider has optimized its cost to acquire and/or service their customers. Forwarders and carriers spend a lot of time and resources to sell into their existing and potential new customer base. A platform can eliminate this time dramatically by eliminating a year in the selling process, creating a real-time procurement process, and reducing the forwarder's unit costs (*Shippers turn to digital booking platforms*, 2019; Toczauer 2019).

"De Nooijer, alongside the key members of AF KLM Cargo team, have presented their vision of digitalization, which comes with a "high-level ambition" to operate in a "transparent and easy to connect way, at a competitive price". He added: "The customer is completely informed, every step of the way, about what is happening with their shipment. It is about the exchange of data, and it is about knowing exactly what is happening. If you are an end consumer and you order a pizza at home, you can see that the driver is going to arrive at your home within ten minutes. So why on earth would you go for less in a business environment where you have shipments worth much more than pizza?" (*AF KLM on a digital journey* 2018)

KLM is a perfect example of a company within the industry taking advantage of a digital approach. As of Oct 2018, they had 460K track and trace visits, and 28% of their business booked directly online. If we take this data and apply it to the costs previously discussed, the results are immense.

The platform's goal is not purely based on rates that are widely misunderstood, although transparency and standardization are the objectives of the platform. Marketplaces eliminate the need for manual requests for rates and, once received, the need to filter through the rates for decision-making. It is extremely challenging and taxing on a team to filter through many different rate proposals that are in different formats and include or exclude different costs. As discussed earlier, the main challenge of a marketplace is the availability of data and how APIs can enable marketplaces to leverage this data to create a neutral collaboration platform that can be used regardless of partners being utilized. A marketplace can be the next-generation TMS by allowing BCOs to work with any provider at any point in time. Simply put, marketplaces can fill a much-needed gap in the industry by aggregating data from fragmenting sources and automating the decision-making process (McKinsey 2016). A common misconception is that these marketplaces are looking to replace forwarders, which could not be further from the truth. Marketplaces are technology companies that do not want to get involved in the operational or human elements of global transport. Carriers that take advantage of marketplaces can have direct access to BCOs that they normally would not and can eliminate their notorious reputation of providing subpar customer service.

One of the important topics discussed in *Platform Revolution: How Networked Markets Are Transforming the Economy – And How to Make Them Work for You* by Geoffrey Parker, Marshall Van Alstyne, and Sangeet Paul Choudary, is platform openness (Parker, Alstyne, and Choudary 2016). This topic is very relevant to any platform manager who is creating the operating guidelines and principles behind a logistics platform. Specifically, we will take a deeper dive into this concept and how it relates to the success or failure of an international logistics marketplace. The level of openness for a platform dictates how the users will function within the platform. If the platform is completely open, then anyone can join, and anyone can interact within the marketplace. This strategy can lead to enormous user growth but has inherent drawbacks that need to be seriously considered. Let's take Craigslist as an example where anyone can post and anyone can interact with the user posting (Kroft and Pope 2014).

This has notoriously created significant issues where users have misused the platform for criminal activity. This was not their goal, but this has been allowed by the platform manager because they have a completely open platform. If we apply this open concept to a logistics platform, we can assume that two things will occur: 1) users will join the demand side that will not have real freight to ship and 2) forwarders that are not vetted or qualified to provide services will join. This creates negative interactions between both sides of the marketplace and will detract from the value that users gain by utilizing a marketplace. On the other hand, if the marketplace is leaning more toward the closed functionality, we do not have a marketplace at all but more of a SaaS approach, which hinders growth and the network effects that make a marketplace valuable. The ideal level of openness for a marketplace is something that needs to be seriously considered as it will control and manage the interactions between the two sides of the platform (Hartmann 2013). The platform manager can choose to restrict one side of the marketplace while remaining open on the other. In other words, it is possible to be closed to one side of the marketplace's customers while remaining open on the other.

As an example, it would be wise to put strict guidelines on the provider side to ensure the providers can provide the services they are quoting on. For example, in the United States, it is required that the forwarder is licensed with the Federal Maritime Commission. If a marketplace decides to be completely open, they will not check these licenses and, in turn, will not create value for their BCOs because they have allowed unlicensed forwarders to provide rates to their customers. This ensures that there are positive interactions in a percentage of transactions and does not inhibit growth on the shipper or BCO side. Another aspect of openness is data sharing and integrations. This decision needs to be managed and monitored as closely as the previous decision on who can join the marketplace. Some marketplaces may choose to remain closed from an integration perspective, and others will open to other platforms by leveraging API integrations allowing easy flow of supply and demand. As discussed earlier, data creates immense value for marketplaces, so the platform manager will need to make sure that the data is not being misused or that the marketplace is not being cut out of the transactions. However, the benefits of opening APIs are significant as they can create scale quickly, lower customer acquisition costs, and enrich the data, which is one of the marketplace's more valuable assets.

The BCO's Perspective

Traditionally, BCOs have looked to regional providers, providers who advertise prominently, and their existing providers for logistics services. Relative to the entire industry, this represents a small sample size of choices. An existing relationship with a provider has typically trumped competing options for several reasons (Sahay and Mohan 2006). Ease of doing business, familiarity with the provider's tracking portal, and having a logistics counterpart who understands your business, combined with the occasional discount, has historically been enough to maintain the business. Furthermore, the difficulty of collecting competing offers and the time required to vet other potential providers have often created a competitive advantage for the incumbent provider. These difficulties in conducting large scale RFQs have prevented the BCO from becoming a more sophisticated buyer and failed to provide senior management with a transparent picture of what the broader industry has to offer.

Another subsequent challenge that small and medium BCOs face is staffing (McKinnon et al. 2017). Often, younger, less experienced professionals begin their careers in the company's customer service or logistics department before moving on to inside sales. The lack of familiarity with the industry and the company's goals, as well as what is often a pure "on-time delivery" focus for these employees, limits the potential for innovation. Many employees end up feeling that managing 2–3 different supply chain management (SCM) portals and comparing prices on the spot is enough to create value in the role. The manual nature of this process and the regular inconsistency in information across providers means that they are likely to correct in this thinking. The additional challenges that come with shipping across borders and preparing documents for customs clearance also push internal SCM teams to rely on small but stable outside providers whose familiarity with these processes creates the impression of outsized value-add that further limits the willingness of SCM departments to hunt for alternative providers.

Time limitations prevent BCOs from expanding and innovating. Examples of these time constraints are; the time needed to educate internal SCM employees, the time needed to conduct an RFQ (Janke 2011), the time lost waiting to hear from new providers, the time lost on back and forth emails clarifying business needs, the time spent updating portals and confirming next steps, and the time spent uploading documents and getting approvals. Many companies have their own logistics manager that oversees running RFPs, which according to Glassdoor, carry of $74K annually, not including benefits and other forms of costs. If technology is not leveraged, the majority of that salary is being spent on aspects of their job that can be automated rather than driving efficiencies elsewhere. From the BCOs perspective, the time lost due to broad process friction is what will ultimately drive changes in the industry and adoption of new solutions. Increasing headcount is just a short-term remedy to the SCM department being spread too thin on manual processes. Ultimately management will be driven to reduce headcount and overhead cost and will look for new solutions to drive these initiatives to the bottom line.

Undoubtedly there will be pushback within SCM departments and the industry. But there are especially two trends that are changing the landscape all together (Kersten, Hackius, and Böhle 2016). First, there is a generational shift where supply chain is not viewed as a clerk's job executing paperwork 24/7. The future workforce grew up on technology. For example, most of the younger generations have never actually been inside a bank (Marous 2016). They do not remember what it was like to hail a cab. They expect groceries to be delivered when and where they want. So, there is absolutely no way they will settle for anything less than technology adoption that makes their life more efficient. Previously, all this technology and industry learning occurred by the BCO working with their forwarders and carriers. The problem with this is that while they are getting a great operations education, the forwarders and carriers are not technologists. The generational shift has already happened and has overcome the first challenge of understanding the tech landscape and, most importantly, how it applies to their challenges. Second, technology has never been as agile and quick to deploy, which means the ROI is easily captured. For example, many SaaS companies are now being built with modularity in mind, so a BCO can take an à la carte approach and pick whatever functionalities their organization requires rather than an all or nothing approach. Previously, IT & Procurement was responsible for selecting what software solutions they will invest in (Jadhav and Sonar 2009). The first challenge is based on awareness, knowledge base, and creating enough branding that the BCO is now a product advocate. The fact that deployment costs are so low for platforms means that the ROI can easily be calculated and used for internal financial support. This makes it easier for the BCO to buy and easier for the SaaS provider to sell.

As these advances in technology and marketplace capabilities come to the fore, BCOs eyeing adoption will find themselves asking a complicated question: how do you put a price on service? Management will see the reduction in provider constraints from marketplaces and the reduction in process friction from new platforms. They will also recognize the resulting cost savings and benefits of decreased headcounts and real-time quotation comparisons. On the other hand, many within the BCOs organization will feel threatened by the seeming loss of responsibility and accountability that will accompany switching to a marketplace. The legacy provider, a trusted counterpart who solved past problems, still has a lot of value in the form of service they will argue. Who will handle disputes? Customs clearance? Broad troubleshooting and terms and conditions? The pressure on white-collar logistics workers to adopt a global and technology-driven mindset will be daunting in many small and medium-sized organizations.

The adoption of marketplaces is inevitable; the value creation in terms of speed and cost savings will be staggering (Jain, Chaudhary, and Patchala 2017), and the resulting transparency in procurement and process will enable businesses to move faster and satisfy more customers. But within BCO organizations, where the ultimate decisions around adoption are made, how can both schools of thought align and make logistics a competitive advantage?

The Future

Based on current trends, API will be widely adopted, enabling real-time data sharing, which will fix many of the challenges faced on both sides and immediately adding to the bottom line of the stakeholders. Marketplaces will serve as the global integrator for global logistics. It will not be done by one forwarder, NVOCC, or carrier, which makes the marketplace's core value unit is the data. Marketplaces are not restrained by pricing, service, or operational capabilities regarding the actual movement of the freight. These platforms will be in a unique position to collect, analyze, and create prescriptive analytics that automates decision-making (McKinsey 2016). Currently, there are sources of rate data ranging from indexes to manually getting spot rates from providers. However, the former is based on port-to-port rate structures, sole reliance on what is reported by index participants, no guarantee freight shipped at the price, and not necessarily a fit for BCOs of all sizes from a volume perspective. An index also does not take into consideration other indicators such as customer satisfaction or performance against KPIs. The latter is time-consuming and does not have the scale to ensure one understands the market and does not guarantee the best providers are part of the process. A well-designed marketplace will be in the best position to capitalize on this by consuming real-time transactions, analyzing the data, and making forward-looking predictions. This is important because these platforms will enable the industry to move past historical data and become proactive rather than reactive.

Earlier, we discussed how freight was historically procured. The BCO sends out an RFP either for contract or spot purposes. The main challenge is the lack of a data-driven decision tool to make instant transactions possible. In many cases, the BCO is reliant on what the carriers are telling them from a market perspective and therefore creates a less than ideal situation between the two parties. This will be history. A marketplace will solve this by first pulling the data in from a BCO, either manually or via integration, and then consuming the data. The platform will then automate the entire procurement process in a few stages (Segev and Gebauer 2001). First, and perhaps the most important, is an immediate understanding of what the market price is for their freight in specific time intervals. The benchmark will include door to door cost rather than port-to-port, which will give a truer overall picture of the cost structure. Now that the data is easily consumable, the platform can make recommendations on the ideal provider portfolio in combination with those that the BCO would like to work with. The forwarders and carriers can then offer a reduction or discount off this benchmark based on volume or specific BCO. The procurement process can be handled end to end all within the platform, including the award to the providers.

This also means added transparency, which is, in many cases, at odds with how the market has been operating. We believe that the real discussions will take place when costs associated with moving freight are shared openly. BCOs are fully aware that it is in their best interest to ensure their providers stay profitable, but the fact is that nearly no one on the BCO side has a true understanding of the costs that their providers incur. This has led to the carriers rolling out surcharges like General Rate Increases, Low Sulfur Bunker Surcharges, or Operational Surcharges to offset their costs (Knowler 2019). The list of surcharges goes on, but in reality, it is all caused by an engrained understanding that the freight procurement is a zero-sum game. This creates friction on both sides as the BCO feels they have no insight into the costs, and the carrier or forwarder is trying their best to maintain margins. It is clear; opacity is plaguing the industry. We hope to see these issues minimalized by using technology and data to drive these discussions.

SimpliShip: Deeper Dive into an Existing Logistics Marketplace

A lot of this chapter has been written by the Co-Founder & CEO of SimpliShip, which is an international air and ocean freight marketplace. SimpliShip started as a bootstrapped (self-funded) logistics technology startup. This creates many opportunities and challenges at the same time as it creates resource scarcity. The most critical aspect of a successful product is that it is solving real challenges that the industry is facing rather than creating tech and searching for problems to solve. This does not take a lot of funding to do. However, in B2B (business to business) as opposed to B2C (business to consumer), there is a requirement of understanding the market and transferring that knowledge into the product. In many cases, your target market will not know how to translate their needs into software requirements.

SimpliShip looks to hire those that have experience in international logistics. The entire founding team, from the leadership team to our developers, is experienced in operations, sales, trade compliance, etc. both on the BCO, provider, and technology side of things. Given the gap between tech and logistics, it is not an easy thing to accomplish, but it does allow the entire team, including developers, to become a sales force. It also saves incredible amounts of time and allows the team to stay lean, especially at the early stages when cash is limited. On average, SimpliShip saves approximately a month of onboarding time by taking this approach. However, it most likely is not completely sustainable if you are growing rapidly. But it does lay down the foundation for hiring as every leader in the company is self-sustainable to a certain extent to onboard and train new hires, bringing them up to speed quickly. If the department the applicant is applying for is customer-facing, we look for deep experience in international logistics.

SimpliShip is a B2B (business to business) marketplace, and that is an important distinction. They are not creating software for consumers who are purchasing products from businesses, and therefore the focus is much different. They are selling into businesses and, more specifically, the operational and IT departments. So, this needs to be engrained into the organizational hiring practices as the hires are expected to be able to understand the challenges and the industry. Discussions with medium to large customers often seem more like a consulting engagement as they want a technology partner that listens, digests, and creates value for them. Our customers are usually experts in their supply chain, but they do not necessarily know best practices. So, this is where SimpliShip works with them hand-in-hand to help engrain these best practices into their organization through technology. This ties into the above focus on people, processes, and technology.

SimpliShip's biggest challenges are twofold. First, we need to help a historically archaic industry revive itself and adopt technology that has existed for a while. For example, APIs were first rolled out in 2000 but have yet to be widely adopted in international logistics. We spoke about data availability earlier, and that will continue to be the biggest challenge on the provider side. The company takes every opportunity to discuss the opportunities to forwarders, NVOs, and carriers.

Startups in this space will most likely echo this as SimpliShip has become the face of this movement. It is taking longer than they would like, but that, in many cases, is not unique to logistics but most likely can be applied to B2B as a whole. Secondly, SimpliShip does the same but on the BCO side of our customer base. In many cases, BCO's are still trying to understand what technology is out there and how it can help them. So, a lot of time is spent engaging in investigative discussions with BCOs on their current issues and then applying technology to solve them. This helps bridge the gap between in-depth operational knowledge that the BCO has and what the currently available software options out there are. The BCO may not have a full understanding of the tech landscape, but they will understand how it makes a real problem go away. SimpliShip, therefore, feel that startups in logistics tech play a really important role in bridging the gap between the BCOs, providers, and technology in general. The real scale for these startups will come when the talent gap is minimalized, which will enable increased inbound customers and decreased customer acquisition costs (CAC). This had led us to SimpliShip's API first strategy. eCommerce is taking over and has been for some time now in every single industry. APIs enable us to be wherever our customers are at any given moment at a very low CAC balancing the costs associated with helping the industry move forward. This makes the product seamless to customers whether they are utilizing it through ERP, TMS, or a sourcing website where products are being bought and sold online.

The good thing about the industry and the product is that it is extremely easy to monitor and measure the return on investment. Product pricing varies by vendor, but if the work is executed correctly, it should be easy to prove savings to your customers. There are three areas of potential savings: 1) the bottom-line hard savings that are acquired through freight procurement, 2) the savings due to resource reallocation, and 3) better and faster data availability.

The Era of Data and the Impact on the Workforce

A logistics professional, whether on the BCO or provider side, has been focused very much on operational and execution challenges, and for good reason. Experienced and young supply chain professionals have had a linear thought process ingrained into their mindset. For example, Sourcing's job is to ensure on-time production. Sourcing then hands off the shipment to logistics, where their goal is to move the shipment and deliver it on time to the customer. This mindset will not only be disrupted but will lead to a new breed of supply chain leaders with much more focus on strategic initiatives. Due to technological advances that we discussed previously, there will be an overwhelming amount of data available, which will force supply chain professionals to adapt quickly to make sense and formulate strategic decisions at an exceedingly fast pace (Kern 2019).

As technology adapts and integrations become seamless with the ability to communicate in real time, the needs of the organization and industry will change, which will create a large and significant talent gap. For example, we already see the organizational integration of SCM operations and the technology departments for large corporations. This change is the first attempt at aligning an organization with the new digital supply chain. But, based on our experience, the lack of qualified candidates for these roles has never been so apparent. Startups are pushing the industry forward while behemoths like Amazon have made SCM a competitive weapon that enables them to service their customers like no one else. But it will take a generational shift in the education system for us to build the leaders of tomorrow. The focus will shift from learning how the departments work together to produce and deliver products or services to how specific technologies such as machine learning (ML), artificial intelligence (AI), and platforms are automating previously mundane and non-value-added processes (Panetta 2018; Kern and Wolff 2019).

Future supply chain leaders will be able to understand the current and future SCM landscape thoroughly but, most importantly, also align the business strategy with technical requirements that will support these initiatives (McKinsey 2017). Given the increase in the importance of the generalist leadership role, it will also force the demand for SCM data experts and developers to increase rapidly. The development side is already growing as we see the advancements in logistics technology rising at a rate unforeseen in our industry. Their job will be to develop solutions to some of the biggest challenges SCM professionals face. It is important to note that successful software vendors will be solving real challenges rather than trying to apply technology to challenges that do not exist or do not properly address these challenges in the first place. Due to the immense amounts of data that is now available, SCM data experts will find themselves on the front lines solving some of the largest challenges in the industry (Heckel 2017). They will be not only be tasked with creating the data structure and process for their companies but also automating the digestion of this data both internally and externally. They will create the baseline for discussions with all key SCM stakeholders and help to engrain the customer's voice into the product development process.

Imagine what is possible when integrators open their data via API? As we discussed previously, one of the most important factors that a platform manager needs to decide on is how open the platform should be. When we apply this same approach to the international logistics industry, we see some intriguing opportunities for both the API providers but also the developers who can create apps on top of the providers. We foresee the growth of this approach from SaaS providers, which in turn leads to the increase in demand for developers in this industry. The key in that the SaaS provider will win over their customers because they have won over the developers.

On the other side of the industry, freight forwarders will be greatly affected by the digital era as follows:

> All this matters because of the potential threat digitization poses to incumbents unless they take steps to harness its potential fully. The transparency the internet brings puts downward pressure on margins at the same time as attackers—often new, technology-savvy companies that adopt innovations with speed—are raising the bar on what consumers expect while reducing the cost to serve those customers and accelerating the pace of digitization. (Dichter et al. 2018)

In the same article, McKinsey identifies that only 6 percent of the world's largest carriers and forwarders have end-to-end booking capabilities (Dichter et al. 2018). There are many reasons that the logistics industry lags in comparison to other industries, but the major reason is culture and defending the old guard mentality. By definition, Industries that are already on their digital path have to go through their operational expenditures (OPEX) and capital expenditures (CAPEX). From a workforce point of view, the transformation for carriers and forwarders will take different paths as carriers have been moving toward a more centralized customer service infrastructure. In contrast, the majority of forwarders have not yet made this move. Unfortunately, the carriers did not think of their digital path simultaneously, which led to the same customer service struggles they had previously. Forwarders have the luxury of looking at this in hindsight and hopefully look at these two workstreams as one integrated journey of reorganization. In either case, there will be a reallocation of resources from sales and customer service to a more tech-enabled business (Maguire 2018). Cutbacks in the typical customer-facing departments will free up cash and resources to invest in their digital aspirations. This will mean that many of these headcounts will be moved into different areas that the current workforce cannot fulfill. So, we will see many of these current employees leaving the industry all together in search of a role they fit while an increase in technology roles will be significant, matching much of what we see happening on the BCO side.

With all these data now readily available and the increasing investment from all parties in digital, we see opportunities to automate using already existing technology such as ML, artificial AI, blockchain, and IoT sensors (Panetta 2018; *Machine Learning on AWS* 2019; Kern and Wolff 2019). These technologies can help in very specific-use cases. For example, IoT platform sensors can be used to truly provide real-time tracking, so all supply chain stakeholders are aligned as opposed to the current EDI technology being utilized (cf. Chapter 3). Further, AI/ML can be utilized to consume large data sets and make decisions that should help drive quicker and more efficient outcomes. An obvious area for a typical BCO is in forecasting, demand, and production planning. Previously, we discussed the customer service challenges and the future reorganization of these departments. According to a Stanford white paper by Hau L. Lee, Haim Mendelson, Sonali Rammohan, and Lauren Blake, there are many specific-use cases for these technologies, such as UPS utilizing Microsoft's Bot Framework to automate delivery updates and customer service interactions. The outcome was that bots handled 200,000 customer conversations in the first eight months (Lee et al. 2018). If we take this example and apply it to the ocean and air freight industry, it is not hard to further understand the enormous opportunity to transform the industry. The international freight industry is much more complicated than parcel and small package, so while we are not advocating a complete shift to bots, we do think there is an opportunity to automate many aspects of customer service.

Another example can be seen when we analyze the operational and vessel planning that plagues carriers as they currently have limited insight into what percent of freight booked will be onboard a vessel. And, vice versa, a BCO may have booked freight that may not be onboarded due to overbookings accepted by the carrier to ensure that their vessels move as full as possible. With platform integrations possible and the increasing investments being made, it is now possible for manufacturing data to be consumed and vessel planning automated (Glave, Joerss and Saxon 2014). For example, if a booking has not been canceled, but the production date inside a BCO's ERP has been updated, the data sharing is all possible via integration, but AI could enable automated booking cancelations and replan accordingly. Another challenge for many international buyers and sellers is the Letter of Credit (LC) process, which utilizes a bank to guarantee payment for services rendered for a fee (Kozolchyk 1965). Blockchain could be utilized to eliminate the requirement for a bank to be part of the transaction leading to more efficient and cheaper workflows.

Lastly, we will continue to see the increase in Chief Supply Chain Officers in the board room (Roh, Krause and Swink 2016). They will be tasked with creating and supporting both horizontal and vertical strategies and focus on translating customer requirements into their products as well as leveraging the entire organization as a competitive weapon. This will force an organizational shift from viewing SCM as a cost center, in which budgets are to be managed and annual savings to be achieved, to perhaps the most important enabler of meeting and hopefully exceeding both short and long terms goals. We will leave you with this quote from Sam Walton: "Communicate everything you possibly can to your partners. The more they know, the more they'll understand. The more they understand, the more they'll care. Once they care, there's no stopping them" (Walmart 2020). Data availability and technological advancements will not only allow an organization to make quick educated decisions but also minimize the naturally occurring bias out of the discussion. With these new requirements and advances in both hard and soft skills, we see the industry evolving at a fast pace. The question is, are we prepared to take advantage of it?

Conclusion

Key Takeaways

- Historically procurement has been handled non-digitally, which has led to the following inefficiencies:
  1) Lack of understanding of the current and future market.
  2) Poor profitability management from a carrier, forwarder, and NVOCC[4] perspective.
  3) The inability for BCOs[5] to ensure they are always optimizing price and service levels.
  4) Decreasing service levels, e.g. slow steaming or canceled port calls.
- The rise of logistics centered marketplaces and platforms for contract and rate management, visibility, transportation management systems, and many others will forever change the international freight landscape by integrating numerous sources of data and automating the decision-making process for all stakeholders.
- We need an industry-wide push for an overhaul of backbone technologies. More specifically, the industry needs to move away from archaic technologies like EDI that slow down the process of data integrations.
- Upgrading platform connectivity tools from EDI to API will be the key to the success of logistics platforms for internal and external use as it enables real-time sharing of valuable data.
- One of the most important factors to consider when creating a marketplace platform or choosing which platform to use is openness. Completely open platforms will attract the most users and build the largest network, but being too open can bring in unwanted parties to the network.
- From the BCOs perspective, the time lost due to broad process friction of manual tasks is what will ultimately drive changes in the industry and adoption of new solutions. Ultimately, management will be driven to reduce headcount and overhead cost and look for new solutions to drive these initiatives to the bottom line.
- As technology adapts and integrations become seamless with the ability to communicate in real time, the needs of the organization and industry will change, which will create a large and significant talent gap.

References

AF KLM on a Digital Journey (2018). *Air Cargo News*. https://www.aircargonews.net/airlines/af-klm-on-a-digital-journey/ (accessed 12 January 2019).

Dichter, A., Rothkopf, M., Bauer, F. et al. (2018). Travel and logistics: data drives the race for customers. https://www.mckinsey.com/~/media/mckinsey/industries/travel transport and logistics/our insights/travel and logistics data drives the race for customers/mck-travel-and-logistics-data-drives-the-race-for-customers.ashx (accessed 12 January 2019).

Glave, T., Joerss, M., and Saxon, S. (2014). The hidden opportunity in container shipping. https://www.mckinsey.com/business-functions/strategy-and-corporate-finance/our-insights/the-hidden-opportunity-in-container-shipping (accessed 12 January 2019).

Hartmann, E. (2013). *B-to-B Electronic Marketplaces: Successful Introduction in the Chemical Industry*. Deutscher Universitätsverlag (Business-to-Business-Marketing). https://books.google.com/books?id=bApJCAAAQBAJ (accessed 12 January 2019).

Hebard, G. (2019). *Interviewed by Cory Margand for Role Of Digitalization In Freight Forwarding*, 5 March.

Heckel, M. (2017). Skills for industry 4.0. http://www.delivered.dhl.com/en/articles/2017/02/skills-for-industry-4-0.html (accessed 8 October 2018).

Hughes, A. (2019). EDI vs. API and the critical role each plays in ecosystem onboarding, CLEO. https://www.cleo.com/blog/edi-vs-api (accessed 12 January 2019).

Huynh, N. and Walton, C.M. (2008). Robust scheduling of truck arrivals at marine container terminals. *Journal of Transportation Engineering* 134 (8): 347–353. https://doi.org/10.1061/(ASCE)0733-947X(2008)134:8(347).

Jadhav, A.S. and Sonar, R.M. (2009). Evaluating and selecting software packages: a review. *Information and Software Technology* 51 (3): 555–563. https://doi.org/10.1016/j.infsof.2008.09.003.

Jain, S., Chaudhary, P.K., and Patchala, S. (2017). Technology-led disruptive freight marketplaces – the future of logistics industry. https://www.tcs.com/content/dam/tcs/pdf/Industries/travel-and-hospitality/Building-Disruptive-Freight-Marketplaces.pdf (accessed 12 January 2019).

Janke, F. (2011). The use of hidden data in electronic business networks: Benchmark and network performance indicators. In: *19th IDIMT*, pp. 341–348. https://idimt.org/wp-content/uploads/proceedings/IDIMT_proceedings_2011.pdf#page=341 (accessed 12 January 2019).

Jung, I. et al. (2018). Big data analysis model for MRO business using artificial intelligence system concept. *International Journal of Engineering and Technology* 7 (3): 134–138.

Kern, J. (2019). *Skills for the Future*. Chengdu: OceanX https://youtu.be/VkRCxuJvUTY (accessed 12 January 2019).

Kern, J. and Wolff, P. (2019). The digital transformation of the automotive supply chain–an empirical analysis with evidence from Germany and China. https://www.innovationpolicyplatform.org/system/files/imce/AutomotiveSupplyChain_GermanyChina_TIPDigitalCaseStudy2019_1.pdf (accessed 12 January 2019).

Kersten, W., Hackius, N., and Böhle, C. (2016). Trends und Strategien in Supply Chain Management und Logistik-Chancen der digitalen Transformation Trends and Strategies in SCM and Logistics-Opportunities of the digital transformation View project Capability Management for Independent Maintenance Servic. www.bvl.de (accessed 14 September 2018).

Knowler, G. (2019). CMA CGM joins Maersk, Hapag-Lloyd in low-sulfur surcharge for short-term contracts, JOC. https://www.joc.com/maritime-news/cma-cgm-joins-maersk-hapag-lloyd-low-sulfur-surcharge-short-term-contracts_20191105.html (accessed 12 January 2019).

Kontainers (2019). Who are Kontainers? https://kontainers.com/videos/kontainers-video-who-are-kontainers.html (accessed 12 January 2019).

Kozolchyk, B. (1965). The legal nature of the irrevocable commercial letter of credit. *The American Journal of Comparative Law* 14 (3): 395. https://doi.org/10.2307/838450.

Kroft, K. and Pope, D.G. (2014). Does online search crowd out traditional search and improve matching efficiency? Evidence from craigslist. *Journal of Labor Economics* 32 (2): 259–303. https://doi.org/10.1086/673374.

Lee, H. L., Mendelson, H., Rammohan, S. et al. (2018). Value chain innovation: the promise of AI. https://www.gsb.stanford.edu/sites/gsb/files/publication-pdf/vcii-white-paper-value-chain-innovation-promise-ai.pdf (accessed 12 January 2019).

Lock, M. (2018). The marketplace model: a proven model for success & why the shipping industry should adopt it, cerasis. https://cerasis.com/marketplace-model/ (accessed 12 January 2019).

Machine Learning on AWS (2019). Amazon. https://aws.amazon.com/machine-learning/ (accessed 12 January 2019).

Maguire, E. (2018). Logistics, supply chain and transportation 2023: change at breakneck speed. Jersey City. http://forbesinfo.forbes.com/l/801473/2019-09-23/27ns/801473/8485/Penske_REPORT_FINAL_DIGITAL.pdf (accessed 12 January 2019).

Marous, J. (2016). Millennials won't wait for banks and credit unions to understand them, the financial brand. https://thefinancialbrand.com/62278/banking-millennial-digital-mobile-marketing/ (accessed 12 January 2019).

McKinnon, A., Flothmann, C., Hoberg, K., et al. (2017). *Logistics Competencies, Skills, and Training: A Global Overview*. Washington, DC: World Bank Group http://documents.worldbank.org/curated/en/551141502878541373/Logistics-competencies-skills-and-training-a-global-overview (accessed 12 January 2019).

McKinsey (2016). The age of analytics: competing in a data-driven world. https://www.mckinsey.com/~/media/McKinsey/Business Functions/McKinsey Analytics/Our Insights/The age of analytics Competing in a data driven world/MGI-The-Age-of-Analytics-Full-report.ashx (accessed 12 January 2019).

McKinsey (2017). Jobs lost, jobs gained: workforce transitions in a time of automation. https://www.mckinsey.com/~/media/mckinsey/featured insights/future of organizations/what the future of work will mean for jobs skills and wages/mgi jobs lost-jobs gained_report_december 2017.ashx (accessed 12 January 2019).

Morley, H.R. (2016). Tension between shippers and carriers over invoice accuracy rising, JOC. https://www.joc.com/maritime-news/container-lines/tension-between-shippers-and-container-lines-over-invoice-accuracy-rises_20160912.html (accessed 12 January 2019).

ORBCOMM (2019). Networks: satellite AIS. https://www.orbcomm.com/en/networks/satellite-ais (accessed 12 January 2019).

Panetta, K. (2018). Gartner predicts 2019 for supply chain operations. *Gartner Predicts*: 1–10. https://www.gartner.com/smarterwithgartner/gartner-predicts-2019-for-supply-chain-operations/ (accessed 12 January 2019).

Parker, G.G., Van Alstyne, M.W., and Choudary, S.P. (2016). *Platform Revolution: How Networked Markets Are Transforming the Economy - and How to Make Them Work for You*, 1e. W. W. Norton & Company.

Perez, S. (2018). Apple's App Store revenue nearly double that of Google Play in first half of 2018. *TechCrunch*. https://techcrunch.com/2018/07/16/apples-app-store-revenue-nearly-double-that-of-google-play-in-first-half-of-2018/ (accessed 19 January 2019).

Petersen, R. (2015). Leaky ships: ocean carriers in the age of profitless shipping. https://www.flexport.com/blog/why-are-ocean-freight-rates-so-low (accessed 12 January 2019).

Richter, W. (2016). China ocean freight indices plunge to record lows, wolf street. https://wolfstreet.com/2016/03/14/china-ocean-freight-indices-plunge-to-record-lows/ (accessed 12 January 2019).

Riedl, J., Chan, T., Schöndorfer, S. et al. (2018). The digital imperative in freight forwarding. https://www.bcg.com/en-in/publications/2018/digital-imperative-freight-forwarding.aspx (accessed 12 January 2019).

Roh, J., Krause, R., and Swink, M. (2016). The appointment of chief supply chain officers to top management teams: a contingency model of firm-level antecedents and consequences. *Journal of Operations Management* 44 (1): 48–61. https://doi.org/10.1016/j.jom.2016.05.001.

Rouse, M. (2015). Transportation management system (TMS), TechTarget. https://searcherp.techtarget.com/definition/transportation-management-system-TMS (accessed 12 January 2019).

Sahay, B.S. and Mohan, R. (2006). Managing 3PL relationships. *International Journal of Integrated Supply Management* 2 (1/2): 69. https://doi.org/10.1504/IJISM.2006.008339.

SAP EDI Document Standard (2018). https://www.edibasics.com/edi-resources/document-standards/sap/ (accessed 16 September 2018).

Segev, A. and Gebauer, J. (2001). B2B procurement and marketplace transformation. *Information Technology and Management* 2 (3): 241–260. https://doi.org/10.1023/A:1011442008251.

Shippers turn to digital booking platforms (2019). *Air Cargo News*. https://www.aircargonews.net/freight-forwarder/shippers-turn-to-digital-booking-platforms/ (accessed 12 January 2019).

Toczauer, C. (2019). Panalpina, DHL GF join growing list of digital freight sales platforms. https://aircargoworld.com/allposts/panalpina-dhl-gf-join-growing-list-of-digital-freight-sales-platforms/ (accessed 12 January 2019).

Trelleborg Marine Systems (2018). Use of big data in the maratime industry. https://www.patersonsimons.com/wp-content/uploads/2018/06/TMS_SmartPort_InsightBee_Report-to-GUIDE_01.02.18.pdf (accessed 12 January 2019).

Walmart (2020). 10 rules for building a business. https://corporate.walmart.com/our-story/history/10-rules-for-building-a-business (accessed 12 January 2019).

Yurong, Y. and He, H. (2000). Data warehousing and the Internet's impact on ERP. *IT Professional* 2 (2): 37–41. https://doi.org/10.1109/6294.839365.

17

The Evolution of the Cold Chain

A Story of Increasing Complexity Amidst a Sea of Traditional Thinking

Alex von Stempel

Freshwater Logistics Ltd, Middlesex, England

Introduction

There are many different products such as seafood, meat, fresh fruit and vegetables, electronics, and pharmaceuticals that can be described as *perishable*. But all of them have one thing in common: the need for temperature control as they are transported and stored. Each category of product is also characterized by essential differences that have spawned unique supply chain processes and technologies in terms of logistical applications.

> According to the International Dictionary of Refrigeration, the cold chain is a series of actions and equipment applied to maintain a product within a specified low-temperature range from harvest/production to consumption (International Institute of Refrigeration 2015–2019). This involves for the purpose of this chapter looking at the element of transportation and logistics of perishables.

While barely a generation ago seafood processing was still an important industry in many geographical regions including Northern Europe and the United States, the migration of this industry sector to Asian countries during the last few decades serves as a good example to illustrate the division of labor in the industrial world, which continues to reverberate across the cold chain.

Cold Chain Transportation

At least 20 million tons of seafood are shipped every year across all transport modes. This may depend partly on the number of catches. However, today, most of the growth is led by aquaculture, which now represents around half of all wild catches (Schrieder 2019).

The Digital Transformation of Logistics: Demystifying Impacts of the Fourth Industrial Revolution, First Edition. Edited by Mac Sullivan and Johannes Kern.
© 2021 by The Institute of Electrical and Electronics Engineers, Inc.
Published 2021 by John Wiley & Sons, Inc.

Most of the seafood is shipped frozen by sea, although there are significant fresh (chilled) volumes transported by air. There is for example a substantial and indeed very lucrative trade in abalones that are harvested in Australia and transported by air to China.

The majority of seafood processing is now taking place in Asia including the preparation of fish fingers that are boxed and typically shipped in reefer containers. In many cases, these are shipped back to the very regions where they came from. Salmon, cod, and shrimp are some of the main types of seafood shipped by sea.

Perishable commodities can be transported by sea, air, road, and even rail. While there is considerable duplication between individual perishable supply chains, it would appear that there is plenty of room for better supply chain coordination in some areas given that there is a degree of duplication and inefficiency among them with container shipping companies and specialized freight forwarders involved in a permanent tug of war.

The container industry, which for a long time has focused largely on the management of physical equipment, i.e. container ships and containers, is a mature industry. It therefore has been slow to incorporate advances in technology developed in other areas. Meanwhile, technology has been characterized by increasing "dematerialization," as poignantly referred to by Steven Pinker in his recent book *Enlightenment Now* (Pinker 2018). Akin to digitization, the process of dematerializing is closely related to the division of labor and the process of specialization. The faster digitization progresses, the starker the contrast with asset-based silos marking out potential fault lines. As a consequence, the very notion of "disruption" carries a critical double meaning.

With an unmistaken reference to the second law of thermodynamics, Pinker states that in an "isolated system" (one that is not interacting with its environment), "entropy never ceases." Closed systems inexorably become less structured, less organized, and less able to accomplish interesting and useful outcomes, until they slide into an equilibrium of gray, tepid, homogenous monotony and stay there (Pinker 2018, p. 15). Applied to the world of logistics, one could say that the world of logistics remains highly inefficient, riven by technical obsolescence.

Perhaps, even more than with dry containers, reefer container operations illustrate the abovementioned diverging trends, especially as there is a much greater need for information and connectivity between product (cargo) and load unit.

Cold Chain Transportation Modes

In the following sections, we will try and shed some light on the underlying dynamics of the cold chain transportation modes and how as a system it must be regarded as being in constant flux.

Ocean Freight

One factor that has fundamentally contributed to the reshaping of global perishable supply chains has been the steady progress of reefer container shipping at the expense of reefer bulk shipping, better known as conventional or "specialized" shipping. The dwindling number of conventional ships that are still being used to transport fresh fruit, frozen fish, or meat indicates the decline of vertical integration of the perishable supply chain at the expense of disintermediation. In the past when there were still national fruit marketing boards, for example, in countries such as South Africa, Morocco, and Israel, these were regarded as a mere extension of the respective export fruit sectors. The advent of the container gradually revolutionized this (Robertson 2019).

The speed of decline of traditional reefer shipping is now largely been held back by the fluctuating price of scrap metal, i.e. what the ship breaking trade is prepared to pay for older ships. According to intelligence provider Alphaliner, the average age of conventional vessels still in operation at the beginning of 2019 was 27 years (Verberckmoes 2019). The seaborne perishable split between reefer containers and conventional ships was 81.8 : 18.2% in 2019. By 2023, maritime research consultancy Drewry Shipping expects the latter to drop down to 14.6% (Dixon 2019). A comparison between the number of reefer containers and dry containers in operation puts the maritime segment of the cold supply chain into perspective. Reefer container shipping represents 7% of total container movements (container global fleet, 42.2 million TEU; reefers, 3.2 million TEU according to Drewry). Toward the end of 2019, reefer box shipments were set to grow at 4–5%.[1]

The ocean freight segment is dominated by a few large shipping lines: Maersk Line, CMA CGM, and MSC (UNCTAD 2019). According to figures released by Alphaliner in January 2019, Maersk had 474 174 reefer plugs out of a total TEU of 4 072 276, followed by MSC with 334 596 reefer plugs[2] (3312.944 total TEU), and CMA with 293 612 reefer plugs (2 664 669 total TEU). Other leading players are Hapag-Lloyd, Cosco, and ONE (Verberckmoes 2019).

Another peculiarity of seaborne perishable shipping is that it is closely associated with the South–North trade lane, which compared with other trade lanes has long been considered to be a "secondary" trade route. Fresh produce, including exotics such as mangoes and papayas and protein from Latin America, is the backbone of shipments between the Southern and Northern hemispheres. Taken together, fresh produce constitutes the largest commodity shipped in reefer containers (Veldman 2019).

There are, for example, considerable volumes of kiwifruit and apples from New Zealand or bananas from the Philippines that complete the picture. Added to this must be the increasing Chinese import demand for frozen meat, typically pork, beef, and white meat (chicken) as well as seafood and other forms of protein including insects, which could become a more important source of the food supply in the future (Ramos-Elorduy 2005). The leading market for cherries in China is Chile, while meat from the Mato Grosso region in Brazil is finding ready markets in China and the Middle East, leading to new shipping services being created to serve these markets.

The rise in demand for halal meat in the Middle East may not be as high as the demand for pork meat in China due to the sheer difference in the number of consumers; however,

---

1 This was before COVID-19 hit the headlines, which is likely to cast severe doubts on any current predictions. For instance, at the beginning of March 2020, Professor Alfred Cheung from JC Food Republic predicted that there was a shortage of 12 000 reefer boxes for the next couple of months.

2 Reefer plugs are electrical power points also known as "reefer points" onboard a vessel or a container terminal.

this denotes the progressive differentiation and evolution of market demand. This is matched with an impressive crisscrossing of matching shipping services.

According to Seabury Consulting, emerging markets "chalked up" 9% growth in perishable trade between 2000 and 2018. Yet, in all but very few cases, reefer container shipments still very much depend on trade volume in the dry box business, which provides the necessary base cargo. If there is not enough inbound or outbound "dry" business, reefers alone would not be able to sustain a trade route.

Airfreight

In terms of intercontinental trades, the volumes of perishable airfreight are much smaller. According to the airline Cargolux, in 2019, perishables shipped by air amounted to 6 710 244 kg, including 1 247 173 kg of pharmaceuticals. By comparison, the global perishable seaborne market in 2018 was 129 million tonnes according to Drewry Shipping (Dixon 2019). Fresh produce and seafood make up the bulk of the remainder within the temperature-controlled food segment. Virtually, all perishable commodities are shipped chilled and as belly freight following finely tuned schedules. The airfreight sector had to recognize a relative decline of dedicated freighters. Another key difference compared with ocean freight is the higher value of perishable cargo transported by air and the dominant role of a small group of freight forwarders in this segment.

While for sea transport cargo serves as the basis of its business model, passenger traffic is the main revenue source for the airlines. Due to the greater length of transit time, the volume and speed of technological development in the airfreight-based perishable supply chain have, until now at least, been slower than in the ocean freight sector.

Technological Differentiation Between Air and Ocean Cold Chain Transportation

The progressive containerization of perishable products (both frozen and fresh) has provided the test bed for technologies designed to extend the shelf life of perishable products. As a temperature-controlled load unit, the reefer container thus acts as a workhorse for intercontinental supply chain integration.

Figure 17.1 typifies the challenges, touchpoints, and quality control points of the maritime-based supply chain for fresh produce.

In a wider sense, the following technical solutions should be regarded as part of the value-added drive for greater traceability and a growing awareness of accountability throughout the perishable supply chain.

Unfortunately, many technology providers active in the cold chain claim that they can provide end-to-end solutions when in fact many of them cannot. Only very few can eradicate all the blind spots that lurk behind every corner and at every crossroad of the supply chain. This is a subject that will be covered later on in more detail.

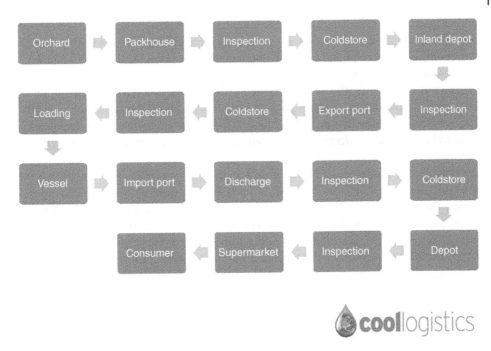

Figure 17.1 Maritime-based supply chain for fresh produce. The Perishable Products Export Control Board (PPECB), South Africa/Cool Logistics Global. *Source*: Cool logistics global. © 2020, Cool logistics global.

Cold Chain Transportation Considerations

Condition Monitoring

There are essentially two types of condition monitoring of perishables, one for monitoring load units, i.e. reefer containers or reefer trucks, sea (container) terminals as a clearly defined operational unit, cold stores, and distribution centers, and the other for monitoring cargo. The further we look downstream at the cold chain, the more different strands emerge, which can be divided into tracking consignments down to pallet level, carton level, or even further down to the individual piece of fruit. The superspecialized demand for this was demonstrated by blockchain company DataHarvest at Cool Logistics Global 2019 in Valencia, Spain (Buhl 2019).

As South Africa-based perishable supply chain expert Andy Connell points out, half the problem of explaining the importance of logistics to farmers is to convince them that the product they sell will never be as fresh as when it was harvested. "The First Mile in reefer shipping is to the fruit condition what the Golden Hour is to outcomes of injured persons. This Golden Hour is that period immediately after the accident. The patient management in this hour is often what determines the outcomes of recovery later (Connell, 2016)."

Other than temperature monitoring, ethylene suppression is the second most important cause of fresh produce deterioration. Ethylene is a hydrocarbon gas, which is produced for the ripening process (respiration and corresponding senescence) (Miller 1947). The third issue is water loss in transit and storage.

There are different methods for controlling the "aging process" of fresh produce. However, the situation becomes complicated by the fact that each fruit and each vegetable emits ethylene at a different rate and that only certain types (right down to different cultivars) can be co-loaded. Different products can create their very own ethylene supply chain reactions.

If, for example, apples, kiwifruit, bananas, and oranges are kept together in a fruit bowl for a period of time, they will ripen each other faster. The same applies to transport and storage. The rest is a procedure of managing the inevitable process of product deterioration. Asparagus, mushrooms, parsley, and peas have a rate of respiration of mg CO_2/kg-hour of >60. By comparison, apples, grapes, and papayas respire at the rate of between 5 and 10 (Mitcham et al. 2020). A way to reduce the risk is to use controlled atmosphere (CA) technology, which can also be applied to meat, though not fresh seafood.

Controlled Atmosphere (CA)

The CA process involves different stages: reduction of oxygen, an increase of carbon dioxide, removal of carbon dioxide, and removal of ethylene. CA for fresh produce was first developed for bulk shipments of bananas. As the fruit arrived from countries such as Ecuador, bananas represented large year-round volumes warranting dedicated banana carriers to devote the entire ship holds to CA treatment. Upon arrival, the fruit was delivered into special banana ripening rooms found typically in every major fruit port in Europe. Today, the technology of reefer units is so advanced to control and pre-ripen the fruit by satellite or other wireless methods. The technology is available, but it is still expensive.

One of the reasons why maritime reefer containers, including 45 ft reefer units, are not more widely used is that retailers typically have no demand for full container loads of any type of fruit. The expansion of the reefer less-than-container load (LCL) business can only provide only a few very limited options. However, long since gone are the days of whole "banana trains" leaving the ports of Antwerp, Bremerhaven, or Hamburg destined for distribution right across Europe. Today, virtually all this is moving in refrigerated trucks, which pick up the fruit from port-based cold stores. The cross-dock, therefore, performs the role of a watershed between maritime import operations and land-based distribution. Of course, as soon as the fruit is transferred, the CA process is broken, which explains the reluctance of wider industry adoption.

A way around this problem is being adopted in China where in some cases imported fruit is still being sold directly from the reefer container to market importers and market traders due to the shortage of cold stores in some parts of the country. However, this situation is rapidly changing. Whereas the risk of temperature shocks at transfer points is relatively small in mature destination countries, most of the problems arise in the countries of origin with high ambient temperatures.

Today, all reefer unit manufacturers pride themselves to be able to offer CA solutions: Carrier Transicold, Maersk Container Industry (MCI, Star Cool), Daikin, and Thermo

King (TK) are the four main suppliers of reefer machinery in the maritime market. Whereas reefer units are manufactured by Carrier, TK, and Daikin from part of bigger refrigeration divisions, MCI remains so far the only example for an integrated reefer solution, providing both reefer containers and reefer machinery units. MCI is also the only manufacturer associated with a shipping line (A.P. Moller Group). CA systems can be divided into active and passive systems. Daikin, for example, offers an "active solution" using nitrogen in order to suppress unwanted ripening.

A particularly interesting form of ethylene control has been proposed by Emeritus Professor Ron Wills of the University of Newcastle, Australia. He suggests that using ethylene scrubbers[3] as soon as horticultural products are harvested could reduce the dependency of temperature control so that fruits such as bananas could in some cases be transported over distances of over 3000 km without refrigeration (Wills 2018).

Modified Atmosphere Packaging (MAP)

Although the difference between modified atmosphere packaging (MAP) and CA is becoming more fluid (especially with the use of active CA), MAP typically relies on the use of special semipermeable plastic film inside a load unit where a delicate balance of different gases is being maintained. However, due to the activity of microbes, including the likely developments of pathogens, the composition of these gases can be subject to change over a period of time.

While the packaging industry continues to develop MAP solutions as well as new forms of intelligent packaging measuring and regulating the immediate environment of "living" produce, all these efforts are directed at one aim: finding better ways to safely extend shelf life. Packaging is typically considered as part of the product, similar to CA, on the cusp between production logistics and transport technology. The use of technologies that come under the remit of postharvest technologies reveal the progressive fusion between perishable product and transport, be it by refrigerated trucks, reefer containers, and other perishable transport devices still under development, such as drones for last-mile delivery.

Evolution of Cold Chain Transportation

Drewry Shipping estimates that the development of smart reefer containers is moving from just above 10% of reefer boxes in 2018 to almost 60% by 2023. Companies active in this field include technology providers such as Emerson, OrbComm, Globetracker, and Traxens. One of the early driving forces behind this has been Maersk Line, which rolled out its reefer monitoring systems in close cooperation with the tech sector (Eurofresh-Distribution 2016).

---

3 Scrubbers refer to the use of a chemical treatment applied to the surface of fresh produce that binds the ethylene receptors on the surface of the fruit rendering the ethylene inactive.

There are so many different companies involved in different segments of the cold chain that it has become difficult to keep track of them. One thinks of all the GPS-based tracking solutions developed for the truck market, which has been extended to include the reefer trucking sector. Others offer solutions that are "mode agnostic."

The technology providers all have one thing in common: they all rely on infrastructure, such as satellites or wireless technologies, which rely on servers. In addition to location and condition monitoring, many new sensor-based applications are being developed for the temperature-sensitive sector, including ETA,[4] power analytics, CO_2 footprint calculations, and freshness indices.

One of the issues associated with both wireless and satellite-based systems is their vulnerability. While data security is a problem not limited to the cold chain, this is also tied up with food safety issues. Hence, there is a constant drive for standardization and interconnectivity between different service providers in order to prevent customers from getting locked into just one supplier (Kern and Wolff 2019). Yet, neither options are hacking proof, which may also explain why there are no fully integrated vertical systems. A more cynical explanation would be the proverbial lack of trust often masquerading as the need to protect commercially sensitive information from competitors. Philosophically speaking, of course, standardization and innovation are in a dialectical conflict. As a result, end-to-end efficiency will remain elusive. Meanwhile, there is also mobile e-commerce, which is a whole new development at the other end of the supply chain, creating its own "Doppler effect," which is beginning to reverberate upstream into supply chain relationships (Amling and Daugherty 2018).

Consumers are becoming savvier and savvier. As they become increasingly technologically enabled, they now demand more frequent deliveries, including home deliveries, naturally all of the freshest sort. This includes fresh (not previously frozen) seafood, meat, or dairy products that are both tasty and safe to eat and preferably in individualized packaging to prolong shelf life right down to individual fridge or freezer level. As consumers do not want to miss out on the finer things in life, flowers are now being delivered in their ecosystem, featuring a guaranteed minimum of seven-day vase life. However, here too without a fully functional cold chain, the risk of customer reclamations would simply be too great. So, the single most critical element in ensuring the maintenance of product quality remains temperature control. Reliable temperature measurement remains the A and O of any perishable supply chain, or in other words, the cold chain is a never-ending task.

Telematics solutions have long since been regarded as a logistical panacea designed to provide end-to-end visibility. However, due to the fragmented responsibilities within the cold chain, the biggest source of potential disruption of the cold chain is, in fact, one of the unintended consequences of measurability.

Traditional temperature charts may have been replaced by electronic versions, which are not only more reliable but also tamperproof. There is currently no commercially viable system that can reliably measure and calculate the effect of the length of the door opening time of a reefer truck or container under high ambient temperature on the expected

---

4 Estimated time of arrival, the time when a ship, aircraft, or cargo is expected to arrive at a certain place.

reduction of shelf life. One of the issues is that basic foodstuffs such as bananas, potatoes, or shellfish do not pay enough. Pharmaceuticals and medical supplies appear to remain the main target for more esoteric supply chain solutions.

Consider the precise location of a pallet inside a container and its location in a reefer truck on the other side of the cross-dock divide. Why should this level of detail be relevant unless you were asked to transport the proverbial crown jewels? Yet, with every new sensor that could be deployed, another gap opens, craving to be filled with a new solution. In fact, in terms of traceability, the sky is the limit, or indeed the depth of the customer's pockets. Not many people would be prepared to pay for the extra bit of security and safety. One of the biggest problems is the mistrust that continues to exist between different actors operating the different segments of the supply chain.

Technology Solutions

A number of digital solutions are gradually being applied to critical areas of the cold chain. However, asset-based priorities and capital-based fiefdoms are proving to be resistant to change.

Blockchain

Blockchain concatenates cryptographically verified transactions into unique, i.e. immutable sequences of lists. Using complex mathematical functions, it arrives at a definitive record of who owns what and when in the supply chain (Iansiti and Lakhani 2017). The technology has been embraced by several perishable shippers, retailers including Carrefour, ocean carriers such as Maersk Line and Hapag-Lloyd, ports such as Rotterdam, and perishable freight forwarders including foodcareplus. To read more about blockchain applications in logistics, see Chapter 6.

The ability to make it harder to mislay goods or tamper with labels while cargo, including perishables, is in transit is an obvious bonus. Blockchain could have real significance in the perishable supply chain, especially as vertical traceability could be of overall benefit (The Economist 2015). Still, one of the biggest stumbling blocks of digitalizing essentially siloed operations is that all too often, decision-makers are fearful to face the risk of exposure. Having nowhere to hide means that transparency is equated with immeasurable risk. The underlying principle of blockchain is the process of randomization. This is psychologically linked to the fear of relinquishing control to automatic supervision. It can, therefore, understandably, fill every transport and logistics manager with a degree of discomfort. By ruling out any lurking variables and establishing watertight cause-and-effect relationships, randomization is often equated with machines taking over the world.

"To build a trusted cold chain blockchain needs to address concerns with a traditional distributed ledger," says Brian Robertson of Emerson Commercial and Residential Solutions, adding that "private ledger could accelerate necessary consensus across the members of the chain" (Robertson 2019). While different competing blockchain solutions are being developed, other forms of machine-to-machine communications are spawning new applications for robots in static storage environments.

IoT and AI

As machine-to-machine communication is progressing in sectors such as artificial limbs (prosthetics), it can be expected that these developments will eventually filter down to the logistics sector. For example, robotizing operations within a confined space such as warehouses can make a lot of sense (Aleksandrova 2019; Rolfsen 2019). One company that has been mapping a route around its extensive cold store operations is US-based Lineage. Having been involved with some of the US' largest retailers and food companies, Lineage owns various patents for automatic warehouse improvements, tracking of vehicles in warehouse environments, and optimization of rack operations.

Don Miller of Globetracker argues that with more sophisticated tags and wireless technology, robots could interact with the same tags used for condition monitoring across the cold chain, using their freshness index to place them in the cold store. "This could be based on directing those products with the shortest shelf life making sure it is the first to arrive on the consumer's shelf."[5] One problem that may be more difficult to master includes having warehouse robots interacting with the external world, with a risk of malfunctioning. Timing the exact arrival of a driverless truck at a cold store, activated through a camera when to unload their cargo, even given accepted tolerances of mutual recognition capabilities, is likely to still be faced with many practical difficulties. Coordinating movements between an independently controlled manned vehicle and a fixed platform are probably possible, but breakdowns might be difficult to manage, notwithstanding the deployment of the "cleverest" algorithms. In any case, it would probably be easier to have a fully automatic interface under single central control.

Gaps to be Addressed

Possibly even more remote is the idea of driverless ships. Operating in a harbor, excluding all wave movement in order to facilitate automatic vessel unloading, seems rather fanciful at this moment in time and would indicate clear limitations of artificial intelligence (AI) in the arena of contiguous transportation systems.

With currently no "one-size-fits-all" vertical solution in place, it is probably inevitable that individual solutions for the different segments of the cold chain are far more likely to attract the interest of technology providers than the illusive search for the proverbial lever of Archimedes through even the most matrix-based concept of concatenation. But again, one should not allow the perfect to become the enemy of the good.

Developing separate technical solutions for individual market segments of the cold chain is most likely to remain the best approach to be pursued by most technology systems providers, simply because it is much easier to develop and offer systems that measure and predict behavior within clearly defined individual segments of the cold chain.

Some systems can be used to measure the condition of a reefer box as it is being picked up by a ship-to-shore crane spreader and lowered on the quay. This would allow measuring exactly the time intervals during which a container is off power. For example, as soon as a

---

5 This was in a direct email communication with the author.

ship comes to rest alongside a quayside, power to reefer boxes tends to be switched off. This may take usually only a couple of minutes, but it may be more. However, when the ambient temperature is 50 °C, this can influence the commodity inside the reefer box. Numerous telematics solution providers are therefore developing different systems for monitoring reefer boxes and target shippers of perishable cargoes separately.

Food Safety as the Key

The standardization of processes measuring food safety using different transport modes such as reefer trucks, cold stores, and reefer containers is still only in its infancy. For example,[6] ATP rules do not cover fresh produce transportation by refrigerated truck.

As already mentioned, it is for practical commercial reasons that reefer containers are used only in very few cases to make door-to-door deliveries. However, as technological solutions are being continuously developed, the need for creating a kind of gateway or "vertical standardization" is becoming more urgent. Food standards in the perishable logistics industry are likely to become stricter. There is, for example, good reason to expect that HACCP rules will evolve to identify and monitor deviations from what is deemed "safe" and crash through the cross-dock as the biggest barrier of the cold chain. HACCP stands for Hazard Analysis and Critical Control Point. It is an internationally recognized system for reducing the risk of safety hazards in food (Pierson 2012).

Conclusion

The world of agricultural production and the world of logistics seem to largely exist in parallel worlds with increasing fragmentation partly driven by technological evolution and partly because of entrenched positions between different transport modes and different realms of responsibility. The most obvious fault line runs between operations on both sides of the cross-dock.

Reluctant to share as little information for free with perishable shippers, reefer box shipping and logistics specialists operating within the arena of perishable freight forwarding or cold store operations will tend to jealously guard what they regard as commercial secrets. This may impede also the progress of the regulatory process and the lack of alignment even in areas where product safety is concerned.

Blockchain technology and advances in Internet of Things (IoT) and AI are likely to help blur the edges that continue to exist between different segments (Sloman 2010; Tjahjono et al. 2017; Panetta 2018; Neher 2020). But since perfection should not be the enemy of the good, the cold chain will continue to evolve in incremental steps.

---

6 ATP is the Agreement on the International Carriage of Perishable Foodstuffs and on the Special Equipment to be used for such Carriage (ATP). It is a 1970 United Nations treaty that establishes standards for the international transport of perishable food between the states that ratify the treaty (United Nations 2016). However, Russia has been unwilling to ratify this (Transfrigoroute 2020).

Legislation governing confidentiality and/or the risk of carriers and (perishable) freight forwarders being accused respectively of market collusion will represent hurdles, but these may vanish in time.

The perishable supply chain contains both the seed of its vulnerability dictated by its exposure to external influences and its regenerative ability in the face of adversity. The process is endless and expanding not unlike the universe itself. Measurability is a tool, but it must not become a weapon. It cannot replace reality. Why is it then that in Aesop's famous fable, the tortoise won the race over the hare?

Key Takeaways

- The cold chain is a subset of both the world of (agricultural) production and the world transport and logistics.
- Fragmentation is a function of technological (industrial) evolution and is eroding entrenched asset-driven positions in the maritime transportation sector in its present form.
- The semipermeability of the cross-dock is both its strength and its weakness, with information acting as a case in point.
- The speed of progress in developing the cold chain from farm to shelf and corresponding regulatory processes could be significantly affected by the effectiveness of blockchain adoption.
- Legislation governing confidentiality and/or the risk of carriers and (perishable) freight forwarders being accused respectively of market collusion will represent hurdles, but these may vanish in time.

References

Aleksandrova, M. (2019). Streamlining your warehouse management with digitalization. https://easternpeak.com/blog/streamlining-your-warehouse-management-with-digitalization/ (accessed 15 April 2020).

Amling, A. and Daugherty, P.J. (2018). Logistics and distribution innovation in China. *International Journal of Physical Distribution and Logistics Management.* 50 (3): 323–332. https://doi.org/10.1108/IJPDLM-07-2018-0273.

Buhl, M. (2019). Digitisation, transformation and supply chain transparency [PowerPoint presentation], *11th Cool Logistics Global*, 17–19 September, Valencia.

Connell, A. (2016). Focusing on the first mile: country and product case Africa [PowerPoint presentation]. *Fruit Logistica*, 3–5 Febraury, Berlin.

Dixon, M. (2019). Perishable reefer shipping [PowerPoint presentation]. *Cool Logistics Global*, 17–19 September, Valencia.

Eurofresh-Distribution (2016). Maersk buys 14 800 'smart' reefer containers. https://www.eurofresh-distribution.com/news/maersk-buys-14800-'smart'-reefer-containers (accessed 15 April 2020).

International Institute of Refrigeration (2019). *International dictionary of refrigeration.* http://dictionary.iifiir.org/index.php (accessed 1 May 2020)

Iansiti, M. and Lakhani, K.R. (2017). The truth about blockchain. *Harvard Business Review* 95 (1): 118–127.

Kern, J. and Wolff, P. (2019). The digital transformation of the automotive supply chain – an empirical analysis with evidence from Germany and China. https://www.innovationpolicyplatform.org/system/files/imce/AutomotiveSupplyChain_GermanyChina_TIPDigitalCaseStudy2019_1.pdf (accessed 15 April 2020).

Miller, E.V. (1947). The story of ethylene. *The Scientific Monthly* 65 (4): 335–342. http://www.jstor.org/stable/19231 (accessed 15 April 2020).

Mitcham, E.J., Crisosto, C.H., and Kader, A.A. (2020). Fruit produce facts english. http://postharvest.ucdavis.edu/Commodity_Resources/Fact_Sheets/Datastores/Fruit_English/?uid=13&ds=798 (accessed 15 April 2020).

Neher, A. (2020). Logistics management in an IoT world. In: *Digitizing the Logistics Industry: Demystifying the Impacts of the Fourth Industrial Revolution* (eds. M. Sullivan and J. Kern). Wiley.

Panetta, K. (2018). Gartner predicts 2019 for supply chain operations. *Gartner Predicts*: 1–10. https://www.gartner.com/smarterwithgartner/gartner-predicts-2019-for-supply-chain-operations/ (accessed 15 April 2020).

Pierson, M.D. (2012). *HACCP: Principles and Applications.* Springer Science & Business Media.

Pinker, S. (2018). *Enlightenment Now: The Case for Reason, Science, Humanism, and Progress.* Penguin Random House.

Ramos-Elorduy, J. (2005). Insects: a hopeful food source. In: *Ecological Implications of Minilivestock*, 263–291. Science Pub.

Robertson, A. (2019). *Intermodal Cool Chain Management: A Practical Guide to the Movement of Temperature-Controlled Cargo.* Kogan Page, Limited. https://books.google.co.uk/books?id=dB3DswEACAAJ (accessed 15 April 2020).

Robertson, B. (2019). Bridging the gap in a sea of data. In: *Asia Fruit Logistica.* Hong Kong: Emerson Commercial & Residential Solutions.

Rolfsen, B. (2019). Amazon's growing robot army keeps warehouses humming, Bloomberg environment. https://news.bloombergenvironment.com/safety/amazons-growing-robot-army-keeps-warehouses-humming (accessed 15 April 2020).

Schrieder, G. (2019). Fresh produce demand and container shipping. In: *Logistics Hub/Fruit Logistica.* Berlin: Maersk Line.

Sloman, A. (2010). What is artificial intelligence? http://www.cs.bham.ac.uk/research/projects/cogaff/misc/aiforschools.html (accessed 15 April 2020).

The Economist (2015). *The Great Chain of Being Sure About Things – Blockchains.* The Economist. https://www.economist.com/briefing/2015/10/31/the-great-chain-of-being-sure-about-things (accessed 28 September 2018).

Tjahjono, B., Esplugues, C., Ares, E. et al. (2017). What does Industry 4.0 mean to supply chain. In: *Procedia Manufacturing*, vol. 13, 1175–1182. Elsevier. https://doi.org/10.1016/J.PROMFG.2017.09.191.

Transfrigoroute, G.J. (2020). Transportation and distribution challenges. In: *Fruit Logistica.* Berlin: Technical Advisory Council.

UNCTAD (2019). Review of maritime transport 2019. https://unctad.org/en/Publications
Library/rmt2019_en.pdf (accessed 15 April 2020).

United Nations (2016). ATP handbook 2016. https://www.unece.org/fileadmin/DAM/trans/
main/wp11/ATP_publication/Handbook-2016e.pdf (accessed 15 April 2020).

Veldman, R. (2019). Global reefer trade growth outlook update [PowerPoint Presentation].
11th Cool Logistics Global, 17–19 September, Valencia.

Verberckmoes, S. (2019). Container carriers target the fruit market. In: *Logistics Hub/Fruit
Logistica*. Berlin: Alphaliner.

Wills, R. (2018). Effect of ethylene on non climacteric produce. In: *Postharvest Partner Meeting*.
Barcelona: Bioconservacion.

Section IV

People

18

People Navigating Digital Transformation

Johannes Kern

Tongji University, Shanghai, China

At a conference on intelligent manufacturing in Beijing, Prof. Dr. Henning Kagermann, the forefather of the Industry 4.0 concept, explained to me that half of all barriers toward the digital transformation are human related. Indeed, his expert opinion was repeatedly confirmed by various studies. In a survey by the consultancy Capgemini, 62% of respondents consider culture as the top hurdle to the digital transformation (Buvat et al. 2017). According to a study by the consultancy McKinsey, the largest internal barriers to digital effectiveness are cultural and behavioral challenges (Goran, LaBerge, and Srinivasan, 2017). Also, research by the supply chain network BVL found that in this context, typical obstacles companies that will have to overcome are corporate culture and ensuring that employees are on board (Heistermann, ten Hompel, and Mallée, 2017). At the same time, managers often underestimate the importance of culture and other soft factors (Buvat et al. 2017). Many either let it go unmanaged or delegate it to the HR function where it then becomes a secondary concern for a company. When managers do not understand the power and dynamics of the human side of business, their digital transformation plans will go off the rails (Groysberg et al. 2018). A company's history and tradition can prevent the development of new mindsets, and in a work environment where humans and machines increasingly complement each other, employees must be willing to go along a transformation journey (Heistermann, ten Hompel, and Mallée 2017). Therefore, this section is dedicated to the people side of the digital transformation.

In the first chapter, Michael Babilon-Teubenbacher outlines key aspects and approaches for helping leaders and their organizations to create sustainable impact in the context of digitalization and digital transformation. He argues that as true digital transformation and digitalization are fundamentally different concepts, they also must be tackled fundamentally different. Based on their business model, players in the logistics industry have to consciously decide how to best benefit from the Fourth Industrial Revolution and move fast, not only technology wise but also in terms of key capabilities such as risk tolerance and vision- and business-model orientation. In this increasingly competitive environment,

The Digital Transformation of Logistics: Demystifying Impacts of the Fourth Industrial Revolution,
First Edition. Edited by Mac Sullivan and Johannes Kern.
© 2021 by The Institute of Electrical and Electronics Engineers, Inc.
Published 2021 by John Wiley & Sons, Inc.

managers will likely have to choose a mix between exploiting existing business and exploring new business models. Here, building a balanced set of methods and capabilities in business transformation as well as change management will become a key differentiator.

The second chapter by Robert Mostert and Johannes Kern addresses organizational culture change (OCC). While this is definitely a challenge, OCC can lead to improved performance if it optimizes the way a firm's culture is aligned with its strategy. The authors conducted expert interviews with CEOs and senior HR managers of multinational enterprises in Singapore, Thailand, Australia, and the United States and triangulated their findings with a case study from the Australian transportation and logistics company "Toll" to understand which practices companies use today to drive OCC efforts. The authors found that the practices that must be aligned with the company strategy are purpose, vision and mission, values and beliefs, recruitment and placement, training and development, rewards and recognition, performance management and feedback, and artifacts. If companies align those with their strategy, OCC is happening.

The third chapter addresses competence management in the digital transformation. Sun Jiayu highlights that while a lot of discussion in the context of the digital transformation is revolving around a variety of technologies, employees, the most crucial factor that makes such technologies actually work, are often ignored. To be prepared for a Logistics 4.0 world, systematically upskilling them is mandatory. Jiayu outlines how skills can be managed through competence models and which ones are required today and will be in the future. The digital transformation is also accompanied by a transformation of the way we are learning, from a passive to an active mode and from a classroom to a digital experience. However, only if people and people's mindset are prepared for the digital transformation, methods and technologies can support the needed change of employee's behavior and their skills.

In the fourth chapter, Cameron Johnson lays out the impact of digitalization on supply chain traceability. As an increasing number of companies across various industries get under pressure to track products from source to consumer, technology that supports traceability such as QR codes or RFID will become more and more common. Johnson shows that employing such technology would not only increase transparency but also, for example, enable manufacturers to capture data on waste generated during various production stages. To increase efficiency while assuring quality, large downstream companies like Boeing and Apple therefore leverage their market power, forcing downstream partners to adopt digitalization tools. Based on his two decades of experience in high-end manufacturing logistics, the author uses the carbon fiber supply chain as an example to demonstrate that despite growth and innovation, digital technology is still underutilized, which creates opportunities for better quality control, global traceability, logistics optimization, regulatory compliance, and waste management.

Business transactions in the freight forwarding industry have traditionally been strongly relationship based. However, technology now has enabled resource pools that can now compete, connect, and collaborate for a fraction of the cost and effort it took a decade ago. In their chapter *The Evolution of Freight Forwarding Sales*, Mac Sullivan, Dennis Wong, and Zheyuan Tang explain that this has spurred a new movement to open digital sales channels for procuring logistics and freight services. As this leads to a more transparent market where customers have abundantly more information available, freight forwarding sales executives must retool and upskill to stay relevant. The authors investigate the challenge through the digital transformation in detail and offer examples how a freight forwarder's salesforce can be futureproofed.

In the final chapter of this section, Colin Cobb and Dyci Sfregola discuss how to manage and select logistics service providers (LSP). In an increasingly volatile, uncertain, complex, and ambiguous market environment, companies want to maintain flexibility while focusing on core competencies, which grant them competitive advantages (Aguezzoul 2014). They therefore employ LSP that can offer customized solutions and provide specialized logistics expertise (Sink and Langley Jr 1997). However, to realize these benefits, a suitable LSP must be selected and the relationship between the buying firm and the service provider managed. The authors of this chapter therefore outline best practices on how to develop strategic alliances with suppliers, which allow to offset financial and operational risk, cultivate profitable relationships, and implement cross-functional, cross-company transparency. While sourcing and managing suppliers can be theoretically accomplished quickly, it requires the right practices executed by competent employees to develop and manage an inclusive, scalable, and streamlined logistics ecosystem that exceeds expectations.

References

Aguezzoul, A. (2014). Third-party logistics selection problem. *Omega* 49: 69–78.

Buvat, J., Solis, B., Crummenerl, C. et al. (2017). The digital culture challenge: closing the employee-leadership gap. https://www.capgemini.com/consulting/wp-content/uploads/sites/30/2017/07/dti_digitalculture_report.pdf (accessed 9 December 2020).

Goran, J., LaBerge, L., and Srinivasan, R. (2017). *Culture for a Digital Age. McKinsey & Company*. https://www.mckinsey.com/business-functions/mckinsey-digital/our-insights/culture-for-a-digital-age (accessed 9 December 2020).

Groysberg, B., Lee, J., Price, J., and Cheng, J. (2018). The leader's guide to corporate culture. *Harvard Business Review* 96 (1): 44–52.

Heistermann, F., ten Hompel, M., and Mallée, T. (2017). Digitisation in logistics. https://www.bvl.de/misc/filePush.php?id=39654&name=BVL17_Position_Paper_Logistics_4.0.pdf (accessed 9 December 2020).

Sink, H.L. and Langley, C.J. Jr. (1997). A managerial framework for the acquisition of third-party logistics services. *Journal of Business Logistics* 18 (2): 163.

19

Change Management Falling Short – the Call for Business Transformation

Michael Teubenbacher

CPC Consulting, Beijing, China, subsidiary of CPC Unternehmensmanagement AG

Introduction

Discourses and studies on digitalization and digital transformation in logistics frequently focus on specific digital solutions or specific parts of the supply chain as these are absolutely crucial and valuable for the development in logistics. However, two aspects are commonly overlooked: first, true transformation requires breaking down existing limitations and thus a comprehensive rethinking of end-to-end logistics in the search for new business models. Second, digital solutions are only half the battle for creating true business value. As long as people and organizations are not truly embracing not only the technology but also the inherent principles, policies, and procedures, the resulting value gap will remain significant.

To fill this gap, a deeper understanding of the concepts around change management and business transformation including related strategies, methods, and tools is an essential foundation. Being familiar with both dramatically increases the likelihood of fully realizing the potentials entailed in digital solutions and Industry 4.0 also and especially in the context of logistics. It is important to state though that knowledge on specific digital solutions and fluency in concepts around change management and business transformation are complementary. In logistics, like in other industries, it takes a combination of both areas of expertise to effectively digitally transform organizations.

This chapter, after mapping out the terminology around change management and business transformation, sets both into the context of "digital" and "logistics." By taking reference to specific examples from logistics and other industries, strategies, methodologies, and key success factors for impactful implementation of digital transformation initiatives are described. Specific emphasis is laid on challenges that are unique to logistics as this industry historically can be seen to be rather at the conservative end when it comes to "digital."

The Digital Transformation of Logistics: Demystifying Impacts of the Fourth Industrial Revolution, First Edition. Edited by Mac Sullivan and Johannes Kern.
© 2021 by The Institute of Electrical and Electronics Engineers, Inc.
Published 2021 by John Wiley & Sons, Inc.

Change Management and Business Transformation – Two Sides of the Same Coin?

Regardless of the industry, the term "digital transformation" has sneaked its way into the standard vocabulary used by organizations. It is spoken either with enthusiasm by executives, with – at least superficial – confidence by middle managers, or with mixed feelings by frontline staff. Meanwhile, it is a highly current topic with, for instance, over 328 million search results on Google (Google 2020). Very little attention is paid to what "(Business) Transformation" at its core refers to and how it differentiates from and plays together with "Change Management." This leads to confusion, misperception, and consequentially the application of less than optimal strategies and approaches in tackling it.

To put it straight, (business) transformation refers to the process of shifting from a current as is state to a desired future target state and thus to the establishment of a *new way of being* (Barrett 2016). It is to *reinvent the organization and discover a new or revised business model based on a vision for the future* by executing a portfolio of initiatives, which are interdependent or intersecting (Ashkenas 2015). It is a *fundamental change in your beliefs of why you perform certain actions* and a modification of beliefs so that natural actions achieve the desired results (Palinkas 2013). Accordingly, success measurement in business transformation is based on assessing how much the future state deviates from the target state, based on the initial vision and organizational strategy definition (Fannin 2018). As business transformation is much more *unpredictable, iterative, and experimental* and entails much higher risk, it rarely builds on considerations around return on investment (ROI) (Ashkenas 2015). This, as we will also explore later in this chapter, poses one of the key challenges for successful business transformation in the efficiency-driven logistics environment.

Change management, on the other hand, subsumes tasks, measures, and activities that aim at effecting a specific comprehensive change for implementing new strategies, structures, systems, processes, or behaviors (Doppler and Lauterburg 2019). It assesses the past, compares it with the present, and determines the ideal future state from the current business state (Chaudron 2019). It is about using *external influences to modify actions to achieve desired results* (Palinkas 2013). Accordingly, success measurement takes the current state as a reference point, thus frequently focusing on optimization and therefore on ROI-related measurements (Fannin 2018).

While business transformation is about defining an *entirely different future*, change management aims at *doing things differently* in a specifically defined context (Barrett 2016; Fannin 2018). Being aware of this distinction is crucial as only the combination of approaches from both fields will lead to success in today's mostly ambidextrous environments where continuous optimization of existing approaches AND business innovation are required for success (O'Reilly 3rd and Tushman 2004). This is also confirmed specifically by CPC's[1] experience in the logistics context.

---

1 CPC is a leading German Change Management consultancy. We are the reliable partner for corporations and medium-sized companies that strive to drive business transformation and implement lasting change. To effect such change, we follow a holistic, customer-centric approach built around the three core competencies: people, projects, and organizations.

In summary, it is crucial to distinguish between business transformation, which refers to a completely new way of being by reinventing the organization and discovering a new or revised business model, and change management, which describes tasks, measures, and activities that are targeted at a specific change like the implementation of a new strategy, of new structures, of new systems, or of new processes and behaviors.

Business Transformation in the Context of "Digital"

In defining the distinction in the context of "digital," it seems that many efforts undertaken under the term "digital transformation" are less of a *business transformation than* a *business optimization*. They do not aim at truly reinventing the business or establishing a new business model, but rather target at optimizing existing structures and processes based on digital technologies. This is commonly referred to as either digitization, simply making analog processes digital, or digitalization, using digital technologies to make processes more efficient, more productive, more profitable, and with greater customer satisfaction in their digital and physical experience with the company without the underlying business model (Brennen and Kreiss 2014).

While not paying further attention to "digitization," but because of the abovementioned very common confusion, we will semantically draw a line between the terms "digitalization" and "digital transformation," which in the remainder of this book are jointly referred to as "digital transformation." This is crucial as required strategies, methods, and key success factors that need to be applied differ. The connection between digitalization, digital transformation, and change management is also outlined in the below graph, putting the pursuit of value into the center of all related activities (Figure 19.1).

This "misleading labeling" does not necessarily create harm by itself. However, a combination of misguided underlying expectations from executives and ineffective application of strategies and approaches frequently does. A global study by Kane et al.

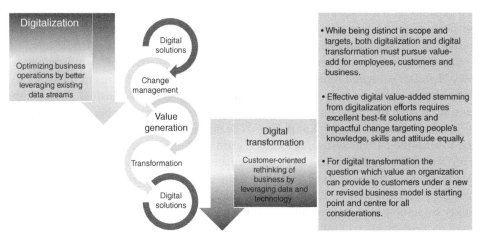

Figure 19.1 Value pursuit in the center of digitalization and digital transformation. Source: CPC Consulting.

found that only one-fourth of organizations had transformed into digital businesses, with the rest either being on transformative journeys or entangled in internal discussions instead of acting (Kane et al. 2015). At the same time, 85% of executives state that digital maturity is critical to organizational success.

Many executives are seeking for digital transformation while driving technology- and solution-focused digitalization. Indeed, oftentimes the fact that digital transformation is in the first place about people, not about technology and solutions, is missed. Without people understanding the larger picture, the general direction in which business is heading based on new digital possibilities and changing business patterns, buy-in, and commitment to supporting initiatives remains low, which might be why in failed transformations, often plenty of plans, directives, and programs, can be found – but no vision (Kotter 1996). For instance, John Kotter, Professor of Leadership, Emeritus at the Harvard Business School, describes a case where "a company gave out four-inch-thick notebooks describing its change effort. In mind-numbing detail, the books spelled out procedures, goals, methods, and deadlines. But nowhere was there a clear and compelling statement of where all this was leading (Kotter, 1996)."

We frequently see this pattern among our customers from a wide variety of industries, including logistics service providers and logistics departments of automotive companies. Driven by cost pressure and burdened by cultural as well as technological legacy, many of them show a strong focus on executing single digitalization initiatives. At the same time, they refrain from creating new or revised sustainable, digitally enabled business models that could be brought to life based on a well-selected portfolio of interrelated initiatives.

Likewise, there are logistic solutions with transformative potential that get implemented, for example, an innovative way of freight space brokerage, but fall considerably short of their potential. Here, individual employees or even entire business units are not aware of the bigger context, which then translates into very limited adoption rates. Without credible communication, and a lot of it, the "hearts and minds of the troops are never captured," and if the underlying vision cannot be communicated to someone "in five minutes or less and get a reaction that signifies both understanding and interest," transformation efforts are likely to fail (Kotter 1996). As they do not have the intended impact, they lead to frustration instead of enhanced performance.

What adds on top is the "unpredictable, iterative, and experimental" and thus also a risky character of digital transformations (Ashkenas 2015). Often, organizations, especially also in logistics, show patterns that aim at minimizing risks instead of fostering learning along transformation trajectories. This often results from having too many managers and not enough leaders (Kotter 1996). The manager's command is to minimize risk and to keep the current system operating (Perrin 2010). In other words, driving digital transformation requires leaders and entrepreneurs who are prepared to take risks, can communicate exceptionally well, and can handle volatility, uncertainty, complexity, and ambiguity (VUCA) well (Bennett and Lemoine 2014). They do so by leveraging their ability to develop and lead progressive timelines, by continuously assessing progress based on a combination of soft and hard factors, and by dynamically leading a portfolio of interlinking initiatives.

We at CPC experienced how effective digital transformation, focused on people and leadership, can be at one of our customers in the telecommunications industry. Besides driving forward digital solutions that will keep the company to stay ahead of the market, top executives of this company showed a strong commitment to investing in current- and next-generation leaders, thus preparing them for the challenges of the digital transformation. They promoted the implementation of a blended learning program for next-generation leaders, which took participants through a 6–8-month learning journey. This learning journey was designed to build up knowledge and skills relevant in the context of digital transformation such as collaboration, empowerment, agile working, ambidexterity, and mindfulness but also digitalization and technology (Kern 2019). By blending different online and offline formats and combining participants in physical and virtual learning groups, the learning journey has also succeeded in triggering the mindset change required in the context of digital transformation. Now, obviously, this is an example from telecommunications. But while the respective digital solutions might be different, the underlying leadership skills required for effective digital transformation as well as the purposefully chosen digital formats for triggering the mindset change are without doubt readily transferable to logistics.

For digital transformation to be impactful, a company needs to have clear future orientation, meaning a clear vision based on a revised, digitally empowered business model. Likewise, communication needs to be done in a language understood across business units and hierarchies. A company also needs to understand their own risk tolerance to strengthen leadership and entrepreneurship based on strategic staffing and retainment of talents who handle VUCA well. Lastly, a shared mindset and belief system must be put in place where reducing doubt, encouraging knowledge building and skill development are emphasized. The mindset and underlying beliefs are the make-or-break factors as building staff with a growth mindset is without alternative if new business models and new ways of interacting with partners are explored. Innovators and early adopters are needed to support change management efforts.

Change Management in the Context of "Digital"

Moving our attention to the change management side of the story, it is surprising to note that despite change management having been around for a couple of decades, still only a few change efforts can be considered successful (Burnes 2011). About 60–80% do not produce the intended results, and some even can be considered as utter failures (Kotter 1996). This is particularly unsettling as change management is bound to target specific, well-defined comprehensive changes such as the implementation of a specific digital solution and thus to generate clear and ideally also clearly measurable business value. The reasons leading to failure or less than optimal outcomes are manifold. Building on those, strategies and approaches will be outlined on how to prevent them or how to at least mitigate potential consequences.

First, there are many cases in which projects with significant change impact are carried out without paying attention to change management at all. Those projects often install very

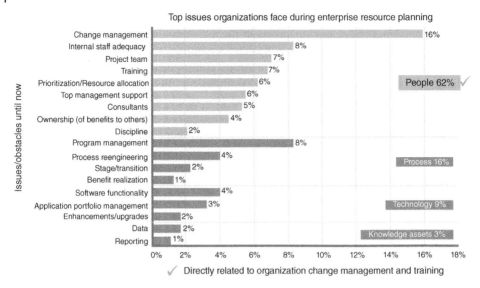

Figure 19.2 Top issues organizations face when implementing enterprise resource planning solutions. *Source*: Benchmarking Partners (2017). © Benchmarking Partners, Inc.

good solutions, new processes, new software, new collaboration models, etc., at organizations under the assumption that adaptation of those solutions is going to happen naturally and will be welcomed with open arms. The tricky thing is that the success of such undertakings is determined by quality at two ends: the quality of the solution and the quality of the implementation. Implementation, as opposed to installation, refers to an actively shaped process in intense exchange with the adapters, the people who need to get used and adapt to the new solution.

The relevance of organizational change management specifically and of people-related topics more generally in the context of project implementation is also highlighted in Figure 19.2, which depicts the top issues organizations face when implementing enterprise resource planning solutions.

Besides those cases where change management is not actively addressed at all, it seems that very often the process character of change is forgotten or ignored. As Kotter puts it, "The most general lesson to be learned from the more successful cases is that the change process goes through a series of phases that, in total, usually require a considerable length of time. Skipping steps creates only the illusion of speed and never produces a satisfying result (Kotter 1996)."

In change, there are no shortcuts, as the full adoption of new solutions takes time as well as varying degrees of efforts by the affected adapters. Especially in changes related to digital solutions, this is frequently not easy to accept as "digital" implicitly is often linked to speed. Accordingly, strategies, particularly around engaging people, are needed, which instead of skipping steps help to make the measures taken more effective and impactful, thus supporting adapters to pass through the different stages of the change curve smoother and ideally quicker.[2] At the beginning, those strategies are related to getting the change purpose clear for creating a sense of urgency. Then, they focus on gathering a critical mass based on a powerful guiding coalition to lead the change, and in later stages, they aim at keeping momentum until the very end to achieve sustainability.

These challenges are also in line with further frequent reasons for failure around change management (Appelbaum et al. 2012). In the initial phase of a change, the main reason for it not taking off is a lack of understanding of why change is necessary. Companies should, as renowned journalist and author Malcolm Gladwell states, leverage the power of context (Gladwell 2006). For instance, bad business results are a good context to create momentum, although they restrict financial room for maneuvering. Good business results give room for maneuvering but tend to be marked by inertia based on the question "Why should we move?"

No matter the circumstances, putting facts about a worsening competitive position on the table and facilitating an intense discussion is always a good starting point for getting things going. As shooting the bearer of bad news is human nature, often outsiders such as Wall Street analysts, customers, and consultants are relied on as messengers "to make the status quo seem more dangerous than launching into the unknown (Kotter 1996)." Questions to be asked in the context of implementing digital solutions might be "Why is the implementation of this solution without alternative?" and "Who could be the most well-suited messenger?"

Aside from having the change purpose clear and sticky, a common reason for failure around change management in early stages is that a one-size-fits-all approach is applied, disregarding that not all groups within an organization have the same influence (Gladwell 2006). According to the "law of the few," single members of an organization have a disproportionate influence in effecting change (Gladwell 2006).

Certainly, management plays a key role here. As much as 75% of a company's management might need to be honestly convinced that business, as usual, is unacceptable (Kotter 1996). In addition, also the theory of the "Diffusion of Innovations" provides guidance on strategies for convincing the right "few" people and having the rest follow (Rogers et al. 2005). Adopters can be divided into innovators (2.5%), early adopters (13.5%), early majority (34%), late majority (34%), and laggards (16%) (Rogers et al. 2005). Focusing on innovators and early adopters helps to create dynamics efficiently. The rest of the organization will then follow sooner or later. For the implementation of solutions in the context of digital, this certainly also implies that leaders must be aware of who are the innovators within the organizations and who are the ones that adopt new solutions early. They further

---

2 The change curve is a model used to understand the stages of personal transition in the context of organizational change. It outlines how people will react to change, so that change leaders can help them make their personal transitions and make sure that they have the help and support they need (Kübler-Ross 1973). It describes four main stages. When a change is first introduced, people's initial response may be shock or denial, as they react to the challenge to the status quo (stage 1). Once the reality of the change starts to hit, people tend to react negatively and move to stage 2, "disruption," where they may fear the impact, feel angry, and actively resist or protest the changes. In stage 3 of the change curve, "exploration," people stop focusing on what they have lost. Instead, they start to let go and accept the change. They begin testing and exploring what the change means and so learn the reality of what's good and not so good and how they must adapt. In stage 4, "rebuilding," they not only accept the change but also start to embrace it: they rebuild their ways of working. Only when people get to this stage can the organization really start to reap the rewards of change (Mind Tools 2018).

must ask themselves whether those innovators and early adopters have the credibility to truly act as influencers for the rest of the organization.

This is one of the key intersections of transformation and change management in the context of digital transformation. As indicated earlier, digital transformation is about creating a "new way of being." Hence, it is of utmost importance to have the right people on board and in the right positions to promote the spirit and the ideas behind it. Those people, in turn, need to be the innovators and early adopters who will help organizations to efficiently reach the tipping point for the introduction of new digital solutions. Having things moving is then a won battle, but not yet a won war. What we regularly observe among our customers is that the first signs of success are a trigger for taking the foot from the accelerator and believing that the change has already achieved a self-sustaining zone. Kotter refers to it as "declaring victory too soon (Kotter 1996)." This results in a possible loss of momentum and even the reversal of hard-won gains.

It should be borne in mind that change sticks when it becomes "the way we do things around here," that is when it seeps into the bloodstream of the corporate body. As long as behaviors are not rooted in social norms and shared values, they will degrade as soon as the pressure for change is removed (Kotter 1996). Here, we also have a striking example from one of our consulting projects in logistics. During the implementation of a new ERP solution at one of our customers, even though the purpose for the change was clear, a strong guiding coalition was in place, and everyone kept pressing consistently for realizing the change, not having the shared value of "data quality" proved a key obstacle to achieving the desired success.

The so-far mentioned reasons for failure have led us at CPC to derive four knowledge areas – distinct field of competency and expertise – which are essential to success in change management. Each of those knowledge areas is supported by corresponding strategies and approaches.

The first area is about developing a clear understanding of the IMPACT that the change is supposed to have. This understanding also has to include a clear understanding of why change has been initiated and how the different adapters throughout the organization will be directly impacted by this change. Mastering this field addresses the risk of not having the change purpose transparent and described in a way so that every person affected by the change can clearly understand it.

The second area refers to gathering a guiding coalition of people who can LEAD the change and helping them to collaborate as a team. This guiding coalition needs to include not only people who lead based on their hierarchical power but also people who lead because of their expertise or personality (e.g. acting as innovators or early adopters of new solutions). Mastering this area helps to create momentum effectively and efficiently by leveraging multipliers.

The third area relates to ENGAGEMENT and ENABLING of adapters – those who will have to adapt to the change. Traditionally, a strong focus is put on enabling based on training measures that help people become familiar with new solutions. We at CPC put a stronger focus on leveraging strong experience-based engagement formats such as small- to large-scale business simulations as they are most effective according to our experience. Mastering this area helps to keep momentum and to develop a strong sense of when to take the foot from the accelerator or when to keep it pressed.

The fourth area is about INSTITUTIONALIZATION, which focuses on making change stick by addressing organizational patterns that hinder change. Whereas the first three areas help to create and maintain momentum, this area focuses on identifying hindering organizational patterns and other obstacles that might prevent change. Our experience shows that frequently, change management projects are pushed ahead with best intentions while at the same time organizational patterns are strongly counteracting the change. Mastering this area helps to make change stick against all odds, thus massively increasing the sustainability and long-term success of change management.

In a nutshell, for change management to be impact – and thus successful – four focus areas and related approaches are crucial. First, a clear understanding of the IMPACT that the intended change will have across adapter groups is required so that this impact can also be actively shaped. Then, execution-focused LEADERSHIP needs to provide clear direction, supported by ENGAGEMENT and ENABLING that draws considerable efforts from the project team, especially as it needs to go beyond standard communication and enable formats to truly encourage active engagement and thus foster commitment. Finally, INSTITUTIONALIZATION is to ensure that change sticks and that it becomes "the way we do things around here." This only works if, besides the implementation work at hand, also hindering patterns (e.g. target setting schemes, budget allocation, recognition schemes) are addressed and adjusted. Even though they may sound simple, the effort for truly excelling at them is considerable. Managing change means leading a process that does not allow for shortcuts. Conflicts are unavoidable; omitting them will fire back – for sure!

Business Transformation and Change Management in the Context of the Fourth Industrial Revolution in Logistics

Logistics trails behind the digital curve compared with most other industry sectors such as media, telco, banking, and retail (BCG 2020). The industry notoriously suffers from fragmentation, which results in poor transparency and thus underutilization of assets, inefficient processes, and outdated customer interfaces (cf. Chapter 4). At the same time, competition is increasing massively. New players are entering the market, many of them with digital roots. Some of them are integrators who take advantage of industry fragmentation and who offer integrated and streamlined end-to-end services, some of them are suppliers who go for digitizing their offerings and operations themselves, and some of them are even customers who turn into competitors (BCG 2020).

Many traditional players in the logistics industry see themselves confronted with the challenge to optimize core operations, thus exploiting existing business while at the same time exploring new opportunities, thus competing effectively with new entrants. At the same time, also in-house logistics units (e.g. spare parts logistics in automotive after-sales) are facing increasing pressures to optimize operations and even to develop new business models (e.g. direct delivery from suppliers to dealers). Without doubt, business transformation AND change management must play a crucial role in logistics.

Neither business transformation nor change management are well-established concepts in this industry. Rather, players in the market concentrate their efforts on technologies and solutions instead of actively shaping the underlying value-generating business transformation. Without a proper picture of which new identity the business could assume, well-developed solutions are left far behind their potential and fail to deliver the expected digital transformation. Without a proper understanding of why a certain new solution is critical for business success or even company survival, new solutions get installed instead of implemented. This leads to a failure in realizing the expected benefits, for example, in terms of ROI.

As logistics is an industry rather on the conservative side, we can assume that a lack of risk tolerance is one of the key challenges it faces on its way toward transforming for the future. We observe that the focus on exploiting existing revenue streams to the max has cannibalized the much-needed exploration capacities and capabilities, where first investment is needed into people as well as into new technologies and where neither the amount nor the time horizon for payoff is directly predictable. Besides that, operational focus also has withdrawn attention from developing an overarching vision of what else the respective business could be, a vision that is supported by business as well as technology units likewise. Lacking such a vision adds little fuel toward a mindset change among employees who would drive the business transformation into a more digital future.

Still, we at CPC see more and more cases where those challenges could be overcome and businesses were transformed to become more digital. The adopted strategies vary. One of CPC's customers is adopting a technology-centric approach where IT supports the business to evolve the business model by implementing innovative solutions in a very focused manner. Other companies such as Maersk take a more radical approach. They are pursuing a strategy of grooming an entire ecosystem of innovative startups and service providers that all work toward helping Maersk to provide an ever more seamless experience to customers.

Whatever the selected strategy is, driving digital transformation from the people and the solution side, equally and in a balanced fashion, is crucial to bundle energies and to make progress along the transformation trajectory. Also, the mentioned examples entail both kinds of initiatives, implementation of new digital solutions and continuous investment into people, new hires, training, and talent programs, to shape the required mindset change and to support the company's vision.

Just as business transformation, we do not experience change management as a consistently established concept in logistics. A strong focus on solutions, such as new systems, new processes, or new technologies, leads to these solutions being installed rather than implemented, thus leading to far less than optimal adoption rates. Frequently, the true reason why change is being initiated and cannot be avoided is not described in a language that is understandable for everyone, nor is the impact on the people who must adapt to the change evaluated appropriately, although it could be easily done by applying a combination of quantitative, for instance, online surveys, and qualitative techniques, such as interviews.

Without understanding the true impact and the circumstances under which people need to make use of new solutions, also communication and training measures are frequently applied in a one-size-fits-all approach that is neither efficient nor effective. As a result, we frequently see people sticking to old patterns such as twisting and tweaking a new ERP system and "enriching" it with paper-based and manual workarounds to be able to run processes as they knew them. They are far away from truly "adapting" and thus from perceiving new solutions as a growth opportunity that will enable them to work more effectively and therefore increase business value.

Accordingly, aside from understanding the actual impact, the way of enabling and engaging those people affected by change makes a significant difference. This is particularly true for key areas in logistics, such as inventory management or trade financing, which are knowledge and experience based. Having experts who need to change their behavior works best if those experts are actively engaged and not just affected by the change. Such engagement can take a variety of formats, ranging from having them co-shape the solution to giving them an active role in driving the change, thus safeguarding their position of expertise.

Also, for nonexperts, experiencing instead of only hearing about a change can make all the difference between people doing because they are told and people adapting to a new solution because they truly understand and accept. In many projects at our customers in the logistics industry, we see a tremendous difference that simulation-based formats can make, be it the simulation of end-to-end processes while implementing a new ERP solution or the focused stimulation of new procedures while implementing a new freight space brokerage solution.

Finally, we notice that digitalization projects in logistics frequently also struggle with significant amounts of organizational legacy. Budget allocation procedures, target setting schemes, decision-making procedures, and established leadership principles are generally all suited to hinder sustainable improvements due to the implementation of new technological solutions. For example, having an established culture of mandatorily "beautifying KPIs" significantly hampers the effective usage of state-of-the-art dashboards in new ERP solutions, thus significantly reducing its potential benefits. Having target setting schemes established that focus strongly on service level fulfillment makes unnecessary the attempts to leverage a new ERP system to decrease inventory levels. Finally, having an organizational setup without clear data governance structures poses a threat to the implementation of any solution that relies on data quality.

Thus, besides clearly understanding the impact of new technological solutions on different adapter groups and besides leaders who guide the organization by using effective enabling and engagement techniques, successful implementation of new digital solutions requires a good understanding of where the organization is coming from and which legacies might hinder change efforts. Considering the increasing number and frequency of changes in the context of digitalization, we are convinced that establishing standardized and effective change management methodologies and well-proven tools and engagement formats as well as building up in-house change management capabilities is an investment that will quickly pay off for logistics providers.

In summary, players must strike a balance between exploiting and optimizing existing business on the one hand and exploring new business models on the other hand. This requires literacy in methods and approaches around business transformation and change management likewise – and especially a balance thereof. From business transformation perspective, mainly low-risk tolerance and operational focus hold back enterprises and prevent them from developing an attractive vision for business. Whatever the selected strategy to tackle this challenge is, striking a balance between solution and people focus will be the make-or-break point. From change management perspective, an almost exclusive focus on excellent solution prevents players from fully realizing the solution's value by also properly implementing it. Understanding the true impact, making change tangible and handling of organizational legacy based on a consistent set of methodologies and tools are key to success.

Conclusion

First, it is crucial to differentiate true digital transformation, the redefinition of business based on digital technologies, and digitalization, the application of digital technologies to optimize operations. As both concepts are significantly different, the equation that "enough properly carried out digitalization" will sooner or later lead to "Digital Transformation" does not hold. "Digital Transformation" means an entirely new organizational existence based on new business models.

Therefore, players in the logistics industry must take a conscious decision on how to best benefit from the 4IR. Many players will find themselves needing the capability to handle ambidexterity, skillfully combining exploitation, optimizing existing operations based on digitalization, and exploration, identifying completely new ways of doing business, thus driving digital transformation.

To tackle both fields, capabilities in business transformation and change management are needed likewise and need to be applied in a balanced fashion. Focusing purely on business transformation for sure helps to convey the big picture and set the ground rules. However, evidence shows that this strategy lacks the execution capabilities required to give underlying initiatives the needed organizational impact. Overemphasizing change management, on the other hand, poses the risk of conveying the impression of running a set of single initiatives without pursuing an overarching transformational goal.

For business transformation, related capabilities are mainly future orientation, risk tolerance, and a mindset and belief orientation. Looking into logistics in the context of the 4IR, we encounter challenges along with all three capabilities. Logistics providers who want to survive a digital transformation need to understand what a new or revised business model might look like in the context of "digital" and that the business might need to become a data broker instead of a physical goods mover or an integrator instead of a warehouse operator. Leaders need to encourage risk tolerance in their efficiency- and risk avoidance-driven environments and balance exploitation focusing on driving efficiency, with exploration focusing on driving innovation. Logistics leaders need to establish a growth mindset among their staff across hierarchy levels by adapting, recruiting, training, and evaluating their talent management strategy.

For the business transformation vision to become impactful, discrete initiatives are required, which are supported by capabilities and approaches around change management for realizing the desired business value. Looking into the implementation of digital solutions in logistics, we see the need for logistics service providers to think about several crucial topics if their solutions should truly realize the intended business value. Leaders need to see change as a process and not as an event, equipping themselves with the needed resources, time wise, capability wise, and cost wise. They need to have a clear vision for the desired impact of the change, developing approaches for measuring the intended outcome and translating the global impact into a proper understanding of the impact for specific adapter groups. To be able to do so, leaders need to comprise the right combination of logistic experts, digital natives, and people handlers who are the formal and informal leaders in the organization and thus can act as an effective guiding coalition that will lead change all the way through. Skillful combination of the right enabling and engagement measures needs to be applied to get people on board so that they will take an active role in the change. Ultimately, policies, procedures, and mechanisms need to be assessed for potentially working against the change, thus preventing it from sticking.

Especially in the context of change management, it is strongly advisable for companies in the logistics industry to consider establishing standardized methodologies, tools, and engagement formats to be prepared for the changes that the 4IR is expected to bring about. There is no doubt that the 4IR will hit logistics and it is showing signs that it is already upon the industry. Fostering capabilities around business transformation and change management is a necessity. Also, strongly supported by CPCs experience from a wide variety of consulting projects, these capabilities will be a differentiator that is as decisive for future success as the new technologies that 4IR will bring.

Key Takeaways

- Business transformation is about defining an *entirely different future*; change management aims at *doing things differently* in a specifically defined context. Being aware of this distinction is crucial as only the combination of approaches from both fields will lead to success in today's mostly ambidextrous environments where both, continuous optimization of existing approaches and business innovation, are required for success.
- Due to fragmentation and poor transparency, logistic trails behind the digital curve as compared with many other sectors, leaving players with underutilized assets, inefficient processes, and outdated customer interfaces. Traditional players in the industry see themselves confronted with the need to optimize core operations and to innovate at the same time that leads to an urgent need to build competencies around business transformation AND change management.
- Particularly, change management capabilities are an investment that will quickly pay off for logistics providers. Understanding the true impact, making change tangible, building a strong guiding coalition, and addressing organizational legacy based on a consistent set of methodologies and tools are the keys to successfully managing change.

References

Appelbaum, S.H., Habashy, S., Malo, J.-L., and Shafiq, H. (2012). Back to the future: revisiting Kotter's 1996 change model. *Journal of Management Development.* 31 (8): 764–782. https://doi.org/10.1108/02621711211253231.

Ashkenas, R. (2015). We still don't know the difference between change and transformation. *Harvard Business Review* (1 May 2020). https://hbr.org/2015/01/we-still-dont-know-the-difference-between-change-and-transformation (accessed 2 May 2020).

Barrett, R. (2016). *A New Psychology of Human Well – Being: An Exploration of the Influence of Ego – Soul Dynamics On Mental and Physical Health*. Lulu Press.

BCG (2020). Digital transformation in the logistics industry. https://www.bcg.com/de-de/industries/transportation-travel-tourism/center-digital-transportation/logistics.aspx (accessed 2 May 2020).

Benchmarking Partners (2017). Top issues organizations face when implementing Enterprise Resource Planning solutions.

Bennett, N. and Lemoine, G.J. (2014). What a difference a word makes: understanding threats to performance in a VUCA world. *Business Horizons* 57 (3): 311–317. https://doi.org/10.1016/j.bushor.2014.01.001.

Brennen, S. and Kreiss, D. (2014). Digitalization and digitization, culture digitally. http://culturedigitally.org/2014/09/digitalization-and-digitization/ (accessed 2 May 2020).

Burnes, B. (2011). Introduction: why does change fail, and what can we do about it? *Journal of Change Management* 11 (4): 445–450. https://doi.org/10.1080/14697017.2011.630507.

Chaudron, M. (2019). Change vs. transformation, rock'n change. https://rocknchange.com/change-vs-transformation/ (accessed 2 May 2020).

Doppler, K. and Lauterburg, C. (2019). *Change Management: Den Unternehmenswandel gestalten*, 14the. Campus Verlag.

Fannin, K. (2018). Organizational change & transformation – 6 critical differences and why they matter. https://www.intelivate.com/team-strategy/transformation-vs-change-6-differences (accessed 2 May 2020).

Gladwell, M. (2006). *The Tipping Point: How Little Things can Make a Big Difference*. Little Brown.

Google (2020). Google trends 'digital transformation. https://trends.google.com/trends/explore?q=digital transformation&geo=US (accessed 2 May 2020).

Kane, G.C., Palmer, D., Phillips, A.N. et al. (2015). Strategy, not technology, drives digital transformation. *MIT Sloan Management Review and Deloitte University Press* 14 (1–25).

Kern, J. (2019). *Skills for the Future*. Chengdu: OceanX https://youtu.be/VkRCxuJvUTY (accessed 2 May 2020).

Kotter, J.P. (1996). Why transformation efforts fail. *Harvard Business Review*: 59–67.

Kübler-Ross, E. (1973). *On Death and Dying*. Routledge.

Mind Tools (2018). The change curve. https://www.mindtools.com/pages/article/newPPM_96.htm (accessed 2 May 2020).

O'Reilly, C.A. 3rd and Tushman, M.L. (2004). The ambidextrous organization. *Harvard Business Review* 82 (4): 74.

Palinkas, J. (2013). The difference between change and transformation. *CIO Insight*. https://www.cioinsight.com/it-management/expert-voices/the-difference-between-change-and-transformation (accessed 2 May 2020).

Perrin, C. (2010). Leader vs. manager: what's the distinction? *Catalyst* 21519390 (39): 2.

Rogers, E.M., Medina, U.E., Rivera, M.A., and Wiley, C.J. (2005). Complex adaptive systems and the diffusion of innovations. *The Innovation Journal: The Public Sector Innovation Journal*. Citeseer 10 (3): 1–26.

20

Organizational Culture Change

Process to Sustainably Improve Performance

Robert Mostert[1] and Johannes Kern[2]

[1] *Toll, Singapore*
[2] *Tongji University, Shanghai, China*

Introduction

During the past decades we have seen several corporate success stories. In the 1980s and 1990s, General Electric (GE) was one of the most respected companies in the world, and their Chairman and Chief Executive Officer Jack Welch one of the most admired captains of the industry. Many books were written about the GE Way (Tichy and Cohen 1997; Ulrich et al. 2002) and about how Jack "straight from the gut" led GE from one success to the next with principles like: you need to be No. 1 or 2 in every market or get out, and every year you should get rid of your bottom performers (Welch and Byrne 2003). During the next decade, the automotive company Toyota was in the limelight and seemed to outperform all its competitors based on their product development and manufacturing excellence (Liker 2001). Problems were instantly brought to the surface as soon as someone became aware of them to eliminate waste and find a solution. During the most recent decade, it was the technology giant Apple that became one of the most admired companies, with Steve Jobs' Think Different marketing campaign, the strategy to kill your own products before somebody else does, and the value to not settle for anything less than excellence (*Have you ever read about Apple's core values?* 2016). What these corporate success stories have in common is that they did not just happen because of great strategies or colorful CEOs, but also because of organizational cultures that were different from their competitors (Slater 1999; Liker and Hoseus 2008).

Organizational culture is the tacit social order of a company, "the way things are done around here," that shapes attitudes and behaviors in wide-ranging and durable ways (Manley 2008; Groysberg et al. 2018). While literally hundreds of definitions for the term culture exist across a variety of disciplines and agreement on specifics is sparse among

The Digital Transformation of Logistics: Demystifying Impacts of the Fourth Industrial Revolution,
First Edition. Edited by Mac Sullivan and Johannes Kern.
© 2021 by The Institute of Electrical and Electronics Engineers, Inc.
Published 2021 by John Wiley & Sons, Inc.

them, some common attributes are typically accepted[1] (Baldwin et al. 2006). Culture is *shared*, a group phenomenon that cannot exist solely within a single person. It resides in joint behaviors, values, and assumptions, commonly experienced through unwritten rules. Culture is *pervasive*, which means it permeates every aspect of a company. It is manifested, for instance, in collective behaviors, physical environments, group rituals, visible symbols, stories, and legends. It is also *enduring*, which means it can direct thoughts and actions of the members of an organization over the long term. Finally, culture is *implicit* (Groysberg et al. 2018). Psychological dispositions, perceptions, and motivations are not visible, which makes it sometimes even difficult for members of an organization to understand their very own culture (Ruhly 1976; Hennestad 1990). Colleen Barrett, former president of Southwest Airlines, summarized that "culture permeates every aspect of our company. It is our essence, our DNA, our present, and our future (Barrett 2008)."

Organizational culture is an important driver of firm performance. Researchers linked features of organizational culture with a range of organizational performance outcomes, such as positive employee attitudes, innovation and financial effectiveness (Hartnell et al. 2011). For instance, in a study conducted among franchise automobile dealerships in the United States, Boyce et al. could demonstrate a causal culture–performance relationship, where the right company culture led to higher customer satisfaction rating and vehicle sales (Boyce et al. 2015). This is also acknowledged by practitioners. In a survey among over 2000 executives and employees working across several countries, 84% agreed that their organization's culture is critical to business success. In another, 88% of American employees believed that a distinct workplace culture is important to business success (McShane and Von Glinow 2017). Lou Gerstner, the former chairman and CEO of IBM, highlighted the importance of company culture when explaining how he turned the company from the brink of bankruptcy to exceptional business success. "Until I came to IBM, I probably would have told you that culture was just one among several important elements in any organization makeup and success - along with vision, strategy, marketing, financials and the like. (. . .) Since then, I came to see that culture isn't just one aspect of the game; it is the game (Gerstner 2003)." However, to be an effective performance driver, company culture has to fit to the strategic initiatives of a company. Culture and strategy must be complementary so that they support the same mission and hence mutually reinforce each other (Ford et al. 2008; Hartnell et al. 2011). This poses especially a challenge once companies adjust their strategy, which is a must in today's Industry 4.0 age, where the market landscape is coined by constant disruption (cf. the chapter by Sullivan in this book) (Schroeck et al. 2019). Harvard Business School professors Nitin Nohria and Michael Beer explain that "not since the Industrial Revolution have the stakes of dealing with change been so high. Most traditional organizations have accepted, in theory at least, that they must either change or die (Beer and Nohria 2000)." This is particularly relevant for the logistics and supply chain management sector, where the digital transformation will lead to sensors placed in everything, networks created everywhere, anything becoming automated, and everything getting analyzed (Alicke et al. 2016). Unsurprisingly, many logistics service providers (LSPs) are

---

1 In their review of definitions for "culture," Baldwin, Faulkner, and Hecht even suggest that the term is "an empty vessel waiting for people - both academicians and everyday communicators - to fill (. . .) with meaning (Baldwin et al. 2006)."

working on updating their strategy, which correspondingly may demand a change in their culture (Langley Jr 2019).

Changing culture is extraordinarily challenging as values, power groups, and deeply rooted structures are difficult to access for influencing, and employees do not respond predictably to change efforts (Alvesson and Sveningsson 2015). However, under certain conditions, with sufficient resources and with the right practices, organizational culture can be changed (Palmer et al. 2006).

Our aim is to understand which practices companies use today to drive their organizational culture change (OCC) efforts.[2] To this, we conducted qualitative expert interviews with 10 CEOs and senior HR managers of multinational enterprises in Singapore, Thailand, Australia, and the United States between January and April 2020. This method is particularly suitable for explorative research into a complex topic (Gläser and Laudel 2009; Bogner et al. 2014). We triangulated our findings with a case study from the Australian transportation and logistics company Toll. Case studies are a useful methodology to answer questions where in-depth research is needed using a holistic lens (Yin 2017). The remaining part of this chapter has been organized in the following way: In the first section, business practices that must be aligned with the company strategy are discussed. This includes purpose, vision, and mission, values and beliefs, recruitment and placement, training and development, rewards and recognition, performance management and feedback, and artifacts. Second section then lays out the case study. Here, the company's background, the reason for its OCC initiative, its new purpose and values, and the way performance management was used to drive changed behavior are explained. The chapter ends with a summary of the findings and a conclusion that OCC actually happens when certain business practices are aligned with a new strategy.

OCC Business Practices

Strategic Matching

Fundamental to success is a clear sense of direction, a purpose and a reason to exist (Munro-Faure and Munro-Faure 1996). This requires a clear strategy, a decision how a company wants to differentiate itself from the competition (Porter 1998). Strategy is the direction and scope of an organization over the long term, which achieves advantage in a changing environment through its configuration of resources and competences with the aim of fulfilling stakeholder expectations (Johnson et al. 2008). Companies cannot be everything to everyone; if a company decides to go north, it cannot go south at the same time (De Flander 2013).

Generically, companies can pursue either a cost leadership or a differentiation strategy (Porter 1998). For LSPs this translates into two business models, one promising value by effective and efficient processes of the service provider and the other promising value by providing know-how and impulses for innovation for the processes of the client (Prockl

---

2 Often, an underlying change management model or strategy is used, cf. the chapter by Teubenbacher in this volume. For a simple comparison of common approaches and models of change management, cf. (Belyh, 2015).

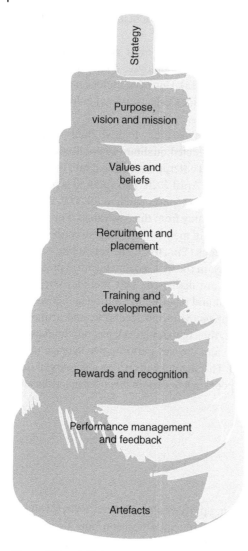

Figure 20.1 OCC through alignment of business practices with strategy.

et al. 2012). However, on an operational level, manifold strategies can be adopted to create value for customers. For example, Marchet et al. identified 30 strategies, such as an increase in asset flexibility by reallocating warehouse staff across multiple sites, exploitation of supply chain synergies, or partnering up with research institutes to introduce innovation (Marchet et al. 2017).

Once a company decides which strategy to pursue, its organization's culture must be aligned to make it work effectively (Bashforth 2019). Managers need to consider the fit between the strategy and organizational culture so that they support the same mission and mutually reinforce each other (Ford et al. 2008; Hartnell et al. 2011). In the spirit of Alfred Chandler's "structure follows strategy," similarly "culture follows strategy (Doppler and Lauterburg 2013)."

This is also why OCC activities are often triggered by strategic changes (Alvesson and Sveningsson 2015). In our interviews we found that OCC projects typically had been launched to address one or more strategic issues. There were situations where the strategy of a company was simply not working anymore or cases where a merger or acquisition made it essential that two cultures were aligned. However, our informants also shared cases where companies embarked upon a change journey when a new CEO in essence simply wanted to make his mark and show that he was different from his predecessor.

Our study substantiated that OCC is happening through a variety of business practices. They are purpose, vision and mission, values and beliefs, recruitment and placement, training and development, rewards and recognition, performance management and feedback, and artifacts. This multiplicity of practices jointly influences a company's ability to change whereby as illustrated in Figure 20.1, like the stick in a stack tower, the strategy is holding and arranging all together (Waterman Jr et al. 1980).

If a company wants to conduct OCC activities, i.e. "moving the stack tower," all elements must move together. Throughout the course of our investigation, we also encountered an unsuccessful change project. Here, in a crisis situation, the "hard part" of change had been done through restructuring and headcount reductions, but when it was time to address softer items, it didn't get done due to a lack of conviction and/or planning.

Purpose, Vision, and Mission

Experts in our interviews stressed that the clarity of a revised purpose, vision, and mission are especially useful to make OCC successful. They deemed the change of those as a helpful ingredient for change. Companies must have a sound *purpose*, a legitimate reason for being there (Schabracq 2007). This raison d'etre should not be about a company internal goal, but instead explain why this company exists and how it is making a difference for someone else. For example, the CEO of REA Group, a digital advertising company that operates Australia's leading property websites, stated that the purpose of his company was "to make the property process simple, efficient, and stress free for people buying and selling a property (Kenny 2014)." The online service LinkedIn exists "to connect the world's professionals to make them more productive and successful (Bresciani 2019)." The BBC explains that its purpose is "to enrich people's lives with programmes and services that inform, educate and entertain (Bresciani 2019)." And the beverage firm Coca-Cola Company sets out "to refresh the world. To inspire moments of optimism and happiness. To create value and make a difference (Bresciani 2019)." All these company purposes clearly communicate why these companies exist and how they are contributing to individuals and to society. Such powerful statements explain to employees how they are making a difference, which inspires them to do good work. They have to feel it (Kenny 2014). Different from the purpose is a company's *vision*, sometimes also referred to as strategic intent. It is the desired future state of a company, its aspiration (Johnson et al. 2008). For instance, the communications equipment provider Ericsson wants to be "the prime driver in an all-communicating world (Kenny 2014)." Finally, the *mission* describes what business the organization is in and where it wants to be in the future (Kenny 2014). For example, the biopharmaceutical company Gilead explains that they "discover, develop and commercialize therapeutics that advance patient care (Bresciani 2019)." The United Nations has a mission of "maintenance of international peace and security (Bresciani 2019)." Google wants to "organize the world's information and make it universally accessible and useful (Bashforth 2019)." And a consulting firm might state that they are "in the business of providing high-standard assistance on performance assessment to middle to senior managers in medium-to-large firms in the finance industry (Kenny 2014)."

Values and Beliefs

Espoused beliefs and values include ideals, goals, values, aspirations, ideologies, and rationalizations (Schein and Schein 2016). Especially values are a critical component, which is why business leaders often want to work with them (Reger 2006). In values-based approaches, the values themselves are often brief and clarified through more detailed explanations or translations in behaviors (Schabracq 2007; Alvesson and Sveningsson 2015). Once created, the values must be communicated. Companies may identify a list of changes needed to embed values, e.g. recrafting their competency

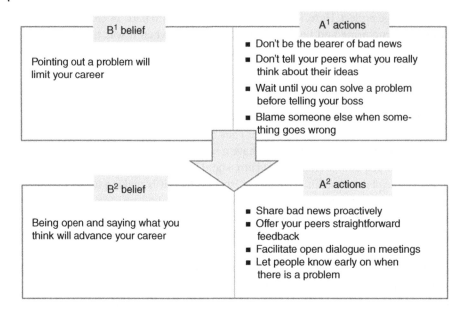

Figure 20.2 Change of beliefs leads to change of actions. *Source*: Connors and Smith (2011). © PENGUIN RANDOM HOUSE, INC.

definitions and subsequently incorporating them in their performance appraisal system and training programs. Companies that work with beliefs realize that they can greatly enhance their success in bringing about OCC by increasing awareness about beliefs. There is a simple yet powerful relationship between the beliefs people hold within an organization and the actions they take. It is based on the conviction that if you change people's beliefs about how they should do their daily work (B^1) and help them to adopt new beliefs (B^2) that you want them to hold, then they will produce the actions (A^2) that you want them to take (Connors and Smith 2011). Figure 20.2 shows an illustrative example.

It will be clear that a shift from B^1 to B^2 is beneficial and necessary to change actions from A^1 to A^2. Both values and beliefs are based on the conviction that successful OCC requires a focus on actions "below the waterline."

A different and perhaps complementary approach to values and beliefs is to focus on habits. Many people might be familiar with the quote from the American writer Will Durant: "We are what we repeatedly do. Excellence, then, is not an act but a habit (Durant 2012)." Habits are response dispositions that are activated automatically by context cues that co-occurred with responses during past performance. Previous studies showed that much of everyday action is characterized by habitual repetition (Neal et al. 2006). While culture is elusive, it is also a series of habits at an individual, team, and organizational level (Randel 2016). When culture needs to be changed, it is important to focus on the tactics to translate the elusive into something tangible and concrete instead of getting lost into shaping mythical elements. Tinkering with lists of cultural values and perfect wordings does not mean much if people cannot relate to it and do not know how to live it. Culture change as a series of habit shifts stands a better chance of success. One small change in a few teams within your organization could lead to thousands of interactions shifting and bring you closer to the culture you want. These changes are not simple, but practical approaches such as this can help to move the needle. The American social scientist Brian Jeffrey Fogg (2019) makes a similar point in that the key to changing behavior is the opposite of what we may think (Fogg 2019). It is not about systems, processes, or individual will power. We often want to change but aim too high or think too big, which makes us fail. He argues that it is about starting small and making success feel good. By creating many improved small habits, we can achieve something big in a sustainable way. Action focused on changing habits could therefore also be helpful in OCC.

Recruitment and Placement

Once the new vision and values or beliefs of an organization are clear, a company can design screening processes that will help to identify the right people. The values become a lens in interview and hiring decisions. It also means that applicants are given a clear image of the values and understand the expectation that joining means that they need to commit to applying the values of the organization in their behavior and decision making (Bashforth 2019). Most companies focus on hiring based on the skills needed for a specific job, but firms with a strong culture hire for how well the person will fit the company. This does not mean that they will ignore a candidate's abilities, but instead recognize that to really contribute over a longer period, a person must feel comfortable in the organization. Candidates might be "right" for a job but wrong for the organization (Dessler 2017). To operationalize the intended target values, companies should define clear culture targets that are specific and achievable. While, for instance, "we value our customers" is ambiguous and poses the risk of inconsistent hiring choices, "we build genuine and positive relationships with customers; we serve our customers with humility; and we act as ambassadors for our rich brand heritage" leaves less room for interpretation (Groysberg et al. 2018). Key abilities are those that help someone grow, change, and develop to meet changing business needs. These abilities could include a willingness to learn and try new things, to be part of a team, and to accept responsibility. Southwest Airlines, for example, hires people not for their first job that they will hold, but for their potential to move up in the organization in the course of their career (O'Reilly and Pfeffer 2000). To hire for cultural fit, the recruiting process should be largely driven by employees who know both the job and the required values best and can accurately convey the expectations required to succeed. The process itself invariably involves multiple steps and enough time so that applicants get a good sense what the culture is and are given plenty of opportunity to bow out (Reger 2006).

Training and Development

A key ingredient of leading an organization's culture is training and development (Liker and Hoseus 2008). For example, Toyota is a learning organization that has gathered a lot of information about how to do each of their jobs but also how people can effectively work together, communicate, and solve problems. They feel that it would therefore be foolish to assume that people coming in the door will simply know everything about how to do the job and how to acclimate to the Toyota culture, so they must be taught both (Liker and Hoseus 2008).

Also, the beauty company L'Oreal has a comprehensive onboarding program that takes several years and includes special training and roundtable discussions, meetings with key insiders, on-the-job learning, individual mentoring, and other experiences such as site visits (Dessler 2017). Southwest Airlines had a corporate university where 25 000 people are trained each year. The emphasis is on doing things better, faster, and cheaper, understanding other people's jobs, delivering outstanding customer service, and keeping the culture alive and well. For instance, all employees begin by attending a program called "You, Southwest, and Success." This gives all newcomers, from pilots to ramp agents, a crash course in Southwest's history, reputation for impeccable service, culture, and how they can contribute. From the beginning, the emphasis is on getting the employees enthused and excited. A continual concern is to help employees avoid complacency. Some trainers are worried that positive press accounts do not help and may lull employees into believing that they are better than they really are (O'Reilly and Pfeffer 2000). One of our interviewees shared a case from his company where during an OCC program, reskilling was deemed important to jointly get ready for the digital age. Therefore, they decided to train their top group on-site at a business school, the next 600 people virtually, and subsequently another group of 600 virtually. They also launched a next-generation program as part of this project, where young people could share their ideas. This led to 40 people being invited to their headquarters in Norway to present their ideas and be stimulated to keep up the good work. A situation where the effect of training can be very accurately measured is in a call center. The below example was shared during another of the interviews. It is about an OCC project in a financial services (FS) organization that wanted to improve customer satisfaction while maintaining performance in terms of efficiency. They focused on skill development of the customer service representatives (CSRs), whose performance is measured in customer satisfaction and numbers of calls/average duration of the calls and requires listening skills, customer orientation, problem solving skills, and efficiency. The CSRs were trained in dealing with complaints first in a classroom, then through role-play, and subsequently initially through listening in on calls, followed by taking live calls themselves while a supervisor listens in and gives them whenever necessary feedback. In such case, the human-centered approach (observing, watching, really understanding what is happening) worked very well. It enabled the company to provide additional training based on actual data that was received (number of calls, customer satisfaction ratings), combined with a supervisor's qualitative observations. As part of this process, CSRs are not just taught to respond to various kinds of behaviors, but also to recognize what emotions unreasonable customers stir up within themselves in order to better deal with these situations. Our interviewee concluded that "if you do this with hundreds of CSR's, then you can directly influence how the organization displays an important value."

Rewards and Recognition

Another important business practice in OCC is to define a compensation plan that advances the company's new strategy. Compensation packages should lead to employee behaviors that the company needs to achieve specific business goals (Kaplan 2005). The traditional approach to pay is that salary increments are based on three factors: job classification, length of service, and performance. All jobs are classified into grades, and (base) salaries of people will fall within the range of their grade. Their performance will determine how fast they grow within their range. People that stay longer in roles will become better at it but often regardless of their performance will grow within their range and in due course get to the maximum of the range. Rewards include traditional pay, incentives, and benefits but also rewards in the form of more challenging tasks (job design), career development, or recognition (Dessler 2017). A common trend in performance-driven cultures is to move away from paying for years in role and to move predominantly to paying for performance. Base salaries are kept moderate, and superior performance will be recognized in the form of bonuses (Ashby 2012). However, employee engagement is impacted heavily by both material and nonmaterial rewards (Scott and McMullen 2010). Studies showed that base pay and benefits are only weakly correlated to employee engagement. Instead, intangible rewards are having high impact on engagement and performance, when combined with base salary and short-term incentives or bonuses (Scott and McMullen 2010; Schuster and Zingheim 2013). The mechanism by which people are incentivized can be further used to deepen the sense of what matters. If team performance is to be valued more, then a bigger percentage of reward could be factored off teamwork and less off pure individual achievement. Annual awards or monthly staff recognition can be based on specific criteria. Those criteria can be defined to be certain values or specific behaviors you want to emphasize and celebrate. A demonstrable celebration of that element to the wider organization is a symbolic confirmation of its importance (Bashforth 2019).

Performance Management and Feedback

Performance management, the continuous process of setting objectives and identifying, measuring, and developing the performance of individuals and teams, is important to ensure that each employee's performance is relevant in terms of a company's new strategic goals (Dessler 2017). For instance, when Dan Vasella became CEO of Novartis, he clearly explained his expectations toward employees and ensured that the company's new values were cascaded through the organization. Clear metrics for gauging performance were set up as he believed that a good performance management system would help keep employees focused on the right things (Lorsch and McTague 2016). Effective performance management is also considered a key success factor for the culture at SAS Institute, a developer of analytics software that is consistently named as one of the best places to work in the United States. David Russo, the company's former SVP HR who also oversaw its growth from 50 to 5000 people, explained the simple philosophy behind SAS's use of performance management. People should be given the tools to do their jobs, and then management should step aside and let them do it while holding them accountable. The company believes in setting high expectations for both performance and conduct, which become self-fulfilling, and in giving people a lot of freedom to determine what they feel works best to meet these objectives (O'Reilly and Pfeffer 2000). Bashforth stresses that how people behave and apply the values in decision making becomes part of performance. It is factored into the overall performance rating for the individual that year, showing that not just what you deliver is important, but also how it is delivered, culturally in line with the organization values (Bashforth 2019). For example, the mountain resort company Vail Resorts relies on a culture that is incorporating joy. Therefore, in their annual review, it is evaluated how well each employee integrates fun into the work environment and displays supporting behaviors, such as being inclusive, welcoming, approachable, and positive (Barsade and O'Neill 2016). A key element that gets attention in any training of performance management skills is the ability to give effective feedback. Connors and Smith refer to this as focused feedback, which should be both appreciative and constructive. Appreciative focused feedback lets people know you value their demonstration of the cultural beliefs. It reinforces the thinking and behavior needed to move the cultural boulder forward. Providing appreciative focused feedback in a timely way not only clarifies the desired cultural behavior but also supplies the repetition needed to reinforce the desired actions (Connors and Smith 2011). Part of performance management is also to determine when people are ready to be promoted. For this it is important to not only look at the results that need to be achieved in a job, but also the intangible characteristics of the person that is being considered for the promotion, which includes factors such as judgment, problem solving ability, technical competence, creativity, and the level of trust and respect generated by the person (Ashby 2012). All company values should play a role[3].

Artifacts

OCC can be supported by artifacts, visible structures/processes, and observed behavior (Alvesson and Sveningsson 2015). They are phenomena that can be seen, heard, or felt. Artifacts can include visible products, such as architecture, language, technology, artistic creations, clothes, manners of address, emotional displays, myths and stories told about the organization, a published lists of values, and observable rituals or ceremonies (Schein and Schein 2016). Examples are prizes that were awarded or won in various competitions or even ordinary objects, such as a bouquet of flowers at a reception area. While artifacts are easy to observe, they are difficult to decipher. That means that observers cannot obviously reconstruct what artifacts mean to a given group (Schein and Schein 2016). Artifacts send a value statement, for example, an open floor environment in an office with glass walls and ping-pong tables might send the message that the company values openness, creativity, and fun (Causey 2017).

Amazon offers to pay full-time associates at its fulfillment centers up to 5000 USD to leave the company. Jeff Bezos, its CEO, explained "that the goal is to encourage folks to take a moment and think about what they really want." And a spokesperson adds that "we want people working at Amazon who want to be here. (. . .) In the long-term, staying somewhere you don't want to be isn't healthy for our employees or for the company (Umoh 2018)." Formal rituals and celebrations are ways in which a company celebrates key events under special circumstances. They may be used when someone celebrates an anniversary or in cases where people are promoted or retire (Trice and Beyer 1993). Some companies celebrate annual dinners, an annual sports day, or events around Chinese New Year or Christmas. Stories and myths are useful to "explain the unexplainable" but also to sustain the company self-image, how to get things done, and how to handle relationships (Schein 1996; Hatch and Schultz 2004; Schein and Schein 2016). For example, a story about the home appliances and consumer electronics manufacturer Haier about its quality

---

3 Dependent on the role, leadership abilities should also be considered as they are key to obtaining the willing cooperation of others to achieve desired results.

mindset goes as follows. In 1985, the company's Chairman Zhang Ruimin received a letter from a customer complaining about Haier's product quality. So he went to the warehouse, ordered a quality inspection, and was shocked to hear that 20% of the refrigerators in stock failed the test. Instead of following the common approach at that time to sell the problematic products to employees at a discount price, Zhang decided to smash them in front of his staff. Considering that the price of a refrigerator at that time was roughly equivalent to two years of their salary, many employees broke out in tears. While this might sound drastic, it became a symbol of what Haier wanted to stand for in terms of quality (Nothhaft 2018). Today, Haier is the world's largest refrigerator company (*World Refrigerator Industry Enters 'Chinese Era,' Haier Takes the Lead* 2018). Likewise, symbols and symbolic action act as triggers and remind members of a culture of its norms and beliefs. They can also be used to indicate status, which might include office décor, IT equipment, clothing, and access to special sections of the office (e.g. the executive washroom). Status symbols provide messaging to use the right behavior toward hierarchy in an organization (Brown 1995). Language is also an important vehicle to convey messages. During one of our expert interviews that focused on the OCC program at a Dutch MNC, terms like "fighting units," "going galactic," and the name "Nexus" for an onboarding tool were used. This seemed to convey an image of excitement but also the need to do battle and to work in a new way, unlike anything that was done before.

OCC at Toll

This case describes the OCC initiative that Toll started in 2018. Data sources were interviews with experts from Toll and archival records in the form of internal and publicly available company-related documents.

Background

The Toll organization has a profound history. Founded in Newcastle, Australia, in 1888 by Albert Toll who began hauling coal by horse and cart, it had developed at the end of his life in 1958 to a fleet of trucks in five locations. It was purchased in 1959 by National Minerals and became part of mining conglomerate Peko Wallsend, which used the Toll business for the transport activities of its entire operations. The company subsequently underwent a name change to Toll-Chadwick, as its new owners sought to integrate its businesses and expand into containerized shipping. By the mid-1980s, Toll-Chadwick had grown into one of Australia's biggest transport operations outside of the capital cities.

In 1986, Toll was bought by its management team, led by Managing Director Paul Little. They were armed with a clear vision of growing the business through the acquisition of small strategically placed transport companies. Toll's growth and expansion accelerated when this team led Toll's 1993 listing on the Australian Stock Exchange (ASX). After its listing Toll responded to increasing customer demand for complete end-to-end logistics solutions by progressively building its reach and service capabilities via a program of expansion and strategic acquisitions both in Australia and internationally. When Toll was listed in 1993, it was valued at A$63 million. By the time Little announced his retirement in late 2010, Toll's worth had grown to almost A$3.8 billion. Little was succeeded as CEO by Brian Kruger, who had joined Toll in 2009 as its CFO. Kruger continued the process that Little had started, and Toll expanded further internationally. In 2014 Toll had become one of Asia Pacific's leading providers of integrated logistics services, employing around 40 000 people across 1200 locations in more than 50 countries, with annual revenues of A$8.7 billion. Early 2015, Toll's Board announced that it had accepted a proposal from Japan Post (JP), one of the world's largest postal and logistics companies, to acquire all of Toll's shares. JP was looking to Toll to spearhead its international logistics operations. On 28 May 2015, Toll formally became a division of JP.

Following the acquisition by JP, in January 2017, John Mullen was appointed as chairman of the board, and Michael Byrne as Toll's CEO. This team was given little time to get settled because they needed to turn around Toll as profits tumbled due to a delayed economic recovery in Australia and a decline in Toll's domestic business as prices had been cut to maintain volumes in a highly competitive market environment. During the next few years, Michael Byrne led a significant restructuring to improve Toll's profitability, which he followed with significant investments in Toll's infrastructure and systems. He also rightly pushed for increased focus on safety and compliance, areas that during the years of expansion had not been given sufficient attention. Unfortunately, the group's results did not improve the way they should following the significant investments that had been made, and what Byrne had not gotten around to is changing the culture of Toll by retaining the good things that were unique to Toll while maintaining the increased levels of discipline and focus that he had tried to institutionalize. He announced his departure late 2019 and Thomas Knudsen, until then CEO of Toll's Forwarding division, as his successor, which became effective on 1 January 2020.

As to values, it was Brian Kruger who had launched "The Toll Way" in 2013, which was refreshed in 2016. It comprised a primary purpose, mission, vision, core beliefs, and values, all of which were captured in "Toll's Code of Practice – Helping you living the Toll Way." The code of practice listed five values: integrity and trust, safety, continuous improvement, teamwork, and openness and transparency. Until mid-2018 the Toll Way had been left largely unquestioned, but at that time a project team started to look at ways in which both Toll's brand and its values could be brought more in line with what the organization needed going forward.

New Purpose, Values, and the Case for Change

In September 2019, a new company purpose and promise were presented during an annual leadership team conference in Torquay, Australia, by Vikram Cardozo, Toll's new CHRO[4] who had started in April 2019. Toll's new purpose became "we move the businesses that move the world." Its promise became "Customers demand the best from us, so we harness the very best of ourselves—our passion for logistics, our dedication to operational excellence, and our drive for innovation—to deliver the seamless solutions they need so they can keep doing, moving and growing."

---

4 Chief human resources officer, in some companies also called chief people officer.

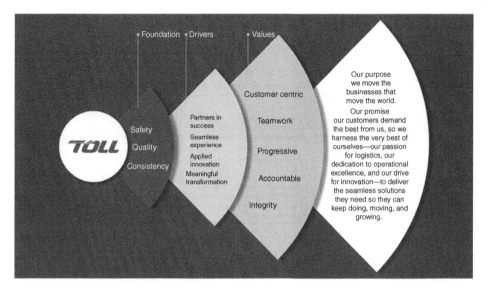

Figure 20.3 Toll purpose and promise. *Source*: Slides Toll Leadership Conference 2019 – Making it Count Our values and behaviors – presented by Vikram Cardozo. © 2019, Toll Holdings Limited.

Based on an extensive series of interviews with customers, employees, and other stakeholders, and with the help of an external agency, Toll had decided to revise both brand and values and did so by developing and/or redefining not only its company's purpose and promise but also its foundational elements, brand drivers, and its values. Figure 20.3 shows how all these elements fit together.

Perhaps as important as these new elements, awareness grew that there was a case for OCC that needed to be made. This was based on three focus areas (or lack thereof) during the prior years, which demanded a response to mobilize the collective energy of Toll's employees to move forward and be successful:

1) The focus on large infrastructure projects, new systems, and safety and compliance, together with strong central control whenever results were disappointing, had led to a stifling of entrepreneurship and initiative in the field and a need for more decentralization.
2) Too much emphasis on cost and EBIT had diminished Toll's customer focus.
3) A lack of appreciation for expertise in areas like customer service, operational excellence, and cash management had created issues that undermined day-to-day operations.

These three elements are illustrated in Figure 20.4.

The choice of Toll's new values fitted perfectly with what was required based on the listed issues, and especially the adoption of the new values "accountability" and "customer" was a vital step in the right direction. The company realized that a strong focus on accountability makes it possible to delegate responsibility and empower managerial levels. Instilling strong customer focus and awareness throughout the whole organization is a second essential element to align activities and drive behavior and essential for any logistics company to be successful.

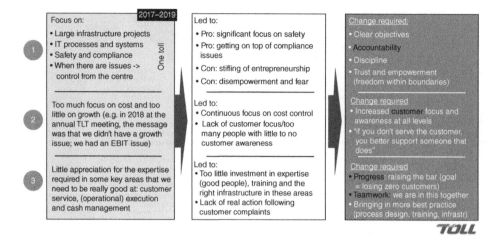

Figure 20.4 Business reasons for Toll's organizational culture change. *Source*: Training materials new performance management system, Toll Group (2020). © Toll Holdings Limited.

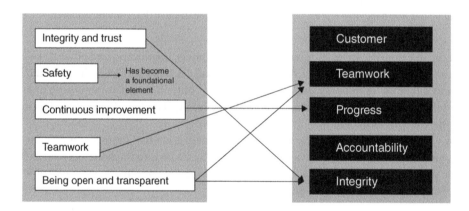

Becoming highly customer focused is essential for our future success

Figure 20.5 Differences and similarities in old and new values.

Toll's new values are (Figure 20.5):

- Customer: We put our customers at the center of everything.
- Teamwork: We work together to succeed and strive to contribute beyond our own responsibilities.
- Progress: We are passionate about progress. We challenge the status quo and pursue simplification and improvement opportunities with urgency.
- Accountability: We are accountable to one another. We take personal responsibility and deliver upon our commitments.
- Integrity: We do the right thing the right way, even when nobody is watching.

Shortly after their introduction to the top team in Torquay, the values were officially launched, and since then they have become a cornerstone and are regularly referred to as an important source of inspiration and as compass that helps to determine how to run Toll's business.

Changing the Way to Manage Performance

With the case for change clear and the values settled, the next step was working the new values into Toll's performance management system. This was done by adopting the principle that an employee's performance on objectives (the "what") and values (the "how") should count equally in determining an overall performance rating. For each value behavioral descriptors as well as indicators that would point at exceed, meet and development needed were developed. Next the value descriptors were translated into the various languages that are in use within Toll, and as pilot Toll arranged to have 750 employees spread across the various businesses, functional areas and geographies try out the new system, which was supported by training of all the managers and pilot participants.

The new rating scales that were developed for both objectives and values were linked to the existing (prior) rating scale to enable a smooth transition. The overview in Figure 20.6 shows how new 3-point rating scales on performance objectives and values were linked to the existing 5-point scale through a 9-block grid.

The pilot showed that employees will take the behavioral/cultural aspects of their roles (the "how") more serious if they are separately assessed on it and if they are convinced that this plays a significant role in determining their overall rating. Toll's performance management process was further improved through more emphasis on calibration to ensure a balanced outcome, consistent use of the system, and overall fairness.

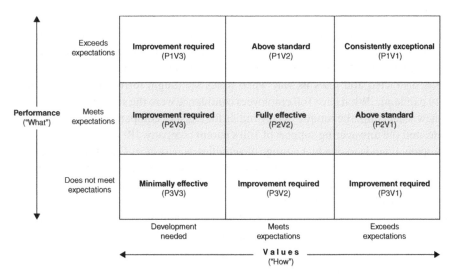

Figure 20.6 9-Block ratings. *Source*: Training materials new performance management system, Toll Group (2020). © Toll Holdings Limited.

Tone from the Top (Leadership)

With the transition of CEOs at Toll and a few additional changes within the top team that happened shortly thereafter, a more personable, down-to-earth style of management was introduced.

A distinct act was that during the difficult period of the COVID-19 pandemic, which had a significant effect on the company's results, Toll's top team led by example. It decided that to help weather the storm, Toll's CEO would take a 40 percent pay cut for a 6-month period. The global top team would take a 30 percent pay cut, Toll's leadership team (c. 100 executives spread across Toll's footprint) committed to a 20% pay cut, and managers and specialists as of a certain grade level would do 10 percent, all on a voluntary basis. This sign of solidarity and willingness to make a personal sacrifice on behalf of the company fitted with the new Toll style of leadership and made an executive declare that upon seeing the headline in the Australian Financial Review "Toll CEO takes 40 pc pay cut," he felt proud to be part of Toll (something that he had not felt for a long time) (Wiggins and Sullivan 2020). What followed were regular video messages where people were thanked for their efforts and confidence was expressed in the organization's ability to get through this difficult period and be successful again. The fact that Toll's CEO and CHRO are now based in Asia, so outside of Australia, is another change that is more than symbolic and a clear sign that JP wants Toll to focus its growth ambitions on Asia.

Lastly, during the difficult COVID months, Toll continued its emphasis on the need for change through a program with the help of Harvard Business School, whereby 100 executives are taken through a modular program that focuses on driving customer focus, enterprise leadership, and leading change.

Next Steps

Changing an organization culture is a 3–5-year process, and at the time of writing, Toll was at the early stages of its journey. This case is, however, particularly worthwhile to study because the initial steps of OCC are of great importance. It can easily happen that an organization gets distracted and loses its way when times are tough, for instance, during the COVID-19 pandemic. What gave Toll employees confidence were the sincerity and capability of its new leadership, the unique opportunities for logistics companies during the coming decade, and the unwavering support of Toll's parent company, JP.

The company intends to conduct a range of activities as part of its OCC program during the next 12 months. It will ensure that all main HR processes are aligned to the new values (the way Toll recruits, onboards, appraises, promotes, remunerates, and trains and develops staff), which will provide consistency of message about what is important for Toll and its employees to be successful. It will launch values awards and appoint value ambassadors to make sure that it reaches all part of the organization. It will share lessons learned and stories that will regularly make the values come alive. It will regularly (annually) measure progress and arrange focus groups to discuss the outcomes. Lastly, it will review symbols, rituals, and artifacts and ensure that both explicit and implicit communication messages are aligned.

Conclusion

This chapter set out to summarize currently used business practices for OCC. Data was gathered through a mixed method approach with qualitative expert interviews among 10 CEOs and senior HR managers of multinational enterprises in Singapore, Thailand, Australia, and the United States, triangulated with a case study at the Australian transportation and logistics company "Toll."

Fundamental to OCC is that there is a clear direction in the form of a compelling strategy. As companies cannot be everything to everyone, they must decide to go north or south. For an LSP this typically means promising value through effective and efficient processes or promising value by providing know-how and impulses for innovation for the processes of the client. Then, the organization's culture must be aligned to make it work effectively – as culture follows strategy. In the end, strategy and culture must "fit" so that they can support the same mission and mutually reinforce each other.

While various approaches and models of culture change are commonly used, OCC actually happens when a range of business practices are aligned with the new strategy. Jay W. Lorsch and Emily McTague from Harvard Business School explain that cultural change is "what you get when you put new processes or structures in place to tackle tough business challenges (Lorsch and McTague 2016)." Throughout our study, we have identified a variety of business practices that, when reworked, lead to new behaviors.

Revised *purpose, vision, and mission* are particularly useful to make OCC successful. The purpose explains why a company exists and how it is making a difference for someone else, the vision states the desired future state of a company, and the mission describes what business the organization is in and where it wants to be. Also, working with *values and beliefs* can greatly enhance success in bringing about OCC as changed beliefs lead to changed actions. In a cultural change initiative, it is important to focus on the tactics to translate the elusive into something tangible and concrete instead of getting lost into shaping mythical elements. Particularly, by creating many improved small habits, success can be achieved in a sustainable way. Through *recruitment and placement* measures, it is ensured that people are identified that fit to the aspired vision and values of the organization. While abilities remain relevant, companies recognized that to ensure that persons will contribute over a long period, they must feel comfortable in the organization. While candidates might be "right" for a certain job, they could be wrong for the organization. Another important practice is *training and development*. Leading companies like Toyota, L'Oreal, and Southwest Airlines use comprehensive onboarding programs to enthuse and excite newcomers and teach them the company values as well as necessary job skills. *Rewards and recognition* can help to advance a company's strategy. For example, compensation packages should encourage employee behaviors that are required to achieve specific business goals. Equally, as employee engagement is impacted not only by material but also nonmaterial rewards, things like annual awards or monthly staff recognition activities should emphasize and celebrate certain values or specific behaviors. *Performance management* is crucial to ensure that each employee's objectives and performance are relevant in terms of a company's strategic goals. How people behave and apply values in decision making must become part of performance, showing that not just what is delivered is important, but also how it is delivered. Finally, OCC can also be supported by *artifacts*, phenomena that can be seen, heard, or felt. Artifacts such as an open floor environment in an office with glass walls and ping-pong tables send a value statement. In addition, stories and myths are useful to explain something that is unexplainable and to sustain the company self-image, how to get things done, and how to handle relationships. When Haier's chairman destroyed substandard refrigerators worth multiple annual salaries in front of his staff, a symbol of what Haier wanted to stand for in terms of quality was set.

Companies that are actively managing their culture are outperforming others that do not. An organization culture that is fitting to the company's strategy leads to improved business results such as positive employee attitudes, innovation, financial effectiveness, and/or higher customer satisfaction, while one that is not fitting can be a significant liability (Groysberg et al. 2018). Albeit difficult, it is possible to change the culture of a company with a long tradition, as shown, for example, in the case of Toll. When the company's CEO visited employees and customers in the beginning of 2020, the OCC initiative was gaining momentum, and he could proudly share that "people genuinely are being very passionate – including our customers – about Toll and our opportunities. And they want us to succeed (Toll Group 2020)."

Key Takeaways

- Organizational culture, the tacit social order of a company or "the way things are done around there," can lead to a lot of benefits such as positive employee attitudes, increased innovation, or higher financial returns.
- However, to lead to this higher performance, a company culture has to be complementary to a firm's strategic initiatives. That means that in the case of strategic changes, the organizational culture may have to be changed as well. This is particularly relevant for the logistics and supply chain management sector, where the digital transformation will lead to drastic changes that demand strategic shifts.
- The most important activity in organizational culture change (OCC) is therefore to match the company's direction and its culture. Like the stick in a stack tower, the strategy is holding and arranging all other business practices, through which OCC is happening, together.
- The clarity of a revised purpose, vision, and mission is especially useful to make OCC successful. A company's *purpose* explains why the company exists and how it is making a difference for someone else. The *vision*, also referred to as "strategic intent," is the desired future state of a company, its aspiration. The *mission* describes what business the organization is in and where it wants to be.
- Espoused beliefs and values include ideals, goals, values, aspirations, ideologies, and rationalizations. Companies that work with them can greatly enhance their success in bringing about OCC as changed values or beliefs lead to changed actions.
- Recruitment and placement measures ensure that people are identified that fit with the new vision and values of the organization. To learn everything about how to do a job and how to acclimate to a company's culture, employees should receive comprehensive training and development.

- Rewards and recognition are also important business practices in OCC. A compensation plan must be defined that advances a company's new strategy by leading to desired employee behaviors. Equally, performance management must incorporate not just what is delivered, but also how, culturally in line with the organization values.
- OCC can be supported by artifacts that are visible structures/processes and observed behavior. They send a value statement, such as an open floor environment in an office with glass walls and ping-pong tables that might send the message that a certain company values openness, creativity, and fun.
- A good example of OCC can be found within the Australian Logistics company "Toll," which is currently undergoing OCC. Based on important reasons, they launched an OCC initiative where it developed a new purpose and values and cemented it by changing the way performance management was conducted.
- Ultimately, OCC occurs when the mentioned business practices, aligned according to the company's strategy, are adequately put in place and processed.

References

Alicke, K., Rachor, J., and Seyfert, A. (2016). *Supply Chain 4.0 – the next-generation digital supply chain*. McKinsey Supply Chain Management Practice https://www.mckinsey.com/~/media/McKinsey/Business%20Functions/Operations/Our%20Insights/Supply%20Chain%2040%20%20the%20next%20generation%20digital%20supply%20chain/08b1ba29ff4595ebea03e9987344dcbc.pdf (accessed 1 June 2020).

Alvesson, M. and Sveningsson, S. (2015). *Changing Organizational Culture: Cultural Change Work in Progress*. Routledge.

Ashby, F.C. (2012). *Revitalize Your Corporate Culture*. Taylor & Francis https://books.google.com/books?id=PyD5Puui-PIC (accessed 1 June 2020).

Baldwin, J.R., Faulkner, S.L., and Hecht, M.L. (2006). A moving target: the illusive definition of culture. In: *Redefining Culture*, 27–50. Routledge.

Barrett, C. (2008). Talking southwest culture. *Southwest Airlines Spirit Magazine* 5 (6).

Barsade, S. and O'Neill, O.A. (2016). Manage your emotional culture. *Harvard Business Review* 94 (1): 58–66.

Bashforth, K. (2019). *Culture Shift: A Practical Guide to Managing Organizational Culture*. Bloomsbury Publishing https://books.google.com/books?id=UpukDwAAQBAJ (accessed 1 June 2020).

Beer, M. and Nohria, N. (2000). Cracking the code of change. *HBR's 10 must reads on change*. *Harvard Business Review* 78 (3): 133–141.

Belyh, A. (2015). Major approaches & models of change management. https://www.cleverism.com/major-approaches-models-of-change-management (accessed 1 June 2020).

Bogner, A., Littig, B., and Menz, W. (2014). *Interviews mit Experten, Qualitative Sozialforschung*. Wiesbaden: Springer VS http://www.worldcat.org/oclc/886627996 (accessed 1 June 2020).

Boyce, A.S., Nieminen, L.R.G., Gillespie, M.A. et al. (2015). Which comes first, organizational culture or performance? A longitudinal study of causal priority with automobile dealerships. *Journal of Organizational Behavior* 36 (3): 339–359. https://doi.org/10.1002/job.1985.

Bresciani, A. (2019). 51 mission statement examples from the world's best companies. https://www.alessiobresciani.com/foresight-strategy/51-mission-statement-examples-from-the-worlds-best-companies (accessed 1 June 2020).

Brown, A. (1995). *Organisational Culture*. Pitman https://books.google.com/books?id=9paPQgAACAAJ (accessed 1 June 2020.

Causey, C. (2017). Company culture as artifacts. https://www.agilepartnering.com/2017/05/26/company-culture-as-artifacts/ (accessed 1 June 2020).

Connors, R. and Smith, T. (2011). *Change the Culture, Change the Game: The Breakthrough Strategy for Energizing Your Organization and Creating Accountability for Results*. Penguin.

De Flander, J. (2013). The execution shortcut. Performance factory. https://books.google.com/books?id=dhX9nQEACAAJ (accessed 1 June 2020).

Dessler, G. (2017). *Human Resource Management*. Pearson Higher Education (Always learning) https://books.google.com/books?id=qMUIswEACAAJ (accessed 1 June 2020).

Doppler, K. and Lauterburg, C. (2013). *Managing Corporate Change*. Berlin Heidelberg: Springer https://books.google.com/books?id=1W3mCAAAQBAJ (accessed 1 June 2020).

Durant, W. (2012). *Story of Philosophy*. Simon & Schuster https://books.google.com/books?id=suLI7RoaBEEC (accessed 1 June 2020).

Fogg, B.J. (2019). *Tiny Habits: The Small Changes That Change Everything*. HMH Books https://books.google.com/books?id=s7OCDwAAQBAJ (accessed 1 June 2020).

Ford, R.C., Wilderom, C.P.M., and Caparella, J. (2008). Strategically crafting a customer-focused culture: an inductive case study. *Journal of Strategy and Management* 1 (2): 143–167. https://doi.org/10.1108/17554250810926348.

Gerstner, L.V. (2003). *Who Says Elephants Can't Dance?: Leading a Great Enterprise through Dramatic Change*. HarperCollins https://books.google.com/books?id=tCoJsXNiZ_UC.

Gläser, J. and Laudel, G. (2009). *Experteninterviews und qualitative Inhaltsanalyse*. Wiesbaden: Springer-Verlag.

Groysberg, B. et al. (2018). The leader's guide to corporate culture. *Harvard Business Review* 96 (1): 44–52.

Hartnell, C.A., Ou, A.Y., and Kinicki, A. (2011). Organizational culture and organizational effectiveness: a meta-analytic investigation of the competing values framework's theoretical suppositions. *Journal of Applied Psychology*. American Psychological Association 96 (4): 677.

Hatch, M.J. and Schultz, M. (2004). *Organizational Identity: A Reader*. Oxford University Press on Demand.

Have you ever read about Apple's core values? (2016). *Think Marketing Magazine*. https://thinkmarketingmagazine.com/apple-core-values/ (accessed 1 June 2020).

Hennestad, B.W. (1990). 'The symbolic impact of double bind leadership: double bind and the dynamics of organizational culture. *Journal of Management Studies* 27 (3): 265–280.

Johnson, G., Scholes, K., and Whittington, R. (2008). *Exploring Corporate Strategy*. Financial Times Prentice Hall (Exploring Corporate Strategy). https://books.google.com/books?id=SmjnLQwlSdsC (accessed 1 June 2020).

Kaplan, S. (2005). Total rewards in action: Developing a total rewards strategy. *Benefits & Compensation Digest* 42 (8): 32–37.

Kenny, G. (2014). Your company's purpose is not its vision, mission, or values. *Harvard Business Review* 3: 285–306.

Langley Jr, C.J. (2019). *2019 third-party logistics study – the state of logistics outsourcing.* https://www.kornferry.com/content/dam/kornferry/docs/article-migration/2019-3PL-Study.pdf (accessed 1 June 2020).

Liker, J. (2001). *The Toyota Way: 14 Management Principles from the World's Greatest Manufacturer.* McGraw-Hill.

Liker, J.K. and Hoseus, M. (2008). *Toyota Culture: The Heart and Soul of the Toyota Way.* McGraw-Hill Education https://books.google.com/books?id=7Y6jjkmqByQC (accessed 1 June 2020).

Lorsch, J.W. and McTague, E. (2016). Culture is not the culprit. *Harvard Business Review* 94 (4): 96–105.

Manley, K. (2008). "The way things are done around here"—developing a culture of effectiveness: a pre-requisite to individual and team effectiveness in critical care'. *Australian Critical Care* 21 (2): 83–85. https://doi.org/10.1016/j.aucc.2008.02.002.

Marchet, G., Melacini, M., Perotti, S., Sassi, C. and Tappia, E. (2017). Value creation models in the 3PL industry: what 3PL providers do to cope with shipper requirements. *International Journal of Physical Distribution & Logistics Management* 47 (6): 472–494. https://doi.org/10.1108/IJPDLM-04-2016-0120.

McShane, S. and Von Glinow, M.A. (2017). *Organizational Behavior.* McGraw-Hill Education https://books.google.com/books?id=zLYrvgAACAAJ (accessed 1 June 2020).

Munro-Faure, M. and Munro-Faure, L. (1996) *The Success Culture: How to Build an Organization with Vision and Purpose.* Pitman Pub. (Financial Times). https://books.google.com/books?id=gDhZAAAAYAAJ (accessed 1 June 2020).

Neal, D.T., Wood, W., and Quinn, J.M. (2006). Habits—a repeat performance. In: *Current Directions in Psychological Science*, vol. 15(4), 198–202. Los Angeles, CA: SAGE Publications.

Nothhaft, C. (2018). Households upgrading appliances: haier. In: *Made for China*, 109–118. Cham: Springer International Publishing https://doi.org/10.1007/978-3-319-61584-4_10.

O'Reilly, C.A. and Pfeffer, J. (2000). *Hidden Value: How Great Companies Achieve Extraordinary Results with Ordinary People.* Harvard Business Press.

Palmer, I., Dunford, R., and Akin, G. (2006). *Managing Organizational Change.* New York: McGraw-Hill.

Porter, M.E. (1998). *Competitive Advantage.* New York: Free Press.

Prockl, G., Pflaum, A., and Kotzab, H. (2012). 3PL factories or lernstatts? Value-creation models for 3PL service providers. *International Journal of Physical Distribution & Logistics Management* 42 (6): 544–561. https://doi.org/10.1108/09600031211250587.

Randel, A. (2016). Changing culture: shift small habits for big wins, medium. https://medium.com/the-ready/changing-culture-shift-small-habits-for-big-wins-c408c45b3e9f (accessed 1 June 2020).

Reger, S.J.M. (2006). *Can Two Rights Make a Wrong?: Insights from IBM's Tangible Culture Approach.* Pearson Education https://books.google.com/books?id=9Qkb1D4j7-0C (accessed 1 June 2020).

Ruhly, S. (1976). *Orientations to Intercultural Communication.* Science Research Associates.

Schabracq, M.J. (2007). *Changing Organizational Culture: The Change Agent's Guidebook.* Wiley https://books.google.com/books?id=wkeBtwEACAAJ (accessed 1 June 2020).

Schein, E.H. (1996). Culture: the missing concept in organization studies. *Administrative Science Quarterly* 41 (2): 229–240. 40th Anniversary Issue.

Schein, E.H. and Schein, P.A. (2016). *Organizational Culture and Leadership*. Wiley (The Jossey-Bass Business & Management Series). https://books.google.com/books?id=l2jpCgAAQBAJ (accessed 1 June 2020).

Schroeck, M., Kwan, A., Kawamura, J., Stefanita, C. and Sharma, D. (2019). Digital industrial transformation, Deloitte LLP. https://www2.deloitte.com/us/en/insights/focus/industry-4-0/digital-industrial-transformation-industrial-internet-of-things.html (accessed 1 June 2020).

Schuster, J.R. and Zingheim, P.K. (2013). Recalibrating best practice. In: *Compensation & Benefits Review*, vol. 45(3), 134–135. Los Angeles, CA: SAGE Publications.

Scott, D. and McMullen, T. (2010). The impact of rewards programs on employee engagement. https://www.worldatwork.org/docs/research-and-surveys/survey-brief-the-impact-of-rewards-programs-on-employee-engagement.pdf (accessed 1 June 2020).

Slater, R. (1999). *The GE Way Fieldbook: Jack Welch's Battle Plan for Corporate Revolution*. McGraw-Hill Education https://books.google.com/books?id=2KUKsu-clz4C (accessed 1 June 2020).

Tichy, N.M. and Cohen, E. (1997). *The Leadership Engine. How Winning Companies Build Leadership At Every Level*. New York, NY: Paperback.

Toll Group. (2020). Thomas Knudsen, Managing Director. https://vimeo.com/387622304 (accessed 1 June 2020).

Trice, H.M. and Beyer, J.M. (1993). *The Cultures of Work Organizations*. Prentice-Hall, Inc.

Ulrich, D., Kerr, S. and Ashkenas, R. (2002) '*How to Implement GEs Revolutionary Method for Busing Bureaucracy and Attacking Organizational Problems Fast*', iTwo Penn Plaza, NY: McGraw-Hill.

Umoh, R. (2018). Why Amazon pays employees $5,000 to quit. https://www.cnbc.com/2018/05/21/why-amazon-pays-employees-5000-to-quit.html (accessed 1 June 2020).

Waterman, R.H. Jr., Peters, T.J., and Phillips, J.R. (1980). Structure is not organization. *Business Horizons* 23 (3): 14–26.

Welch, J. and Byrne, J.A. (2003). *Jack: Straight from the Gut*. Business Plus (accessed 1 June 2020).

Wiggins, J. and Sullivan, R. (2020). Toll group CEO takes 40pc pay cut, Australian Financial Review. https://www.afr.com/companies/transport/toll-group-ceo-takes-40pc-pay-cut-20200331-p54fqc (accessed 1 June 2020).

World Refrigerator Industry Enters 'Chinese Era,' Haier Takes the Lead. (2018). https://www.prnewswire.com/news-releases/world-refrigerator-industry-enters-chinese-era-haier-takes-the-lead-300613167.html (accessed 1 June 2020).

Yin, R.K. (2017). *Case Study Research and Applications: Design and Methods*. SAGE Publications https://books.google.com/books?id=6DwmDwAAQBAJ (accessed 1 June 2020).

21

Competence Management as an Enabler for the Digital Transformation of the Supply Chain

Jiayu Sun

Bosch, Shanghai, China

Introduction

When we talk about digital transformation, we usually talk a lot about a variety of technologies such as big data, Internet of Things (IoT), artificial intelligence, and machine learning. However, the most important factor that makes these technologies work is the people. Without them being well prepared for this huge change, the digital transformation will be nothing more than just a collection of buzzwords. This chapter will focus on how companies can help employees get prepared for these changes and challenges.

Many companies have been using competence models to align their people strategy with their business strategies. With this alignment, companies can help translate the business strategies into an expectation of employee behavior. A study done by Lombardi and Saba showed that 89% of "best-in-class companies," defined as over 80% of employees receiving a rating of "exceed" on the last performance review, had core competences clearly defined for all job roles (Lombardi and Saba 2010).

Effective competence or skills management can drive greater success for the organization and the employees. Well-defined competences can help foster a strong corporate culture, build a more aligned workforce, and help to establish key competitive differentiators. They also help to ensure that the company has consistent performance standards for employees, which increase employee engagement and retention (Halogen Software 2012).

In the context of digital transformation, competence management has become more important than ever. This is because digital transformation leads to plenty of changes and challenges for employees, including new technologies, new processes, and also a new mindset. How well employees can meet and deal with these challenges will influence not only their working attitude but also their performance and, as a result, will have a great impact on the company as a whole. Hence, companies need to provide systematic and up-to-date competence management models to help employees get well prepared for the future.

The Digital Transformation of Logistics: Demystifying Impacts of the Fourth Industrial Revolution,
First Edition. Edited by Mac Sullivan and Johannes Kern.
© 2021 by The Institute of Electrical and Electronics Engineers, Inc.
Published 2021 by John Wiley & Sons, Inc.

Assuming an angle based on my background at Bosch, this chapter will provide an overview about competence management in supply chain functions and how competence management for the digital transformation will also be a learning transformation.

Competence Management in Supply Chain Functions

Fundamentals of Competence Management

The introduction of competence management into the business environment dates back to the early 1970s. David McClelland, a psychologist at Harvard University, brought the concept of "competence" to the human resource area while advising the United States Information Agency (USIA). The USIA was challenged to improve its selection procedures for Foreign Service Information Officers, as they noted that traditional academic scores and school grades were not reliable predictors of job performance. Therefore, McClelland proposed *competency-based* selection. Competences are a combination of explicit and tacit knowledge, behavior, and skills, which give someone the potential to be effective in task performance (Draganidis and Mentzas 2006). According to McClelland's research, he revealed that competencies such as interpersonal sensitivity, cross-cultural positive regards, and management skills differentiated vastly among excellent compared with average information officers (Dubois 1993).

To manage competences, a competence model is needed to identify the competencies that truly impact results. Competence management is considered a set of practices that identify and optimize the skills and competencies required to deliver on the organization's business strategy (Loew 2016). A competence model is a narrative description of the competencies for a specific job category, occupational group, division, department, or analysis units. It can provide identification of the competencies that employees need to develop to be prepared or improve in their job (Draganidis and Mentzas 2006). Typically, competence models include (i) competence names and detailed definitions, (ii) descriptions of activities or behavior, and (iii) a diagram of the model.

Bosch Case Study: A Competence Model Example

A particularly well-developed example comes from the German multinational engineering and technology company Robert Bosch GmbH (Müller et al. 2014). Its competence model is based on the company's mission and serves as basis for the complete recruiting and associate development process. Bosch interprets competences as characteristics, skills, and conduct, which are essential for the successful mastery of current and future tasks. The Bosch competence model consists of four areas of competence, which consist of two competencies each:

- Entrepreneurial competence
  - Result orientation
  - Future orientation
- Leadership competence
  - Leading my self
  - Leading others

- Interpersonal competence
  - Cooperation
  - Communication
- Professional competence
  - Breadth of experience
  - Depth of knowledge

Underlying each competence area are detailed criteria explanations that facilitate company associates' understanding (Robert Bosch GmbH Human Resources Management 2015). In Bosch's practice, competencies are defined for specific target groups by field setting owners whose tasks consist of recording and updating the relevant target groups, communication in own functional area to establish competence and training standards and review, and alignment of training contents/competence standards according to strategy and future suitability. For instance, in the various logistics areas, a setting owner responsible for logistics will discuss and define the competencies for each logistics role with the logistics experts.

Current Competencies Required

In the supply chain field, employees must be sufficiently competent to conduct tasks and actions in a timely and cost-effective manner. This requires a set of basic and operational competencies. Basic competencies in supply chain management include (Ngai et al. 2011) (i) adequate knowledge of the different functions of the supply chain and understanding of the complete business environment; (ii) the ability to communicate effectively with different supply chain parties, working in collaborative environments, and implementing inter-organizational projects; and (iii) the ability to integrate the supply chains of different organizations or business units and identify ways to enhance the efficiency and effectiveness of responses to market situations.

Required operational competencies are capacity planning, demand management, order processing, master production scheduling, inventory management and optimization, materials replenishment planning, logistics, warehousing, and distribution as well as knowledge of continuous improvement processes or methods (Shapiro et al. 2004; Muddassir 2015). For example, at Bosch, competencies defined for a Logistics Returnable Planner include basic competencies such as understanding of logistics concepts, the Bosch Production System (BPS), supply chain design and production network planning, project management, and cost accounting, as well as operational competencies such as inventory management processes, transportation processes, or the continuous improvement process (CIP).

Future Competences Needed

Digital transformations have led to changes in a variety of areas, as well as in the logistics job market. A study of Germany's digital association Bitkom found that in the past years, 10% of jobs have disappeared, while 21% of new jobs have appeared. In supply chain management, for example, jobs like "warehousemen" or "pre-sorters" do not exist anymore, but new jobs such as "data mining specialists" and "robot coordinator" are appearing (Dirks 2016). Cisco Senior Vice President Angel Mendez explained that he was impressed

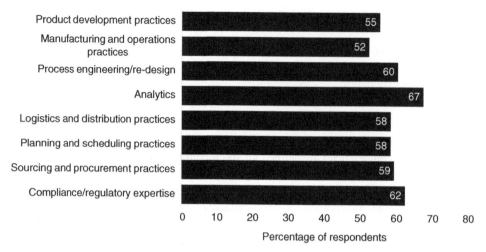

Figure 21.1 Which technical competencies of company's employees become more important in the future? *Source*: Based on Marchese and Dollar (2015). © John Wiley & Sons.

by how the supply chain profession is evolving, where "increasingly this type of role – certainly our supply chain operation positions here, and in many, many companies – is at the core of bringing new capabilities to market quickly, and reacting to competitive pressure and market dynamics (Marchese and Dollar 2015)."

With the appearance of new types of jobs, competencies required along the supply chain will also change. In a survey of the consultancy Deloitte, 67% of supply chain executives think that data analytics will become the most important competence to a company's supply chain organization over the next five years. The second most up-and-coming competence was compliance and regulatory expertise with 62% of participants and the third one process engineering/re-design with 60% (Marchese and Dollar 2015). The results are shown in Figure 21.1.

The supply chain advisory firm Mortson Enterprises has forecasted that while many of the basic competencies will remain, a variety of new competencies are going to be required for supply chain employees in the future (Mortson Enterprises Inc. 2019).

Strategic planning and big thinking includes an understanding about technology trends, such as IoT (cf. Chapter 3) and their possible impact on the supply chain. As research by the Global Supply Chain Institute at the University of Tennessee showed, "the unique nature of the end-to-end supply chain must be clearly understood to establish how to deploy collaboration initiatives to accelerate value creation (Burnette and Dittmann 2018)." Therefore, *real-time end-to-end supply chain management expertise* including understanding how value is created, which collaboration interfaces are involved, and being able to assume a holistic perspective across the supply chain are needed (Burnette and Dittmann 2018). New technologies such as automated guided vehicles (AGVs), drones, and IoT combined with new ways to store, process, and analyze big data will transform the way how supply chains will function in the future (cf. Chapter 25). *Technological expertise* and a *data analytics mentality* are therefore required, including an understanding on how these technologies will advance a supply chain and how they can be applied. Players in the logistics and supply chain domain will need to change technology wise and in terms of key capabilities that require conscious decisions between exploiting existing business and exploring new business models. To be able to implement these changes, methods for *business transformation* as well as *change management* must be understood (cf. Chapter 19). This is closely linked to *holistic, global leadership* skills, and *control tower leadership*. Supply chain employees must be able to work seamlessly within and across companies, cultures, functions, and geographies (Mortson Enterprises Inc. 2019). They need a lead in an open, agile style, removing barriers for their teams, integrating other's opinions, allowing experiments, and fostering internetworked communication across business silos (Kern 2019). Finally, *risk management expertise*, including fast problem-solving and decision-making, is needed to be able to quickly respond to turbulence and uncertainty in today's marketplace (Christopher and Lee 2004).

Management consulting firm Korn Ferry identified *learning agility* as one of the most important competencies for a supply chain leader whose capabilities are like those demanded by general management roles. This war for talent is further compounded as the additional digital imperative comes in, to drive performance as well as revenues through the supply chain, which needs to be done from the very top. Most supply chain leaders are not proficient in information technology (IT) and therefore need to have this expertise existing within their teams (Korn Ferry Institute 2017).

In the short term, automation and digitization will create more roles for data scientists, analytics engineers, and IT and big data specialists, who bring new knowledge to a supply chain organization. But to produce long-lasting results, "end-to-end experts", with well-rounded capabilities, including understanding supply chain realities, end-to-end expertise, cross-functional communication, and leadership skills, are in strong demand as they tend to meet tomorrow's requirements more. In preparing for digital transformation, the essential task is to prepare people for the most precious human role: managing the supply chain from end to end across internal boundaries (McKinsey 2017).

To address this required skill shift and obtain new competencies, also a new form of competence management will be required.

Digital Transformation Competence Management: A Learning Transformation

Fundamentals of Learning

Learning is one of the main competence management foci for organizations that are currently subject to the digital transformation, as, during a transformation, people need training and support to stay engaged and help develop and adopt new ways of working (McKinsey 2017). It is crucial to transform the learning experience from skill training to knowledge-based learning in the supply chain area as particularly in this field associates struggle to find time and suitable methods to train their employees effectively. Bowersox et al. highlight the importance of shifting training to knowledge-based learning, which means that skill development must be placed in the context of the overall process in terms of objectives, dynamics, and measurements (Bowersox et al. 2000). It is also important to build the competence of key managers and planners in understanding the risks and benefits of supply chain integration and managing the relationship between supply chain partners.

Learning in organizations was habitually synonymous with classroom courses and formal education, but learning is often driven by individuals' curiosity and interest, which is hard to embrace in a formal learning setting (Wahlund and Wahlberg 2018). Learning transformation can help to improve the effectiveness of a company's learning programs, the efficiency of the learning operations, and alignment of the learning and business strategies. According to a report from Deloitte, companies decreased their spending on learning and increased their return on an investment after going through learning transformation (Deloitte 2014). Learning is a set of personal and interpersonal activities, deeply rooted in a specific social and cultural context. When this context changes, also the way how people learn changes. Technology cannot but influence how people learn and therefore what makes for effective learning (Beetham and Sharpe 2013).

Transitioning from Passive Learning to Active Learning

Passive learning, where learners were passive receivers of knowledge from a trainer, used to be mainstream in organizations. Passive learners were considered empty vessels or sponges that should be filled. They took notes and internalized the knowledge, while they did not receive feedback from the trainer. This style of learning is trainer-centered and contrasts with the learner-centered mode active learning.

Passive learning is not simply the result of an academic model. Passive learners may quietly take in information and knowledge without typically engaging with the information they received. They may not interact with others, share insights, or contribute to the conversation. According to the leading company Lominger, some 60% of people are passive learners, only 10% are active learners, and the remaining 30% are blocked, which means they are closed off from even hearing about or trying to learn something new (Chief Learning Officer 2013). However, different research studies have shown that using active learning as a strategy has improved learners' performances. For instance, Hoellwarth and Moelter showed that when instructors transformed their classes from traditional instruction to active learning, students' learning improved 38%, from around 12% to over 50%, measured by the Force Concept Inventory, the measurement standard for student physics learning (Hoellwarth and Moelter 2011).

In contrast to passive learning, active learning indicates that the trainer strives to create a learning environment in which the learner can learn to restructure the new information and their prior knowledge into new knowledge about the content and to practice using it (MacManus 2001). Active learning engages learners mainly in two aspects – doing things and reflecting on the things they are doing. Barnes suggests seven principles of active learning: *purposive*, the relevance of the task to the learners' concerns; *reflective*, learners' reflection on the meaning of what is learned; *negotiated*, negotiation of goals and methods of learning between learners and trainers; *critical*, learners appreciate different ways and means of learning the content; *complex*, learners compare learning tasks with complexities existing in real life and making reflective analysis; *situation-driven,* the need of the situation is taken into account to implement learning tasks; and e*ngaged,* real-life tasks are reflected in the activities conducted for learning (Barnes 1989).

Transitioning from Classroom Training to Digital Learning

In the 1990s, the Internet brought a revolution to the traditional classroom training, which became even more comprehensive with the introduction of Web 2.0 (Wang and Heffernan 2009). From passively sitting in a classroom to clicking, swiping, and touching on a digital device, learning transformation enables adapting learning to new audiences and changing business requirements. This transformation focuses on updating training delivery methods and media. Using new digital technologies, learning becomes more effective and more interactive.

Digital learning, or e-Learning, has been referred to as "technology-enhanced learning." It describes a set of technology-mediated methods that can be applied to support student learning and can include elements of assessment, tutoring, and instruction (Wheeler 2012). Digital learning can improve the learning experience by giving learners controls over time, place, path, and pace. It means that learners can choose whenever and wherever they want to learn, decide their way of learning, and decide their own pace. These enhanced learning experiences can help achieve better learning results and improve the competences of the learners. Companies like Bosch currently employ a variety of learning formats, such as classroom training, live online training, web conference, web-based training, learning video, community learning, virtual classroom, mobile apps, manual guide, and shop floor training.

Future Learning Trends

The digital learning landscape keeps embracing new technologies while at the same time facing new challenges. In digital learning areas, four key areas have been identified:

Micro-learning

Due to the information boom, people tend to have shorter and shorter attention spans to concentrate on learning content. Here, micro-learning helps to transform learning contents into fragmented and bite-size learning nuggets, which are easier for people to learn anywhere and anytime. Micro-learning is a digital learning experience designed to be effective, engaging, and enjoyable. It is a transition from common models of learning toward micro perspectives in the process of learning.

Eliminating the digital gap is and will remain one of the most significant educational challenges for the coming decades, and micro-learning can contribute to meeting it (Chrisholm 2006). Investments need to be made in designing low-threshold tools and methods that can engage learners with low-level skills – both basic and digital. This can provide an attractive and encouraging format that is relevant to learners' daily lives and can provide rapid, visible affirmation and recognition of what they have learned.

Micro-learning has been considered by many organizations as an important digital learning solution. For example, Bosch China started a micro-learning award competition campaign in 2016. It was designed to facilitate learning and sharing through user-generated content and digital learning solutions by continuous explorations to develop user-oriented learning contents, enable digital transformation, and support business growth. The competition received positive feedback from participants and acknowledgment by the whole company. Blake Cai, a learning and leadership development director at Bosch China, highlighted that as one of the learning formats, micro-learning embodies learning culture and learning agility at the organization and shapes a learning company (Cai 2020).

Mobile Learning

Mobile learning refers to learning processes supported by the use of mobile and wireless information and communication technologies that have as a fundamental characteristic the learners' mobility, who may or may not be physically or geographically distant from one another and also from formal educational spaces, such as classrooms, instruction, and training rooms or the workplace (Ferreira et al. 2013). The new generation of learners uses these mobile devices as part of their daily routine.

Mobile learning is one of the key current trends in the digital learning area, where new practices and methods to employ mobile learning are now being shaped. Kukulska-Hulme states that increasingly learners will use mobility and awareness of their immediate context as starting points for keeping social contact alive (who is nearby?), accessing fresh content (what resources are available here?), getting local information (what is interesting here?), and becoming visible as creators and producers of content (what am I able to contribute?) (Kukulska-Hulme 2010).

The effective use of mobile learning depends mainly on what it means to its users. A culture of listening to learners will be needed to find out about their current practices with mobile technologies and seeking to extend them or channel them in the right direction. Bosch China has launched an online learning platform called "myTransform" in 2019, aimed at driving the digital transformation of Bosch, that responds well to the rapid changes in the era and business environment by grappling with the issue of building a mobile learning platform that fits the new era and new dynamic. The rollout of "myTransform" means that mobile learning in Bosch China has entered a new era, in a way more economical, more relaxing, more convenient, and much closer to business (Cai 2020).

AR/VR

Augmented reality (AR) and virtual reality (VR) are both growing rapidly as important digital learning formats. AR's most vital advantage is its "unique ability to create immersive hybrid learning environments that combine digital and physical objects, thereby facilitating with developing processing skills like critical thinking, problem-solving, and communicating through interdependent collaborative exercises (Dunleavy et al. 2009)." Virtual reality provides an innovative educational instrument that enables students to assess the value of their solutions requiring them to apply relevant knowledge and understanding to a particular real-life complex problem (Abulrub et al. 2011).

While AR and VR offer unique advantages, such as its combination of virtual and real objects in a real setting, there are also some challenges to be considered. For instance, there are certain hardware (mobile devices, tablets, etc.) prerequisites for the use of this technology, such as processing speed and ergonomics. These challenges are relatively minor and should not prohibit the use of this technology in the learning area (Akçayır and Akçayır 2015). The current technical problems will likely be resolved by new developments that will make AR and VR applications even more useful in the learning area in the future.

Blended Learning

Despite many advantages of digital learning, interaction with people and the exchange of experiences are still regarded as important factors for the participants of face-to-face training, which cannot be digitalized completely. Hence, traditional face-to-face training will not be replaced fully by digital learning, but digital technology will rather enrich and transform the traditional learning experience. New learning formats such as blended learning combine traditional face-to-face training and digital formats with the help of collaborative tools, such as learning communities, working groups, impulse talk sessions, etc. This kind of learning format will keep developing over time as it can be very flexible and compatible and thus be seen as a mix of all possible digital learning formats.

Still, blended learning faces some technical, organizational, and instructional challenges. For example, participants need to be able to use the technology successfully with support from the facilitator, whose role needs also be redefined by immersing them in a blended learning program so they can fully understand the participant experience. While designing the blended learning programs, it needs to be ensured that all the elements of the blend are well coordinated (Hofmann 2011). Bosch has started the blended learning approach in October 2016 with the project "virtual classroom." Learning activities were held in small groups with high levels of interaction, both in 2D and in 3D learning environments. The learning format has encouraged learning regardless of time and location while at the same time saved travel efforts and time (Pape 2017).

Conclusion

Competence management can help to prepare employees and companies better for digital transformation. With an appropriate and up-to-date competence model, a company can manage competences more systematically. Most of the competence models used in organizations usually include occupation-specific knowledge and skills, as well as the soft skills to collaborate and work effectively with people. Each organization has its own definition of required competencies, while in the supply chain the most needed competencies include knowledge of different functions of the supply chain, understanding of the business environment, communication with different supply chain parties, collaboration, and integration supply chain to enhance the efficiency and effectiveness. Competences considered as critical for the future include data analytics along with process engineering, strategic planning, or risk management.

As a major approach of competence management in digital transformation, learning has experienced a transformation from passive to active and from classroom to digital, and the trends we see in the digital learning area keep updating consistently. However, it has to be taken into account that the first step of competence management in digital transformation must be a mindset change. Indeed, setting up a competence model, transforming learning from passive learning to active learning and from classroom training to digital learning, competence management nowadays has used a lot of new methods and new digital technologies to facilitate. But after all, either methods or digital technologies by themselves do not change the essence of the digital transformation. It is a matter of people and people's mindset. Only if both are prepared for the digital transformation, methods, and technologies can support the required change of associate's behavior and their skills.

Key Takeaways

- Competence management can help to prepare employees and companies for the digital transformation. An appropriate and up-to-date competence model allows companies to systematically manage competences. However, organizations often have own definitions of required competencies.
- In the supply chain field, the most needed competencies include knowledge of different functions of the supply chain, understanding of the business environment, communication with different supply chain parties, collaboration, and integration supply chain to enhance the efficiency and effectiveness.
- Competences considered as critical for the future include data analytics along with process engineering, strategic planning, or risk management.
- The first step of competence management in digital transformation must be a mindset change. Only if people and people's mindsets are prepared for the digital transformation, methods, and technologies can support the required change of associate's behavior and their skills.

References

Abulrub, A.-H., Attridge, A., and Williams, M. (2011). *Virtual Reality in Engineering Education: The Future of Creative Learning*. Amman, Jordan: IEEE.

Akçayır, M. and Akçayır, G. (2015). Advantages and challenges associated with augmented reality for education: A. *Educational Research Review* 20: 1–11.

Barnes, D. (1989). *Active Learning*. Leeds: Leeds University TVEI Support Project.

Beetham, H. and Sharpe, R. (2013). *Rethinking Pedagogy for a Digital Age: Designing for 21st Century Learning*. New York: Routledge.

Bowersox, D., Closs, D., and Stank, T. (2000). Ten mega-trends that will revolutionize supply chain logistics. *Journal of Business Logistics* 21 (2): 1–15.

Burnette, M. and Dittmann, P. (2018). *End-to-end Supply Chain Collaboration Best Practices*. Knoxville: Haslam College of Business.

Cai, B. (2020). Blake's activity. https://www.linkedin.com/in/blakecai/detail/recent-activity/ (accessed 02 January 2020).

Chief Learning Officer (2013). Engage passive learners. https://www.chieflearningofficer. com/2013/01/10/engage-passive-learners/. (accessed 23 August 2019).

Chrisholm, L. (2006). *Micro-Learning in the Lifelong Learning Context*. Innsbruck: Innsbruck University Press.

Christopher, M. and Lee, H. (2004). Mitigating supply chain risk through improved confidence. *International Journal of Physical Distribution & Logistics Management* 34 (5): 388–396.

Deloitte (2014). Learning transformation: optimizing your learning operations to reduce costs or fund investments.

Dirks, T. (2016). *Neue Arbeit – wie die Digitalisierung unsere Jobs verändert*. Berlin: Bitkom.

Draganidis & Mentzas (2006). Competency based management: a review of systems and approaches. *Information Management & Computer Security* 14 (1): 51–64. https://doi. org/10.1108/09685220610648373.

Dubois, D. (1993). *Competency-Based Performance: A Strategy for Organizational Change*. Boston: HRD Press.

Dunleavy, M., Dede, C., and Mitchell, R. (2009). Affordances and limitations of immersive participatory augmented reality simulations for teaching and learning. *Journal of Science Education and Technology* 18 (1): 7–22.

Ferreira, J.B., Klein, A.Z., Freitas, A. et al. (2013). Mobile learning: definition, uses and challenges. In: *Increasing Student Engagement and Retention Using Mobile Applications* (eds. P. Blessinger, C. Wankel and L.A. Wankel). Emerald Group Publishing Limited.

Halogen Software (2012). The importance of being competent at competency management. https://www.hrgrapevine.com/content/article/2013-12-12-the-importance-of-being-competent-at-competency-management (accessed 14 October 2019).

Hoellwarth, C. and Moelter, M. (2011). The implications of a robust curriculum in introductory mechanics. *American Journal of Physics* 79 (5): 540–545.

Hofmann, J. (2011). Soapbox: top 10 challenges of blended learning [Online]. http://www.trainingmag.com/article/soapbox-top-10-challenges-blended-learning (accessed 09 January 2020).

Kern, J. (2019). Skills for the future. https://youtu.be/VkRCxuJvUTY (accessed 14 October 2019).

Korn Ferry Institute (2017). The supply chain digital disruption. https://www.kornferry.com/content/dam/kornferry/docs/article-migration/SupplyChainWinter2018.pdf (accessed 30 October 2019).

Kukulska-Hulme, A. (2010). Learning cultures on the move: where are we heading? *Journal of Educational Technology and Society* 13 (4): 4–14.

Loew, L. (2016). Competency management: challenges and benefits. https://trainingmag.com/competency-management-challenges-and-benefits/ (accessed 14 October 2019).

Lombardi, M. and Saba, J. (2010). *Talent Assessment Strategies: A Decision Guide for Organizational Performance*. Aberdeen Group.

MacManus, D. (2001). The two paradigms of education and the peer review of teaching. *Journal of Geoscience Education* 49: 425.

Marchese, K. and Dollar, B. (2015). *Supply Chain Talent of the Future. Findings from the Third Annual Supply Chain Survey*. New York: Deloitte.

McKinsey (2017). Digital supply chains: do you have the skills to run them?. https://www.mckinsey.com/business-functions/operations/our-insights/digital-supply-chains-do-you-have-the-skills-to-run-them#. (accessed 30 October 2019).

Mortson Enterprises Inc. (2019). Quantum leap to the top 10 supply chain skills of the future!. https://supplychaingamechanger.com/supply-chain-skills-of-the-future-quantum-leap/ (accessed 20 August 2019).

Muddassir, A. (2015). 8 must have supply chain competencies. http://www.scmdojo.com/8-must-have-supply-chain-competencies/ (accessed 21 August 2019).

Müller, H.-M., Thomas, A., and Müller, J.-P. (2014). Globalising how? The route towards international HR development. In: *Strategic Human Resource Development: A Journey in Eight Stages* (ed. M.T. Meifert), 315–328. Berlin: Springer Science & Business Media.

Ngai, E., Chau, D., and Chan, T. (2011). Information technology, operational, and management competencies for supply chain agility: findings from case studies. *The Journal of Strategic Information Systems* 20 (3): 232–249.

Pape, K. (2017). CLP028 corporate learning 2025 MOOCathon – Woche 8 start session. https://colearn.de/clp028-corporate-learning-2025-moocathon-woche-8-start-session/ (accessed 09 January 2020).

Robert Bosch GmbH Human Resources Management (2015). Roadmap for your career. http://www.bosch-career.ch/media/public/de/documents/wachsen_bosch/aufstiegsmoeglichkeiten/mitarbeiterentwicklung/Working_Together_Toward_Professional_Success.pdf (accessed 20 August 2019).

Shapiro, B.P., Rangan, V.K., and Sviokla, J. (2004). Staple yourself to an order. *Harvard Business Review* 82 (7/8).

Wahlund, F. and Wahlberg, V. (2018). *Digitalization & Competence Management: A Study of Digitalization and Competence Management Within the Telecommunications Industry*. Gothenburg: Chalmers University of Technology.

Wang, S. and Heffernan, N. (2009). Mobile 2.0 and mobile language learning. In: *Handbook of Research on Web 2.0 and Second Language Learning* (ed. M. Thomas), 472–490. IGI Global.

Wheeler, S. (2012). e-Learning and digital learning. In: *Encyclopedia of the Sciences of Learning* (ed. N.M. Seel). Boston: Springer.

22

Impacts of Digitalization on Traceability

A Case Study of the Carbon Fiber Supply Chain

Cameron Johnson

Tidal Wave Solutions, Shanghai, China

Introduction

Carbon fiber is everywhere. If you see a Ferrari, fly in a Boeing 787, generate energy with a windblade, or play hockey with a CCM stick, you might not realize, but they all rely on carbon fiber to function properly. Carbon fiber applications span over 20 industries, including aerospace and defense, automotive, energy, marine, consumer electronics, and sporting goods. The global carbon fiber market was valued at $2766 million in 2015 and is expected to grow at a CAGR of 11.7% to reach $5991 million by 2025.

Major carbon fiber production centers are the United States, Europe, and Japan, with capacity growing in Korea, China, and Turkey. Toray, based in Japan, has a 35% global share of the market. The automotive, wind energy, and pressure vessel industries are the primary drivers of growth from 2020, whereas aerospace was the primary driver previously. The United States and Europe are the main consumers due to aerospace and automotive demand, but Japan, China, and Korea are developing new supply chains of raw materials to finished products and are the new markets and users (Cook et al. 2017).

Product demand is driven by the following:

- Demand in aerospace in North America and Europe
- Rising demand for lightweight vehicles coupled with government regulations regarding automotive pollution
- Demand in sports and leisure applications, especially in the Asia Pacific region (Research 2017)

The global business landscape is changing rapidly. Digitalization is driving transformation, first in parts of a business, and then the overall efficiency of the company as a whole. For companies in Asia that are surprisingly lagging in digitalization, this means opportunities to implement a system that improves compliance, drives waste efficiency and material traceability, and reduces cost.

The Digital Transformation of Logistics: Demystifying Impacts of the Fourth Industrial Revolution,
First Edition. Edited by Mac Sullivan and Johannes Kern.
© 2021 by The Institute of Electrical and Electronics Engineers, Inc.
Published 2021 by John Wiley & Sons, Inc.

As a result of this growth, leading firms are contemplating implementing solutions to digitalize their quality systems and other business functions. These services will enable firms to better utilize internal resources; drive production and supply chain efficiency, quality control, risk management, and compliance; and increase profitability.

Carbon Fiber, The Material of the Future

Carbon fiber is an extremely strong material that is also very lightweight. It is ten times as strong and 25% the weight of steel (*What is Carbon Fiber?|ZOLTEK*, 2019). In addition to this, it also has high stiffness, high tensile strength, and low thermal expansion. Due to these attributes, carbon fiber is popular in many industries such as aerospace, automotive, military, and recreational applications (*What is Carbon Fiber|Innovative Composite Engineering*, 2020).

Global Supply Chains: A Fragmented Picture

The carbon fiber supply chain is global, interlinked, and regulated, but at the same time, it is fragmented, with a substantial gray market and many middlemen (Schofield 2018). It is roughly categorized into three stages shown in Figure 22.1.

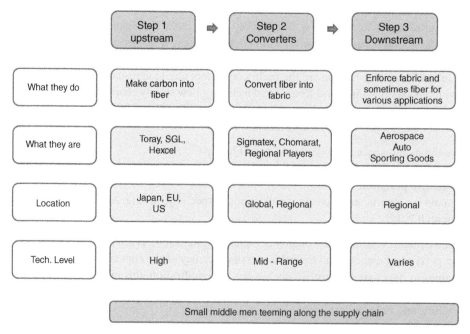

Figure 22.1 Diagram of the global carbon fiber supply chain. *Source*: Based on Schofield (2018). © John Wiley & Sons.

Producers. The first stage is relatively consolidated with 10 major players. Three companies control 71.9% of the global producer market. The two largest companies in step 1 are Toray and SGL. Toray is the largest producer of carbon fiber in the world with an annual production capacity 46 200 metric tons. SGL produces 15 000 metric tons. These companies provide the majority of product to aerospace and automotive industries. (*Carbon fiber manufacturers by production capacity 2017\Statista,* 2017)

Converters. The middle stages of the supply chain are complicated and fragmented. Raw fiber is sold to middlemen, converters, processors, and end users that process material into various supply chains. A **converter** converts carbon fiber into a textile, takes fiber, and using various textile machines weaves it into a fabric or preform that can be used in the downstream supply chain. Once the fiber is converted, it is sold to either a processing company or an end user. A **processor** takes material and, after impregnating the fabric with resin, will then sell it downstream, often to an end user. Another sales channel is to an **end user** who takes the raw fiber and/or fabric and puts it directly into internal processes to produce a part. One such example is Cobra in Thailand, which makes the majority of the world's kite boards.

Middlemen. Middlemen exist due to their specific technology (customized machinery for low production runs), ability to be in areas where large producers do not want to be, and can offer customized products to meet customer needs. These companies will often use different types of carbon fiber depending on price, customer demand, and location. These companies form an integral part of the supply chain to help facilitate fiber usage in areas and with customers that larger players are unable to or do not want to.

End users. End users use raw fiber or processed fiber that is impregnated with resin, depending on how a finished part is produced. Demand is volatile as raw material prices are effected by currency conversion, oil prices, and limited capacity for some fiber.

Several Constraints Characterize the Global Supply Chain

Due to the numerous downstream applications using carbon fiber, demand forecasting on a global and synchronized basis is quite difficult. Carbon market demand forecast is based on an estimate of growth per industry and country on a yearly basis. In addition to forecasting concerns, the industry is plagued by more constraints such as capacity and strict regulations.

Capacity. Carbon fiber producers have a finite number of carbon fiber production lines, and the investment of additional production is significantly high, often amounting to over $100 million per line. When market demand is high and supply is tight, special programs that use carbon fiber in the aerospace and automotive industries are given priority over other projects and uses. This allows fiber companies to maintain market share in these industries and keep prices higher than if sold to other industries.

Regulation. Certain types of carbon fiber, such as T800 or T1000, are not allowed to be sold to countries such as China or Iran due to concerns over military and nuclear use. This keeps prices higher as it is limited to certain countries and/or company use.

Profitability. It is also a challenge, particularly in East Asia where prices are historically lower than other parts of the world. The East Asia market seems to be more willing to accept a cheaper product that is "good enough," especially when the majority of the applications are for the non-aerospace industries in the region.

Case Studies on Lack of Material Traceability and Digitalization

From the author's years of experience in the global carbon fiber industry, he identified several challenges and opportunities to modernize the global supply chain, which can be shown through three case studies:

1) An aerospace company sourcing carbon fiber shows quality control concerns
2) Traceability and its effect on the supply chain
3) A global carbon fiber converter's internal procedure points to an opportunity in digitalization

From these examples, one can observe a clear lack of digitalization in the carbon fiber value chain, which inhibits its ability to be efficient, control waste management, and drive compliance. Also, from these case studies, the following conclusion can be constructed: digitalization of logistics and supply chains, especially for small and medium enterprises (SMEs), can revolutionize an industry player's ability to drive these aspects of their business, control risk, and contribute to profitability. Moreover, the application of digital solutions will enable the whole ecosystem to function as a seamless global system, which may not be desired by the industry's gray market and numerous noncompliant players.

Case Study 1: An Aerospace Company Losing a Qualified Supplier

The first case study looks at an aerospace company that developed China's first electric airplane (Sivani 2015). The prototype was made and flown with great fanfare and was shown on a national news channel as a feat of local engineering and performance. When it was time for the second round of production, one of this company's key suppliers received an order of $4000\,m^2$ for a carbon fiber aerospace quality level material. This supplier first sent a small portion to the Chinese aerospace company for quality review and comparison with the original material used to make the first plane.

This supplier had previously completed advanced aerospace quality audits and certifications and was a supplier of Boeing, Airbus, and Embraer. The product sent to the aerospace company was similar to the material it supplies to these other companies. However, when tested, the material failed the needed standard in weight, performance, and quality. With these results, the customer required an explanation from the supplier as they were told this supplier is "aerospace certified." The supplier performed an extensive audit of the documentation and quality results, which included sending additional material for testing. Serious quality challenges continued to occur.

Confused and frustrated, the supplier visited the customer to investigate these discrepancies. While at the customer's facility, it became clear they did not have any standard operating procedure (SOP) for quality procedures when testing or using the material. This is common in some cases due to the business environment and lack of resources (Hussein and Cheng 2016). This is juxtaposed to an aerospace company that has SOPs, a quality management system (QMS), and AS9100 quality certifications (Aerospace Basic Quality System Standard) for its material and products (Bulaeva 2015; Oschman 2019). Boeing by contrast, before discussing a supplier relationship, requires its suppliers to have a certified and audited QMS and also does not conflict with Boeing's purchasing requirements (Livingston and Marino 2019).

Quality Management

During the production of the prototype, the purchasing manager contacted suppliers through a local online platform and purchased material for testing. There was no standard given to these suppliers, which goes against the aerospace industry practice of giving specifications to be made to (Boeing 2020; Federal Aviation Administration 2013). They bought from several suppliers and comingled material, and the documentation chain was broken. The material was untraceable, meaning there was no accompanying documentation or other information to tell where it came from (Caridi et al. 2014).

Quality Control

Quality control in aerospace is paramount to all other areas (*Aviation Quality Control – Aerospace AS9120|UPS – United States* 2020; Freeman 2018). If there is a defective part or material, an airplane can crash. If there are issues with a plane, one of the first areas reviewed is performance and maintenance. In 10 years from 1994 to 2004, 42% of fatal airline accidents in the United States were traced to maintenance problems (Han 2018). In this case, the aerospace company's production documentation and details were written by hand, and there was no standard SOP followed. If the plane crashed, it might be difficult to determine if the material or supplier could be at fault. Boeing's recent issues in 2019 and 2020 are not related to material sourcing and handling, but rather to engineering challenges.

Standards

In the carbon fiber supply chain, variations in SOPs, QMS, and certifications are challenging because there are multiple standards a company can use: International Standard Organization (ISO), AS9100, company standards such as Boeing's, and also internal company standards that have evolved. This creates problems identifying a supply chain SOP for material traceability, quality issues, and documentation errors. If a quality system based on either ISO or AS standards (Cority IQS 2014) had been adopted by this aerospace company, it could easily find if the standard was met. This, in turn, would save significant time and money and drive productivity (Nanda 2005). It would also ensure consistency when scaling up the material used in production and future replicability (Fawaz 2016).

Case Study 2: Traceability, Which Affects Efficiency and Compliance

In the carbon fiber supply chain, traceability is part of the SOP and QMS systems, and it is important to find issues, or root cause, with products and materials. Digitalization could enable more effective traceability, allow real-time data analysis that an analog system cannot provide, and improve data storage and retrieval. Traceability means the materials used are for the designated production steps and fulfill customer requirements (Conaway 2019). By using a digitalized traceability system, a supply chain can better track where issues start to occur and facilitate faster problem-solving by identifying issues in real time. Apple, as an example, uses a digitalized report called the "Conflict Minerals Reporting Template (CMRT)" that suppliers must fill out to ensure no materials come from conflict zones. The report forces suppliers to keep up-to-date information to ensure compliance and specifications are being met and no conflict-zone materials are used anywhere in the supply chain (Adams 2018). Expanding this platform into tracking material and products beyond conflict zones could enhance Apple's own traceability throughout the value chain. To implement this system would be less cumbersome as most suppliers and auditors are already using it in some form (van Iwaarden and van der Wiele 2006).

Based on the author's observations of the carbon fiber supply chain in Asia, much of the QMS and production paperwork is done by hand. Even companies with a qualified aerospace SOP and QMS are challenged by mountains of manual entry and paperwork that need to be completed for each project. One example of a company that uses digitalization throughout its supply chain to reduce this burden is 3M. 3M uses a digitalized tag and barcodes trees from harvest to end use. This allows the company to track 85% of the global paper supply to the mill (Banker 2017). By using this system to ensure sustainable sourcing, 3M gains productivity, enhances quality control, and increases compliance.

Moving to the carbon fiber supply chain, fiber is produced in batches of spools (also referred to as bobbins) (Warren 2016). Each spool has a "label" identifying the production lot number and other information related to the production of the fiber. This is important for traceability and in sensitive industries like aerospace, a requirement for an SOP and QMS. Using this information, which is passed throughout the supply chain, a company can trace where the material, or part, came from and the processes it went through.

If a label is altered, it can be difficult to trace. For example, for taxation and customs purposes, when fiber enters a country, its label would determine the tax and customs category. If a spool enters country X, a duty of 15% and a VAT tax of 17% would be paid. But if the label is modified, the taxation rate could change drastically (Dumas 2019; Durga et al. 2016). QR codes and RFID have potential use here, but there are security concerns, and it is not a full-proof system, especially in Asia where the creation and usage of fake QR codes are rampant (Ahuja 2014; Mulln 2017; Yoshida and Matsuda 2019).

The inside label of a carbon fiber spool with traceable data. Taken by the author 28 March 2018.

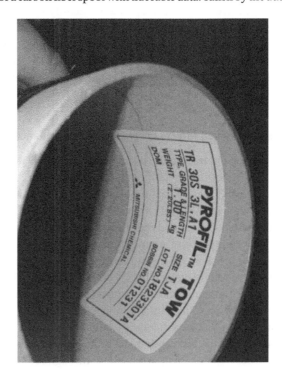

For compliance, traceability is important to follow carbon fiber licensing and end use requirements. Not filing for, or following, proper licensing could lead to jail (Department of Justice 2019); ZTE is one example of a company not following these restrictions in its business with Iran. Once caught, they were then sanctioned, and their senior executives were punished (Department of Justice 2017). Because of its application in the aerospace and defense industries, carbon fiber is considered "dual use," meaning the fiber can be used for both military and commercial applications. In almost every country, carbon fiber is regulated and controlled. Most fiber, regardless of production location, requires an export license. There is also an end-use requirement or import license needed for the material (Kurland 2014; Toray 2019). This creates an opportunity for vendors with digitalization technology and compliance specialties to provide services to those companies that must conform to these rigorous standards.

For example, if a Chinese company imports carbon fiber from Japan, there are several steps to go through before that can occur. The first is that a license application must be made to the carbon fiber supplier, citing the specific usage (automotive, sporting goods, etc.), location, potential customers, price, and quantity (Economic Cooperation Bureau 2011).The second step is the supplier puts forward an application to METI (Ministry of Economy 2014) for approval. Once METI reviews and approves the application and license, shipments to the customer can begin. This process can take 6–12 weeks and is done using paper documentation that is then reviewed. South Korea has a similar process.

Complications with licensing can arise for various reasons. The end user can be put on a blacklist; a trade war can occur, such as between South Korea and Japan, limiting export and imports of goods (Bremmer 2019). Another challenge is companies shifting suppliers because of lead time in issuing a license, especially from Japan or the United States. Other countries, such as Turkey, Taiwan, and China, have fewer restrictions, and license approval is quicker, sometimes only 1–2 weeks. A 4PL or 3PL logistics provider (Moses 2020) in the United States can help the customer apply for an export license through the Electronic Export Information (EEI) system. It can also help manage compliance and screening by checking shipments against current regulations, although the company sending the product is ultimately responsible for its end use. An EEI cannot be used for carbon fiber exports, as the shipper must apply for and obtain approval of a license independently before any carbon fiber can be shipped.

Some countries also limit the amount of carbon fiber a supplier can send to a customer. For example, the Japanese government, through METI, determines how much carbon fiber each company can use every year and has an allowance system. This creates a problem when a company orders more than the allotted limit. Several companies in China have had this challenge due to their market growth, and METI, not trusting the growth in their capacity, restricted additional supply. Once this occurs, a company is reviewed again and must include updated revenue evidence, production capacity, customer information, and reasons why demand has increased (new project, customer, etc.). It also requires a new set of license applications as the company status has changed. Once these new applications are made, METI will review this information and determine if a new amount of carbon fiber can be allotted to the customer (Economic Cooperation Bureau 2011). A digitalized ERP system could improve this process and efficiency, but there can be cultural and managerial challenges to implementation (Sheu et al. 2004).

Case Study 3: A Global Carbon Fiber Converter's Internal Waste System Could Use an Upgrade

One of the largest headaches of the carbon fiber industry is waste (Carnes and Ringenbach 1980). Waste is often tracked on handwritten documents, especially in OEMs and suppliers, based on the author's observation of the typical SME in Asia. Using digitalization would allow companies to track real-time waste, improve production processes, and drive productivity. In other words, the system would flag discrepancies as they occur and allow a timely response instead of catching the problem after it arises and material or product becomes defective.

To resolve these traceability challenges, a global carbon fiber converter created a system that interwove compliance, quality control, waste management, and financial tools into a single process.

Interwoven Processes for Traceability, Created by the Author

This process began with putting all carbon fiber "waste" that accumulated during production manually into a cardboard box. The box would then be weighed on a scale and the weight recorded by hand on paper. The weight would then be calculated by price/kg. For example, if there are 23.3 kg of carbon waste, each kilogram of carbon costs $32, and the wastage cost during production would be $745.60. The finance team would then add this to the total cost of the project. This affects the cost to the supply chain as this cost is passed onto the customer. Waste tracking and recycling, if more efficient, could improve supply chain management by lowering cost and improving utilization.

This process is primarily done by hand and written on paper and then entered into Excel, manually. A digitalized method similar to 3M's bar code system would create greater efficiency and permit real-time tracking and financial calculation, allowing operators and management to be more agile, and adjust production if issues occur (Ćwikła 2013). In 3M's case, from the logger to mill staff, handheld devices are used that provide real inventory information, or information flow, to track material through the entire supply chain.

Tracking Waste

At the carbon fiber converter, different fibers, on different machines, produced different waste rates. Tracking waste raises manufacturing efficiency and drives more precise pricing mechanisms. An Internet of Things (IoT) system could help identify problems and ultimately resolve this by automatically calculating machine feedback and notify operators if an issue occurs. For more information on the use of IoT, please see Chapter 3.

At this converter company, waste tracking is a four-step process, all done by hand. Waste comes off a machine and is placed in a box and weighed. The waste is manually written down on a form (OCR is not used as data is written on paper forms, and each machine has one form to track waste from that particular machine. These forms are kept by the quality control department as a compliance measure). The financial team then inputs this data into an Excel sheet and uses a formula to calculate the scrap cost for the job, per the example given above.

This process uses extensive documentation, all on paper, over an extended period. After completion, documents are physically stored in filing cabinets that are added to each year to accommodate the increasing amount of stored paperwork. Quality samples are also stored this way, without a specific system to track them. An ongoing trend in some industries is a cloud-based ERP system, which could enable data sharing beyond the physical boundary of a single factory. Companies are hesitant to do this because of cost, concerns on data security, and gaps in transferred data (Johansson et al. 2015).

A digitalized system could relay real-time information once the waste is weighed, giving staff and management a heads-up if there is more wastage than "normal," and also provide a baseline per machine. The reason this wasn't previously done is the implementation cost combined with forcing a change in staff behavior has been prohibitive. One such system costs $75/month, with a $1500+ installation, service charge, and other fees (ScrapRight Reviews and Pricing 2019). This cost combined with the challenge of needing a product in different languages, some of which are not supported, makes using such a program unrealistic for the business at this time. With this, the user would still need to "input" the data into the system and still need a hard copy for compliance, thus negating its advantage versus manually writing it on paper.

For example, when first tracked, operators found different fiber types, and machines had different waste rates. Part of this can be attributed to older versus newer machines, where older machines are less productive over time. Newer machines run faster and at greater efficiency (Cooper and Haltiwanger 1993). From the author's experience of converting carbon fiber, older machines have a scrap rate of 7–9%, against newer machines that have a 3–5% rate. How the machine is set up and the type of fiber used also factors into the waste rate (some fiber has more waste than others depending on the setup and customer need). This process helped the converter better understand pricing, as it could set a revised selling price based on the waste rate for certain products and customers. The converter also adds the scrap cost to the total landing cost. It is a lose–lose situation for all involved.

Compliance

Digitalization could make it easier to follow regulations and compliance. Given the sensitivity around military applications of materials and parts related to carbon fiber, several countries, in addition to export and import licenses, also require usage to be tracked and verified through documentation (Power 2019). For export control documents, this is required to ensure the material end use stated in the license is followed. When importing, it is used to ensure materials are taxed correctly throughout the supply chain.

In some price-sensitive industries such as auto parts or sporting goods, if the material is cheap or doesn't come with paperwork, it sometimes can be accepted and used in production. This specifically happens when a factory is not given a material specification for a product, uses cheaper materials to save cost, but still gives the customer a feeling that it is made of carbon and therefore is a high-quality good (Elton-Walters 2016; Barton 2018). In reality, the product could have a carbon fiber layer on the outside, but the core is made of another material such as fiberglass (Crawford 2016). It takes sizable buyers with a clear awareness of the material, production, and higher standards for visibility to change this industry practice. A digital system could also ensure the producer used the correct material and components specified by the customer for the product.

This poses a risk to businesses and suppliers and adds a burden to ensure product quality. If poor-quality material or production is used, it could affect the entire supply chain and the various industries supplied. It also doesn't allow the government to collect the correct tax revenue (Durga et al. 2016). With the US–China Trade War, Chinese exporters often bear the brunt of tariffs, with some companies declaring suppliers must pay all tariffs (Mauldin and Nassauer 2019). This is squeezing already thin margins (Nicita 2019; Lee 2019) and has led to layoffs and bankruptcy of some companies (Lee 2019; Curran 2019). As such, improving waste reduction or recycling becomes even more important. Less scrap and waste could make a difference in the margin, which is typically in single digits.

Digitalization Benefits in the Global Carbon Industry

Part of the challenge in the industry is the lack of technology in both traceability and production of material and parts throughout the supply chain. Some companies, such as Toray that is a producer, use digitalization to track each spool of material. Other companies, such as the aerospace supplier described above, use little to no technology. The supply chain and logistics portion of the downstream carbon fiber industry also struggle with a lack of technology use, as well as inefficiencies along the way.

Implementation Challenges

Manual Entry

Manual entry is easier than implementing an unknown system, particularly for SMEs. A Fortune 500 company can spend time and money to gradually upgrade and deal with challenges from implementing a digitalized system (Rahman 2015; Buonanno et al. 2005). To upgrade a workflow that disrupts, even temporarily, an existing process and migrate from a manual to digital form takes effort and time. Digitalization of workflow and a QMS is costly, increases the complexity of the business and management workload, is rigid, and is time consuming. It can also be prohibitive to make minor changes to these systems as needed. As such, it is easier to proceed with the current process than to make a change (Pandey et al. 2012).

Cost

The carbon fiber industry is fragmented. Many small players earn thin margins and do not have the cash flow to make a significant technology investment and upgrade. Therefore, from a financial perspective, the industry is not making it a priority to invest in an upgrade despite the potential efficiency benefits (Stegkemper 2016).

Training and Staffing

With a new system, a company will either need to hire new workflow or IT staff or else train existing staff with new skills to maintain the digital system (Xue et al. 2013). In a Chinese company, turnover is usually high, on average 20% (Callegarin 2017), and companies fear training an employee who then leaves. This creates a culture where training takes a lower priority and staff are not proficient in these systems. Hai Di Lao is one company that bucked this trend by having a decentralized organizational structure (as opposed to hierarchical in most Chinese companies). It also encourages employees to innovate and, if the innovation is effective, to train other staff to improve efficiency and customer satisfaction (Chen and Zeng 2018). This approach drives customer service innovation and training the company is known for and contributes to lowering staff turnover, which is only 10% (Gu 2016).

Motivation

When a company produces carbon products for multiple customers, it will produce a specification for those companies that pay a price premium. For those that do not pay a price premium, a "gray market" is in place to continue providing products but at a lower price point and quality level. By deliberately not having a digitalized SOP, it provides the opportunity to bypass regulations, compliance, and lower cost (Durga et al. 2016). With this "gray market," companies run the risk of issues with customers, but at that point, a supplier can say it is the customer's problem because it was purchased at a lower price and quality expectation (Friedman 1998). Without digitalization, the opportunity for substituting materials, paying less tax, and bypassing regulations can be used (Stegkemper 2016).

Digitization Opportunities

From the author's experience, there is some digitalization of the supply chain such as the electronic signing of documents by a customs authority, but the majority of paperwork throughout the supply chain is still done by hand on paper. An opportunity could arise with downstream suppliers. At present, downstream users' levels of digitization vary greatly. Boeing or BMW uses extensive technology and digitalization (although Boeing has 737 Max issues, this is due to engineering, not material or the supply chain). However, smaller part producers in sporting goods, cost-conscious auto parts, and even aerospace firms use handwritten documentation systems extensively.

As early adopters, Boeing (*Digitalization Takes Off at The Boeing Company: Building a Next Generation Supply Chain – Technology and Operations Management* 2017), BMW (Boeriu 2015), and Apple (Adams 2018) are encouraging and, in some cases, forcing suppliers to take note and implement digitalization strategies. As this change picks up speed through global supply chains, especially those using similar materials and products, these firms will have to start digitalizing their businesses. This will drive industry and company development and initiate greater efficiency and growth in supply chains.

One possible solution, especially for small players, is a cloud-based ERP system. Initial investment is low and more palatable, and the ongoing maintenance is outsourced at a reasonable standard of quality. It could be more cost-effective rather than developing an in-house system and IT team for maintenance. Such a system could be developed enabling various entities from upstream to downstream suppliers and OEMs to share real-time data, thus greatly improving productivity, traceability, and transparency along the value chain. Many suppliers are not aware of the benefits this could bring, and those that do often shy away because of the cost and risks of implementation.

Conclusion

The carbon fiber industry is fragmented and dominated by SMEs. A traditional QMS relies on redundancy, but a digitalized QMS reduces redundancy and drives efficiency (Ibrahim 2019). A digitalized carbon fiber supply chain will help companies in a multitude of industries become more efficient in quality control, waste management, traceability, and compliance. Adhering to a rigorous program for quality control, which is implemented and applied by all employees and reviewed periodically, will result in reduced product delivery lead times, improved utilization of materials, increased employee productivity, and greater customer satisfaction. Besides, cost savings and enhanced safety for end users will also be a long-term contribution. To achieve this, it will take a change of end-user demand, potential consolidation, tighter regulations, or stricter implementation of current rules and regulations.

Key Takeaways

- Many industries have strict compliance and quality assurance requirements that create a need for detailed product information to move through the supply chain. Digitalization in supply chains, made possible by technology like QR codes and RFID, is the means by which this traceability is possible.
- Improvements to the traceability within the carbon fiber supply chain also enabled manufacturers to capture data on waste being generated during production using the same technologies implemented for traceability.
- Much of the move toward digitalization in supply chains particularly for players that are further upstream are being driven by their downstream partners. These often larger companies will push supply chains into the digital age and force competitors to digitize in the hope of keeping up.

References

Adams, K. (2018). Apple Inc.|2018 conflict minerals report. https://sec.report/Document/0001193125-20-026822/ (accessed 15 February 2020).

Ahuja, S. (2014). QR codes and security concerns. *International Journal of Computer Science and Information Technologies* 5 (3): 3878–3879. http://www.ijcsit.com/docs/Volume%205/vol5issue03/ijcsit20140503268.pdf (accessed 15 March 2020).

Aviation Quality Control – Aerospace AS9120|UPS – United States (2020). https://www.ups.com/us/en/services/aerospace/quality-management.page (accessed 15 February 2020).

Banker, S. (2017). Digitization and the 3M supply chain. *Forbes*. https://www.forbes.com/sites/stevebanker/2017/05/03/digitization-and-the-3m-supply-chain/#4f8a6ea041c1 (accessed 6 November 2019).

Barton, E. (2018). Carbon fiber fails|outside online. *Outside*. https://www.outsideonline.com/2311816/carbon-fiber-bike-accidents-lawsuits (accessed 15 March 2020).

Boeing (2020). Boeing: boeing services: parts support. *Boeing.com*. https://www.boeing.com/commercial/services/parts-solutions/parts/ (accessed 15 February 2020).

Boeriu, H. (2015). BMW uses digitalization to improve processes in production. *BMW Blog*. https://www.bmwblog.com/2015/08/11/bmw-uses-digitalization-to-improve-processes-in-production/ (accessed 15 November 2019).

Bremmer, I. (2019). The Japan–South Korea trade war is worrying for the world|time. *Time*. https://time.com/5691631/japan-south-korea-trade-war/ (accessed 15 February 2020).

Bulaeva, V. (2015). Final report QMS according to AS9100-C. *Windesheim University of Applied Sciences*. https://www.theseus.fi/bitstream/handle/10024/88898/Bulaeva_Valeriya.pdf?sequence=1&isAllowed=y (accessed 15 March 2020).

Buonanno, G., Faverio, P., Pigni, F. et al. (2005). Factors affecting ERP system adoption: a comparative analysis between SMEs and large companies. *Journal of Enterprise Information Management* 18 (4): 384–426. https://doi.org/10.1108/17410390510609572.

Callegarin, M. (2017). High employee turnover in China is not a matter of money|NEW POINT de VIEW. *New Point de View*. https://www.newpointdeview.com/pro-en/intercultural-compentence/high-employee-turnover-in-china-not-matter-of-money/ (accessed 17 February 2020).

Caridi, M., Moretto, A., Perego, A. et al. (2014). The benefits of supply chain visibility: a value assessment model. *International Journal of Production Economics* 151: 1–19. https://doi.org/10.1016/j.ijpe.2013.12.025.

Carnes, R.A. and Ringenbach, L.A. (1980). Disposal of hazardous waste: proceedings of sixth annual research symposium, research symposium disposal of hazardous waste proceedings of the 6th annual research symposium. https://nepis.epa.gov/Exe/ZyNET.exe/2000B1NV.txt?ZyActionD=ZyDocument&Client=EPA&Index=1976Thru 1980&Docs=&Query=&Time=&EndTime=&SearchMethod=1&TocRestrict=n&Toc=&TocEntry=&QField=&QFieldYear=&QFieldMonth=&QFieldDay=&UseQField=&IntQFieldOp=0&ExtQFieldOp= (accessed 14 March 2020).

Chen, H. and Zeng, Y. (2018). Employee motivation management of Hai Di Lao from the perspective of confucianism. *Chinese Business Review*. http://www.davidpublisher.com/Public/uploads/Contribute/5ceba8f55b8d8.pdf (accessed 21 March 2020).

Conaway, I. (2019). 6 reasons why manufacturing traceability is essential now|MECCO blog. *Mecco Blog.* https://www.mecco.com/blog-6-reasons-why-manufacturing-traceability-is-essential-now (accessed 15 February 2020).

Cook, J.J. Booth, S., Cook, J. et al. (2017). Carbon fiber manufacturing facility siting and policy considerations: international comparison carbon fiber manufacturing facility siting and policy considerations: international comparison (June).

Cooper, R. and Haltiwanger, J. (1993). The aggregate implications of machine replacement: Theory and evidence. *American Economic Review* 83 (3): 360–382. https://doi.org/10.3386/w3552.

Cority IQS (2014). ISO 9001 vs AS 9100 for the aerospace and defense industry – IQS blog|IQS blog. *Cority IQS.* https://www.iqs.com/blog/posts/iso-9001-vs-9100-aerospace-defense-industry/ (accessed 15 February 2020).

Crawford, K. (2016). Curious carbon fiber questions with composite specialists. *Front Street Media.* https://frontstreet.media/2016/09/12/curious-carbon-fiber-questions-with-composite-specialists/ (accessed 17 February 2020).

Curran, E. (2019). How the U.S.–China trade war got to this point – the Washington post. *Bloomberg.* https://www.washingtonpost.com/business/how-the-us-chinatradewargot-to-this-point/2019/11/08/1a48866e-01ed-11ea-8341-cc3dce52e7de_story.html (accessed 15 March 2020).

Ćwikła, G. (2013). Methods of manufacturing data acquisition for production management – a review|Scientific.Net. *Advanced Materials Research.* https://doi.org/10.4028/www.scientific.net/amr.837.618 (accessed 15 March 2020).

Department of Justice (2017). ZTE corporation agrees to plead guilty and pay over $430.4 million for violating U.S. sanctions by sending U.S. – origin items to Iran|OPA|Department of Justice. *Department of Justice.* https://www.justice.gov/opa/pr/zte-corporation-agrees-plead-guilty-and-pay-over-4304-million-violating-us-sanctions-sending (accessed 11 March 2020).

Department of Justice (2019). Iranian businessman pleads guilty to conspiracy to violate U.S. sanctions by exporting carbon fiber from the United States to Iran|OPA|Department of Justice, *Department of Justice.* https://www.justice.gov/opa/pr/iranian-businessman-pleads-guilty-conspiracy-violate-us-sanctions-exporting-carbon-fiber (accessed 17 November 2019).

nm13 (2017). Digitalization takes off at the boeing company: building a next generation supply chain – technology and operations management. *Harvard Business School.* https://digital.hbs.edu/platform-rctom/submission/digitalization-takes-off-at-the-boeing-company-building-a-next-generation-supply-chain/ (accessed 15 November 2019).

Dumas, B. (2019). China accused of mislabeling billions in goods for export, dodging US tariffs – The Blaze. *The Blaze.* https://www.theblaze.com/news/china-accused-of-mislabeling-billions-in-goods-for-export-dodging-us-tariffs (accessed 15 November 2019).

Durga, K., Aggarwal, P., and Kumar, A. (2016). E-Commerce frauds using parallel imports. *Journal of Management and IT* 7 (1): 12–15. http://iitmjp.ac.in/wp-content/uploads/2017/06/IT-Conference-2016.pdf#page=14 (accessed 15 March 2020).

Economic Cooperation Bureau (2011). Japan's export control system contents 1. Necessity of export control 2. *Japan's Export Control System.*

Elton-Walters, J. (2016). Cheap Chinese carbon imports: are they worth the risk? – Cycling Weekly. *Cycling Weekly.* https://www.cyclingweekly.com/news/product-news/

cheap-chinese-carbon-imports-are-they-worth-the-risk-173977#zsOZweB4SVejTJgh.99 (accessed 15 March 2020).

Fawaz, Z. (2016). Quality control and testing methods for advanced composite materials in aerospace engineering. In: *Advanced Composite Materials for Aerospace Engineering* (eds. S. Rana and R. Fangueiro), 429–451. Elsevier https://doi.org/10.1016/b978-0-08-100037-3.00015-8.

Federal Aviation Administration (2013). Standard parts. *Federal Aviation Administration.* https://www.faa.gov/aircraft/safety/programs/sups/standard_parts/media/standard_parts.pdf (accessed 15 February 2020).

Freeman, G. (2018). Quality in the aerospace industry. *Intelex.* https://blog.intelex.com/2018/08/10/quality-aerospace-industry/ (accessed 23 March 2020).

Friedman, L.M. (1998). Business and legal strategies for combating grey-market imports. *SMU.edu.*

Gu, B. (2016). Effects of psychological capital on employee turnover intentions. *Journal of Global Tourism Research* 1 (1): 21–28. http://www.union-services.com/istr/jgtr data/1_21.pdf (accessed 21 March 2020).

Han, S. (2018). What are the main causes of airplane accidents? – Everglades University. *Everglades University.* https://www.evergladesuniversity.edu/major-causes-of-airplane-accidents/ (accessed 6 November 2019).

Hussein, A. and Cheng, K. (2016). Development of the supply chain oriented quality assurance system for aerospace manufacturing SMEs and its implementation perspectives. *Chinese Journal of Mechanical Engineering (English Edition)* 29 (6): 1067–1073. https://doi.org/10.3901/CJME.2016.0907.108.

Ibrahim, R. (2019). Digital quality management systems: benefits and challenges. *International Journal for Quality Research* 1 (2): 163–172. http://www.cqm.rs/2019/papers_iqc/15.pdf (accessed 21 March 2020.

van Iwaarden, J. and van der Wiele, T. (2006). Innovative quality management cases – Google Books. *Google Books.* https://books.google.com/books?id=2jGHW57-pgIC&pg=PT109&dq=suppliers+using+existing+quality+control+systems&hl=en&newbks=1&newbks_redir=0&sa=X&ved=2ahUKEwib5c_DtJDoAhWC9Z4KHf0xD44Q6AEwAHoECAAQAg#v=onepage&q=suppliers using existing quality control systems (accessed 11 March 2020).

Johansson, B., Alajbegovic, A., Alexopoulo, V. et al. (2015). Cloud ERP adoption opportunities and concerns: the role of organizational size. In: *Proceedings of the Annual Hawaii International Conference on System Sciences* (eds. S. Abbas et al.), 4211–4219. IEEE Computer Society https://doi.org/10.1109/HICSS.2015.504, https://ieeexplore.ieee.org/stamp/stamp.jsp?tp=&arnumber=7069658.

Kurland, K. J. (2014). End-use monitoring and effective export compliance. moz-extension://f1e5d712-ca7a-4f0a-b8f1-fb26980ecf1d/enhanced-reader.html?openApp&pdf=https%3A%2F%2Fwww.bis.doc.gov%2Findex.php%2Fdocuments%2Fpdfs%2F1020-diversion-end-use-monitoring-concerns-and-best-practices%2Ffile (accessed 17 February 2020).

Lee, Y.N. (2019). US–China trade war: who pays for Donald Trump's tariffs?. *CNBC.* https://www.cnbc.com/2019/06/03/us-china-trade-war-who-pays-for-donald-trumps-tariffs.html (accessed 15 March 2020).

Livingston, B.A. and Marino, A.P. (2019). Boeing quality management system requirements for suppliers. *Boeing.* moz-extension://f1e5d712-ca7a-4f0a-b8f1-fb26980ecf1d/enhanced-reader.

html?openApp&pdf=http%3A%2F%2Fwww.boeingsuppliers.com%2Fquality%2FD6-82479.
pdf (accessed 15 March 2020).

Mauldin, W. and Nassauer, S. (2019). *Target tells its suppliers to handle tariffs costs – WSJ. Wall
Street Journal* https://www.wsj.com/articles/target-tells-its-suppliers-to-handle-tariffs-
costs-11567634705 (accessed 15 March 2020).

Ministry of Economy, Trade and Industry (2014). Electricity system reform/METI Ministry of
Economy. *Trade and Industry.* https://www.meti.go.jp/english/index.html (accessed 3
November 2019).

Moses, A. (2020). Difference between 4PL vs 3PL logistics – Penske logistics. *Penske Website.*
https://www.penskelogistics.com/solutions/supply-chain-management/lead-logistics-
provider/4pl-vs-3pl-differences/ (accessed 10 May 2020).

Mulln, K. (2017). Clever scammers are replacing mobike QR codes and stealing funds.
Technode. https://technode.com/2017/03/09/clever-scammers-are-replacing-mobike-qr-
codes-and-stealing-funds/ (accessed 15 March 2020).

Nanda, V. (2005). Quality management system handbook for product development
companies – Vivek Nanda. *Google Books.* https://books.google.com/books?id=guizsuAAyR4
C&pg=PA19&dq=suppliers+using+existing+quality+control+systems&hl=en&newbks=1
&newbks_redir=0&sa=X&ved=2ahUKEwib5c_DtJDoAhWC9Z4KHf0xD44Q6AEwBHoEC
AUQAg#v=onepage&q=suppliers using existing quality control systems& (accessed 15
March 2020).

Nicita, A. (2019). Trade and trade diversion effects of United States tariffs on
China – UNCTAD research paper no. 37. *United Nations Conference on Trade and
Development.*

Oschman, J.J. (2019). A conceptual framework implementing an AS9100 quality management
system for the aerospace industry. *South African Journal of Industrial Engineering* 30 (2):
1–16. https://doi.org/10.7166/30-2-1930.

Pandey, M.S., Jaiswal, D.M., and Purohit, D.G. (2012). *Challenges involved in implementation of
ERP on demand solution: cloud computing. International Journal of Computer Science Issues*
9 (4), *No 2*: 481–489.

Power, M. (2019). Trade compliance – protecting supply chains from the risks of global
trade – Supply Professional. *Supply Professional.* https://www.supplypro.ca/features/
trade-compliance-protecting-supply-chains-from-the-risks-of-global-trade/ (accessed 15
February 2020).

Rahman, S. (2015). ERP adoption in small and medium sized enterprises in Bangladesh.
Academia.edu. https://www.academia.edu/19853402/ERP_Adoption_in_Small_and_
Medium_Sized_Enterprises_in_Bangladesh (accessed 15 March 2020).

Research, G.V. (2017). Carbon fiber market size & share|Industry growth report,
2018–2025. *Grand View Research.* https://www.grandviewresearch.com/industry-analysis/
carbon-fiber-market-analysis (accessed 9 September 2019).

Schofield, D. (2018). Composites bucking the market trend in M&A: composites world.
Composites World. https://www.compositesworld.com/columns/composites-bucking-the-
market-trend-in-ma (accessed 11 March 2020).

ScrapRight Reviews and Pricing (2019). Capterra. https://www.capterra.com/p/130981/
ScrapRight/ (accessed 15 February 2020).

Sheu, C., Chae, B., and Yang, C.L. (2004). National differences and ERP implementation: issues and challenges. *Omega* 32 (5): 361–371. https://doi.org/10.1016/j.omega.2004.02.001.

Sivani, S. (2015). RX1E – world's first electric passenger plane made in China – The Green Optimistic. *The Green Optimistic*. https://www.greenoptimistic.com/rx1e-worlds-first-electric-passenger-plane-made-china-20150625/ (accessed 4 November 2019).

Stegkemper, B. (2016). Study digitalization of the supply chain successful management of aerospace supply chain networks challenges and solutions.

Toray (2019). What you should know before using TORAYCA®. *Toray*. http://www.torayca.com/en/notice/index.html (accessed 15 November 2019).

Warren, D.C. (2016). Carbon fiber precursors and conversion. *Oak Ridge National Laboratory*. https://www.energy.gov/sites/prod/files/2016/09/f33/fcto_h2_storage_700bar_workshop_3_warren.pdf (accessed 17 February 2020).

Xue, L., Zhang, C., Ling, H. et al. (2013). Risk mitigation in supply chain digitization: system modularity and information technology governance. *Journal of Management Information Systems* 30 (1): 325–352. https://doi.org/10.2753/MIS0742-1222300110.

Yoshida, K. and Matsuda, N. (2019). QR code scams strike China, from merchants to traffic tickets. *Nikkei Asian Review*. https://asia.nikkei.com/Economy/QR-code-scams-strike-China-from-merchants-to-traffic-tickets (accessed 15 March 2020).

23

The Evolution of Freight Forwarding Sales

Mac Sullivan[1], Dennis Wong[2], and Zheyuan Tang[1]

[1] *NNR Global Logistics, Dallas, TX, USA*
[2] *Flexport, San Francisco, CA, USA*

Introduction

As international trade has grown, the use of freight forwarders to facilitate international shipments has grown as they served as a core intermediary in the global logistic network. Freight forwarding has been a very customer-intensive service industry that creates economic value by not only facilitating the timely delivery of goods but also arranging other formalities, such as finding space with a carrier, creating and distributing documentation, and lastly ensuring that the goods get through customs (Shang and Lu 2012). Recently, freight forwarding has fallen behind other industries as it is still very reliant on manual processes, including emails, phone calls, and even faxes in some cases, which are all prone to error and time-consuming (Riedl and Chan 2019). Traditionally, freight forwarding has very much been a relationship-based business-to-business (B2B) service industry with little sway being held by online forums, market reviews, and sales channels as noted in Chapter 11. Technology has now enabled a global resource pool to compete, connect, and collaborate on a variety of business processes for a fraction of the cost and effort it took 10 years ago.

The Rise of the Digital Freight Forwarder as a New Entrant

Digital freight forwarders utilize platforms to offer a range of logistics services with the value proposition of frictionless customer experience in a single-user online portal (Riedl and Chan 2019). The rise of the digital freight forwarder is progressing more quickly than most expected but perhaps not as fast as others thought. For the most part, relationships between many large beneficial cargo owners (BCO) and their freight forwarders have been built through face-to-face meetings, phone calls, and emails. This idea that a personal relationship helps strengthen a buyer–seller relationship in the supply chain world has been investigated and is still open for further research (Butt 2019). That being said, without an alternative to this relationship building, freight forwarding sales executives are under threat from nearly all of what Michael Porter called his five competitive forces; digital freight forwarders like Flexport and entry of e-commerce companies like Alibaba and Amazon (new entrants/increased rivalry), shipping lines selling directly like Maersk (substitution via intermediation), and the rise of marketplaces like Freightos or Uber Freight (a lower barrier of entry/increased buyer's bargaining power). All of these forces are making traditional freight forwarders reevaluate the way they sell and their underlying value proposition. The threat of a new entrant, like a digital freight forwarder, can hurt the profitability of incumbents as they have to either hold down their costs to retain customers or in this case make significant investments in technology to keep up with the market (Porter 2008). The effects of these new entrants have less to do with market share erosion than the fear, uncertainty, and doubt (FUD) in the boardrooms of many traditional freight forwarders.

Companies like Freightos, a leading international freight marketplace, have yet to put a dent in the likes of the top 25 freight forwarders as trust in digital marketplaces is still lacking (Leuschner et al. 2014; Armstrong and Associates 2019). This lack of trust is driven by the risk that either party faces as a BCO give their cargo to an unknown freight forwarder or the freight forwarder accepts the freight of an unknown BCO. The role of indirect procurement in a multinational BCO, the individual that often decides on which freight forwarding to use, has long been held by professionals that place a significant amount of emphasis and weight on the level of trust that they have with their appointed suppliers (Alkhatib et al. 2015). These individuals are seeing the sharing economy grow around them, which will change buying behavior, and the technology of the market will inevitably catch up to the point where they will begin to adopt new forms of digital procurement (Glas and Kleemann 2016).

Market Shifts

Amazon Effect

E-commerce grew by 16% in 2019 and double digits in the previous four years (Young 2019). BCO customers are demanding more visibility, more timely quotes, better billing, and a more intuitive interface. In terms of visibility, customers are shocked when they can have a 5-dollar coffee mug delivered, for free, with real-time turn-by-turn visibility from a distribution point to their door, but can lose visibility of millions of dollars' worth of goods for days at a time as it is in transit from overseas. The Journal of Commerce reported in 2019, "Amazon has surpassed major global players such as DHL and Kuehne + Nagel in 3PL revenue" (Field 2019). Amazon and its Chinese rival Alibaba are not only amassing a captive consumer base but also creating sophisticated logistics networks backed by world-leading data scientists and aggressive expansion plans. In 2016, Amazon filed for its license as a non-vessel-operating common carrier (NVOCC), granting the e-commerce giant the ability to create its own bill of ladings. CEO of Flexport, Ryan Peterson, commented, "It's here that automation, something no traditional freight forwarding company can do even one percent as well as Amazon can, becomes the key competitive advantage over legacy freight forwarders" (Staff 2016). Alibaba, in turn, decided to not only start to offer its suppliers the ability to get a line of credit on local freight forwarding charges but also be able to book directly with some of the largest shipping lines in the world. There is a product offering to which Freightos CEO, Zvi Scheiber, commented, "For Alibaba, this is a direct challenge to global retailers like Amazon. Beyond drones and futuristic supermarkets, Amazon opted to get licensed as a forwarder (NVOCC). Alibaba one-upped them by going directly to the world's largest ocean liner. Point, Alibaba" (Luo 2017).

Instant Visibility and Pricing

Returning to the demands of freight forwarding service buyers in terms of visibility, it is important to note that some providers have paid special attention to the location and condition of shipments out of necessity (Wei and Wang 2010). Pharmaceutical, cold chain, aerospace, and automotive customers have always demanded their providers to provide accurate data on the condition and status of their goods, but they often supported this through higher volumes or higher margins as noted by Wang and Kern as well as von Stempel in this volume. For other industry verticals who simply want their freight moved, the best they can expect from their freight forwarder is basic milestones about when the shipment has completed one step along the journey. Freight forwarders are frantically working to close this gap. Data aggregators, who provide more complete visibility for shipments, like Fourkites, Project44, or Overhaul, are taking advantage of the lack of connectivity between the ports, the trucking companies, and the air and ocean carriers (Johnson 2020).

In terms of quotation turnaround, freight forwarding companies have turned to more and more sophisticated rate management systems, such as Catapult, CargoSphere, and WebCargo, which enable global sales and operations to filter negotiated carrier rates (Johnson 2018). Many of the top 20 freight forwarders and pretty much all of the digital freight forwarders have then taken the step to put a nice interactive user interface (UI) on top of their rate management tool, with some subscribing to or building their own transportation management system (TMS) (Forand 2014). This enables internal customers to be able to query rates on their own time as well as external customers if they are connected to an online portal with an added margin on top. A number of companies, such as the United Kingdom's Kontainers, are making their mark serving this niche market by providing a customizable, white-labeled user interface where customers can go online, grab rates, and, in some cases, book a shipment without ever having to communicate with a freight forwarding representative (KONTAINERS 2019; KONTAINERS 2020). Freightos, the Israeli startup previously mentioned, now offers the ability for a BCO to go and rate shop on their marketplace without even needing to log in.

Many freight forwarders complained that the Shanghai Container Freight Index was decreasing their ability to make a margin by exposing freight rates to the open market, the abovementioned marketplaces and portals are going to take this to a new level (Yang 2016). Buyer's enticement will be too much to ignore, even if you risk upsetting your current key supplier partners (Harris 2000). That being said, a study by Freightos in 2019 shows only 18% of quotations done by a mystery shopper from top freight forwarders received an instant quotation, while as high as 60% of quotation requests did not receive any response (Freightos 2019) (Figure 23.1).

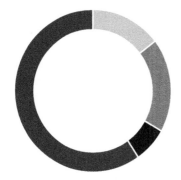

60% - received no response

18% - received instant quotation

15% - manual quotation

7% - received quotation in <24 hours

Figure 23.1 Procurement attempts via forwarders and ocean carriers. *Source*: Based on Freightos (2019). © John Wiley & Sons.

With good instant pricing mechanisms, not only are freight forwarding companies going to save time and resources, but also this will be another way for new startups and technology companies to usurp market share from established players. An inevitable shift occurs: current freight forwarding companies must both protect themselves and expand into potential customer markets. Freight forwarding companies are hiring more data analysts, computer programmers, and software architects than ever before. Instant pricing is going to automate a large part of the traditional sales process, and the sales teams must adapt to create alternative value.

Redefining Customer Service

Freight forwarders have often been tasked with aggregating tracking from different carrier vendors, and the demand for this service is only increasing as customers expect more than just facilitating the movement of freight. Automated solutions like screen scraping, where a computer bot is trained to go out to the website of the carrier vendors and pull back milestones, or the simple retrieval of a milestone from a tracking database can now be pushed to the front-end customer-facing portal. For questions regarding a quote, certain logistics term, or even shipment locations, companies like RPA Labs are creating interactive chatbots that can take and respond to questions all hours of the day. More than 20% of top freight forwarders are offering live chat functions on their online sales portals (Freightos 2019). These tools are providing a frictionless customer experience as the customer goes to quote and troubleshoot their shipments. Automation via chatbots and online rate quoting will push sales teams to upskill and adapt so that they can leverage technology to find new ways to capture more business.

The Evolution of Freight Forwarding Sales

The Traditional Sales Representative

A majority of freight forwarding sales representatives worked in the logistics industry in operations before taking a sales role, so they know the ins and outs of the shipping process and the lingo (Kaschek 2014). The sales process and much of the revenue for freight forwarding companies were driven mostly by the individual sales reps who spent much of their time visiting the customer in person, building trust and developing a relationship with the logistics manager or indirect procurement team, keeping up with market competitive offers, and going through the rigorous exercise of monthly ad hoc, spot quotations or an annual request for quotations (RFQs) (Kaschek 2014). In a digital economy, where mobile technology and quick response times are expected, it is hard for freight forwarding companies to imagine scaling this type of selling. Sales managers are having to upskill their salesforce through training on different tools like LinkedIn, CRMs, and data mining tools like Panjiva or Datamyne (Ahearne et al. 2005).

While interviewing an industry veteran that started as a freight forwarding sales in the mid-nineties, he shared that he got into the logistics business during the golden age, where global trade was growing by double digits annually and profit margins for both the BCO and the freight forwarder were strong. He notes, "Initially logistics was not an industry that attracted young talent, but suddenly this low-end industry became high demand with double-digit growth with needed sophistication for international expertise." He continued to say that the logistics industry has traditionally been a tightly knit large social network, where many people knew each other, and it was often said that "customers follow the sales rep, not the service provider" (Anonymous 2020). The landscape is evolving though, and salespeople are having to learn about the effect that digitalization is having on their job (Elste 2016).

Traditional Sales Reps Evolving

As Bessen states, "Innovative technology is displacing workers to new jobs rather than replacing them entirely"(Bessen 2015). For the logistics salesperson, platforms and chatbots will create the need for these salespeople to upskill and become supply chain optimization consultants as shared by Sun in this volume. A successful salesforce will evolve from a product pusher and "value communicator" to a "value creator"(Rackham 2020). There is now a better chance for a customer to directly purchase services from a freight forwarder earlier in the sales cycle in this market if they have the digital brand power backed by known service quality and pricing (Bienstock et al. 1997; Chumpitaz Caceres and Paparoidamis 2007). Digital marketing is showing the potential to positively impact customer acquisition and shorten their sales cycle (Kaschek 2014). Customers today are no longer looking for the sales rep that takes them out for dinner and a baseball game as compliance rules have changed and cost-cutting targets are higher (Josephson 2014). BCOs want a better experience using a provider that can move their cargo at the right price, the right transit time, and the right visibility, alert them of any changes as soon as possible, and give them additional tools with as little hassle as possible. Customers need their logistics partner to help them stay informed about the market and any potential threats to their supply chain. If they can shop and procure freight online, they are now in need of an experienced professional that can deal with understanding the intricacies of international trade and optimizing their transportation network, without the threat of simply being upsold. This demand is leading companies, like Maersk, to offer supply chain and logistics consultations free of charge to simply have the opportunity to get to know their clients (A.P. Moller – Maersk 2020).

Rise of Technology and Its Effects on Sales

From the book *Challenger Sales* by Matthew Dixon and Brent Adamson, new-age sales are one that challenges the customer, where a salesperson isn't afraid to take control and sales are won through teaching and not persuading (Dixon and Adamson 2011). The true technological advances will provide full transparency in supply chain cost, inventory management, product profitability, and cash flow across digital platforms. Some logistics players are entering into this market doing just that. A large China-based online retailer with their own logistics arm was interviewed and was asked what differentiates them from the competition. He mentioned, "Our strengths are not global network nor shipment visibility, our values are from upstream data information such as inventory data control and even before the manufacturing process such as management of raw material before entering the factory" (Anonymous 2020). The ability to read and understand data is becoming a more critical tool in logistics sales. A growing market, where customers demand quality information and the need for service providers to keep track of all of this data being generated from moving a shipment, has led to a rise in the development and adoption of enterprise resource planning (ERP) systems (Furtună and Bărbulescu 2011).

Connectivity

The COVID-19 situation in 2020 exposed further connectivity opportunities, especially in the inefficiencies of the freight forwarding industry and the need for more remote business applications. Sales will also need adjustments as face-to-face opportunities cease to exist. Before the pandemic, applications such as Slack, Zoom, and Microsoft Teams have already gained a foothold in many companies' internal communications; the pandemic has only accelerated that shift. Microsoft CEO Satya Nadella mentioned Microsoft Teams now has 75 million daily active users in an April 2020 earnings call and is expecting to release a consumer version of the software later in the year (Zaveri 2020). In China, WeChat Enterprise and DingTalk by Alibaba are rapidly evolving workplace and consumer habits, allowing for more direct seller–buyer engagement, easier control of company data, and ultimately increasing consumer expectation for what business in 2020 and beyond looks like.

To make sales in the future, sales teams must expand their communication channels and have an uninterrupted flow of data. Potential customers will want to do much more from afar before they are willing to meet in person. Not only are habits changing, but also potential customers are more likely to be thousands of miles apart than ever. It is important sales teams sharpen their remote conferencing and sales skills as more and more businesses are conducted online. In those few minutes, every detail could make or break a sale. Whether its salesperson-related, such as experience with the tech platform or way they appear on the screen, or environment-related, such as lighting and sound quality, a more diverse set of factors could now impact the sales experience.

Companies also prefer WeChat Enterprise to the stand-alone WeChat in China because the companies become the owner of customer's data and transactions. If a salesperson leaves, the customer's data is still retained by the company and is able to be used by the next salesperson. This continuity in customer understanding will allow for closer engagement between vendors and customers, as well as more loyalty and a higher retention rate. Customers are also expecting to receive updates more frequently than a few emails throughout the day. Staying connected to customers not only is a chance to know them better but also lowers the chance of them going to a competitor. If a company can find or actively solve a problem before anyone else, they will win.

Freight Forwarding Sales Executives at Risk of Losing Their Jobs?

Peter Thiel author of *Zero to One* said, "If your product requires advertising or salespeople to sell it, it's not good enough: technology is primarily about product development, not distribution," but he reiterates, "in business, sales is a vital necessity, innovative products are worthless unless they are sold" (Thiel and Masters 2014). The freight forwarding sales role is not coming to extinction but more of an evolution from a transactional relationship-driven to a consultant business-driven role. Fernando Villar from Maersk Head of E-commerce and M&A Integrations says, "The evolution of digitalization in logistic sales is shifting rapidly from physical transactional sell to value-oriented sell." Villar gives some insights on today's traditional sales versus digital platform ratio, "Today about 98% of sales are still via traditional salesperson versus 2% coming from our digital platform, but in the coming years we anticipate this will shift closer to 70% digital platform" (Villar 2020). The digital platforms will replace human sales with an online inquiry and digital upselling such as insurance, warehousing, and customs. Villar also mentioned, "Potentially in the next 2–3 years, there will only be 3–4 companies you can deal with, but the race is wide open today, it's a win or lose mode, but no one knows who has the right formula" (Villar 2020). Carriers are investing in digitalization, traditional forwarders are reinventing themselves, cargo owners have their own logistics arm, and digital freight forwarders are making an impact to change the industry.

Relationships between sales executives and clients will transition to an advantage, but away from a solution. As fast as technologies are changing, the people are still what make decisions. Business leaders remain reluctant in sharing sensitive business information with others where personal relationship is absent. In many ways, the lack of a personal relationship can generate severe negative outcomes such as lack of trust, longer resolution time for issues, and serious divergence in future plans (Butt 2019). Retaining relationships as the world marches into a digitalized age will allow these companies a head start that could ultimately help them defend against new entrants to the industry.

Analytical and Creativity as Valued Skills

A human touch in logistics is still needed, while we can see port-to-port service as a commoditized product; freight forwarding has changed where professionals that can help customers with supply chain will become more sought after as shown in Chapter 11. With the Amazon effect, more companies are entering the international trade as the barrier to entry is much lower. E-commerce is becoming a big part of the supply chain: an estimated 30% of cargo moved from Asia is via an e-commerce retail channel, and this figure will continue to grow (US International Trade Administration 2019; Young 2019). With complex global regulations and niche local value-added service expertise such as warehousing, fulfillment, and customs clearance, a seasoned logistics professional is even more needed to provide consultative advice, and they need to be back by an operational team that has the ability to execute. An interview with an industry veteran mentioned, "Artificial intelligence will aid in interpreting large volumes of data; however, it will not build trust, develop relationships, have conversations, change minds, or share a vision" (Anonymous 2020). While data analytics can be a powerful tool, there is still a need for a human to put this insight into perspective and pitch a client on the value proposition for the foreseeable future (McKinsey 2016).

Future Job Market

The Boston Consulting Group estimates that the automation of back-office and operational tasks will help logistics companies cut their cost per shipment by up to 40% (Riedl and Chan 2019). By automating a job, it doesn't necessarily mean that the worker should be let go. The automation should improve productivity, which should drive operational expenses down and therefore the cost of the service down, which should in turn increase demand and the end call for an upskilling for the worker (Bessen 2015). What Atkinson and others argue is that this doesn't necessarily mean that employment will drop off; if anything, this could open up new sectors and jobs. Companies that invest in process innovation should be able to increase productivity, lower costs to their customers, and create new more fulfilling jobs (Atkinson et al. 1999). To take this one step further, we can follow the line of reasoning of Say's Law, which states that if automation presents more productivity, therefore lower costs, it can provide higher wages for employees that remain. Say noted, "As each of us can only purchase the productions of others with his own productions – as the value we can buy is equal to the value we can produce, the more men can produce, the more they will purchase"(Wood and Kates 2000). This would signal that we are not all going to be out of a job as we are going to be more productive and this will spur economic activity that will lead to more products being transported.

The job market today is already much more data and analytically oriented than the last decade. In a study of seven large logistics companies, data and software positions make up the bulk of their openings. Roles such as analysts and consultants are also more in demand than carrier relations or operations, signaling a major shift from to value creators and influencers. Individuals are no longer looked at as an extension of the company's products or brand, but as creators that bring unique expertise and insight to specific aspects of operations. There is still room for relationship building here, as customers will prefer to work with salespeople that understand their region, ship route, or company background. Successful salespeople in the future will need to combine those skills with data and analytical skills to effectively give customers the new answers they are looking for (Figure 23.2).

Sales Enablement in a Digital Age

Sales enablement is the process of providing the sales organization with the information and tools that help salespeople sell more effectively. Sales enablement has been adopted by the tech industry in the last decade. For many companies, logistics products still need to be sold by a salesperson, and to maximize individual sales, you need to provide them with all necessary information, tools, and training to help them succeed. CRM platforms and communication platforms like WeChat Enterprise and Microsoft Teams have also made it easier to tackle customers with a group. Sales could no longer be an individualistic role, but rather a combination of various positions all support multiple clients. The communications and collaborations between internal employees are now needed more than ever.

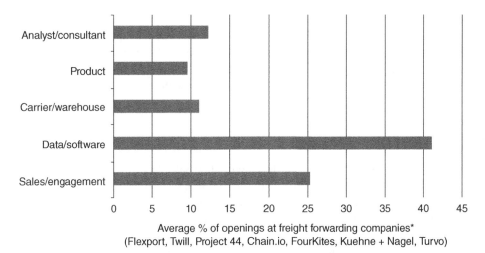

Average % of openings at freight forwarding companies*
(Flexport, Twill, Project 44, Chain.io, FourKites, Kuehne + Nagel, Turvo)

Figure 23.2 Breakdown of openings at freight forwarding companies, based on companies' June 2020 job postings.

Information

In international logistics, it is increasingly more important to figure out what the customer values in terms of price, speed, or quality to best match their expectations (Andreasson and Liu 2010). If a service provider has access to their customer's data, then they can foster longer-term collaborative partnerships. In many ways, sales enablement comes back to giving sales teams access to tools and training that can provide them with this information (Kaschek 2014). Pre-meeting research has never been more important. One of the authors of this chapter vividly remembers while working for a flooring company a meeting where the salesman from the logistics service provider CH Robinson was able to tell the amount of volume that was shipped with them on which lanes, as well as the information of his competitor and even which carriers, were used. The sales representative also offered guidance on warehouse locations, potential supplier bases in alternative lower-risk origins, and services to manage purchase orders.

Digital forwarders look for sales staff that are coming from outside of the industry who can bring a new perspective and new value proposition (Kersten et al. 2017; Riedl and Chan 2019). An industry veteran-turned logistics headhunter states, "Successful new-age digital sales candidates can identify client personas and effectively leverage data and social platforms" (Hextall 2020). Rolodex and Excel sheets are being replaced by customer relationship management (CRM) systems. Customer relationship management, as a structured practice, is increasingly more important to companies as industries like logistics are extremely competitive, and services that used to be unique and somewhat rare to find are now commodities (Duru et al. 2013). The use of a CRM system reduces the cost of customer relationship management and has the potential to increase customer satisfaction (Xu and Walton 2005). Integration between a CRM and an ERP is needed to help a logistics company figure out the desired value of a customer and mitigate any disconnection felt as the emotional bond will be harder to communicate in digital sales (Danielle and Fredrick 2019). It is key to choose the right customer relationship management (CRM) tool that fits your company (Buttle and Maklan 2019). There are CRM systems such as Microsoft Dynamics, Salesforce, and Hubspot that offer a wide range of services (D'Angelo 2020). The success of a CRM tool lies with the users knowing how to use the system and the tool being integrated with core systems like your ERP.

Training

Most freight forwarders lack a focus on developing their sales team with proper training. There are countless training programs in traditional freight forwarding companies that focus on operations and leadership with management training programs, talent programs, and executive training programs, but very few would invest in sales. Companies like Flexport have a training program called Flexport Academy for all employees from the start of their first day. The academy is designed to help all employees understand the freight forwarding business and the vocabulary of the industry (DeLacey 2019). The two-week program first focuses on basics and then will branch off to dedicated fields such as sales. Within the sales training portion, the individual will be introduced to the CRM tool, sales process, and sales techniques and will have a full array of sales resources to help them get started. The academy ends with shadowing sessions with experienced sales colleagues, which will help get the new team member up to speed with their role. As sales roles are making the shift to using more digital solutions to win business, training programs like these will be extremely important since tech-savviness will vary between members of a sales team, and some will require instruction to realize the benefits of using a digital tool such as a CRM (Kodwani and Prashar 2019).

Talent

Human capital must be either purchased by investing in those with on-the-job experience or grown through investment in knowledge training those without experience (Naughton 2007). Digital forwarders are attracting sales talent from tech industries, finance, and even healthcare because logistics has become attractive for young talents. An interview with a logistics recruiter said,

> Candidates want a company that encourages innovation and creativity with an open, collaborative work environment that embraces diversity and inclusion. A company's online presence and social media strategy is also a big draw as Generation Z aspires to work with cutting-edge technology (Hextall, 2020).

Since the essence of selling freight forwarding is changing, companies like Flexport are using this as an opportunity to invest in a different acumen set that prioritizes skills, not experience. Hiring people from outside the logistics industry helps to bring a new element of thinking and broaden the mindset of the organization compared to hiring traditional logistics sales representatives. Tegmark states that the highest levels of creativity from all forms of industries were not motivated by profit, but rather by intrinsic behavioral

Figure 23.3 Top skills in demand.
Source: Van Nuys 2019. © Amanda
Van Nuys.

Skills companies
need most in 2020

Top 5 soft skils	Top 5 hard skils
1. Creativity	1. Blockchain
2. Persuasion	2. Cloud computing
3. Collaboration	3. Analytical reasoning
4. Adaptability	4. Artificial intelligence
5. Emotional intelligence	5. UX design

*data from linkedIn of over 660 + million professionals and 20 + million jobs

characteristics like curiosity, peer appreciation, and a drive to innovate (Tegmark 2017). When employees are allowed to fulfill what they believe is their full potential and are given permission to be creative instead of being pressed into a certain mold, it is beneficial to not only the companies but also other employees in the workplace.

Companies are looking more and more for the ability to actively learn new skills post-interview. Companies are beyond expecting changes; they are anticipating them and preparing for them. Having employees that are creative, curious, and independent and have high emotional intelligence is just as important as ever. Hard skills can be taught through training and certification courses, but soft skills are much harder to change and develop. Executives are putting a high emphasis on soft skills because they are what makes up a company's direction and culture. Not all employees will be asked to use artificial intelligence applications every day, but they will need to be creative, curious, and independent (Figure 23.3).

New Cultures and Skills

We will need employees and individuals that are creative and outside-the-box thinkers. In an industry where creativity is not necessarily valued and sometimes discouraged, creating a culture of innovation is daunting. Gordon Ritter, founder and general partner at Emergence Capital, who is heavily investing in AI companies, says, "The only way for you to have value as a human is to think originally." Ritter says, "If you don't stand out in some way, you fit into the established models, which means the AI system is already done with you" (Hesseldahl 2018). It is the responsibility of business leaders and managers to foster a culture of innovation where employees are encouraged to share ideas and try new things. Employees will need to be enabled to be innovative in their work, and leadership must create an environment for this innovation to happen. For many freight forwarders who have not valued this type of thinking in the past, pushing employees to come up with creative solutions will require a huge shift in company culture.

Reengineering value propositions, execution strategies, and sales models will be constrained by the limitations of today's managers and will lead to conservatism and a retreat to traditional ways. The current industry has been built by the current management, which means there is a lot of room for the next generation to emerge. Digital forwarders try to

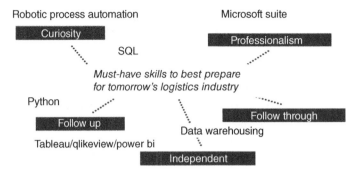

Figure 23.4 Skills to make any candidate valuable moving forward.

avoid hiring traditional sales because of the mindset and habits of selling transactional freight. Flexport is known for hiring only one of five people from the logistics industry, and this formula has shown to be successful when a product can be differentiated in the market. Successful sales in the future, just like the past, will be a combination of people and technology. As technology changes, people must too. It is crucial to remember that the mission of a salesperson is to make the customers' experience better. In the future, sales teams must deliver what the customers want, and technology will be a great way to help themselves do so (Figure 23.4).

The logistics industry is filled with established players. However, it is also in dire need of new blood. To prepare for tomorrow's logistics industry, candidates and salespeople should familiarize themselves with the makeups of the modern communication systems. Programming and data languages like Python and SQL, which can be easily introduced with a few hours on YouTube, can significantly lower training costs and increase competitiveness in the future. Understanding how data warehousing and programs like Tableau work will result in higher value creation and allow for more flexibility in future career paths. This is a great time to learn the skills of improving the logistics industry and carve out a space that previously did not exist.

In a recent conversation with a recruiter within the logistics industry, the recruiter mentioned that when asked about what type of company you would be interested in working for, 6 out of 10 candidates mentioned a modernized digital forwarding company (Hextall 2020). These candidates know how manually driven the industry is, and there is room for startups that have ambitions to bring technology to automate the operations process (Glave et al. 2014; McKinnon et al. 2017).

Conclusion

Freight forwarding sales has increasingly become more difficult over the past 30 years as competition became fiercer, the internet offered more transparency, and customers began to pay more attention to logistics cost as it grew to be a percentage of landed cost. These factors are diminished in scope by what is happening today as online marketplaces gain tractions, digital forwarders aggressively attract attention and further investigation by larger clients, carriers and e-commerce giants enter the market, and perhaps, most importantly, the demographic shift in decision-makers to younger, more technologically savvy individuals plays out. Automation of documentation, customer service, quoting, and billing are enabling a new generation of logistics providers that will quickly be able to disrupt the status quo of the market as they can begin to offer services at a discount in areas where labor costs and quarterly earnings reports have typically left traditional forwarders either sacrificing their already dismal returns or having to investigate new forms of operational excellence by pushing down pressure to the commercial rank and file. It is imperative for those who have a couple of decades in this industry left to investigate how technology is going to change the value proposition that you offer today and whether it is a sustainable means of survival.

Key Takeaways

- There is still a role for the commercial freight forwarding executive, but it will no longer be driven by margins on international freight, it will be more value-driven and consultative.
- It is important not to underestimate the potential impact of disintermediation as suppliers are increasingly more capable of going direct to the cargo owners.
- Marketplaces will have an increasingly larger role in freight forwarding sales as user demographics shift and trust in the capacity of buying freight online.
- Incumbent freight forwarding companies will have to make strategic investments in software to satisfy customer demand, and many of them are not equipped with the talent or resources at this point.

References

A.P. Moller – Maersk (2020). Supply chain development, corporate website. https://www.maersk.com/solutions/supply-chain/development (accessed 10 Jan 2020).

Ahearne, M., Jelinek, R., and Rapp, A. (2005). Moving beyond the direct effect of SFA adoption on salesperson performance: training and support as key moderating factors. *Industrial Marketing Management* 34 (4): 379–388. https://doi.org/10.1016/j.indmarman.2004.09.020.

Alkhatib, S.F., Darlington, R., and Nguyen, T.T. (2015). Logistics service providers (LSPs) evaluation and selection. *Strategic Outsourcing: An International Journal* 8 (1): 102–134.

Andreasson, L. and Liu, S. (2010). European RoRo short-sea shipping – what can ship operators do to unleash its potential? *Management*.

Anonymous (2020). Role of technology for a Chinese retailer.

Armstrong & Associates (2019). A&A's top 50 global third-party logistics providers (3PLs) list. https://www.3plogistics.com/3pl-market-info-resources/3pl-market-information/aas-top-50-global-third-party-logistics-providers-3pls-list/ (accessed 05 February 2020).

Atkinson, A.B. et al. (1999). *The Economic Consequences of Rolling Back the Welfare State*. Cambridge, MA: MIT Press (Munich lectures in economics). Available at: https://books.google.com/books?id=C_yWKPaULG0C.

Bessen, J. (2015). Toil and technology: Innovative technology is displacing workers to new jobs rather than replacing them entirely. *Finance and Development* 52 (1): 16–19.

Bienstock, C.C., Mentzer, J.T., and Bird, M.M. (1997). Measuring physical distribution service quality. *Journal of the Academy of Marketing Science* 25 (1): 31.

Butt, A.S. (2019). Absence of personal relationship in a buyer–supplier relationship: case of buyers and suppliers of logistics services provider in Australia. *Heliyon* 5 (6): e01799. https://doi.org/10.1016/j.heliyon.2019.e01799.

Buttle, F. and Maklan, S. (2019). *Customer Relationship Management: Concepts and Technologies*. Routledge.

Chumpitaz Caceres, R. and Paparoidamis, N.G. (2007). Service quality, relationship satisfaction, trust, commitment and business-to-business loyalty. *European Journal of Marketing* 41 (7/8): 836–867. https://doi.org/10.1108/03090560710752429 M4 - Citavi.

D'Angelo, M. (2020). The best CRM software of 2020.

Danielle, C. and Fredrick, K. (2019). Investigative analysis of CRM system implementation and the challenges attached to its integration into the existing ERP system from an organization perspective.

DeLacey, P. (2019). Flexport academy: how new hires set sail with flexport. http://www.flexport.com/blog/flexport-academy-how-new-hires-set-sail-with-flexport/ (accessed 7 December 2019).

Dixon, M. and Adamson, B. (2011). *The Challenger Sale: Taking Control of the Customer Conversation*. Penguin Publishing Group.

Duru, O., Bulut, E., Huang, S., and Yoshida, S. (2013). Shipping performance assessment and the role of key performance indicators (KPIs): "quality function deployment" for transforming shipowner's expectation. SSRN *Electronic Journal*. doi: 10.2139/ssrn.2195984.

Elste, R. (2016). Paradigmenwechsel im Vertrieb – Konsequenzen neuer Technologien für das Kundenmanagement. In: *Digitalisierung im Vertrieb* (eds. L. Binckebanck and R. Elste), 3–27. Wiesbaden: Springer Fachmedien Wiesbaden. https://doi.org/10.1007/978-3-658-05054-2_1.

Field, A. (2019). Top 50 logistics providers: amazon debuts atop global 3PL rankings. *Journal of Commerce*. https://www.joc.com/international-logistics/logistics-providers/amazon-debuts-atop-global-3pl-rankings_20190420.html (accessed 8 January 2020).

Forand, P. (2014). Pros and Cons of using a TMS platform for managing freight costs. https://www.re-transfreight.com/blog/pros-cons-using-tms-platform-for-managing-freight-costs (accessed 02 March 2020).

Freightos (2019). The State of Online Freight Sales 2019. https://www.freightos.com/wp-content/uploads/Research/201912 Mystery Shopping - If You Build It. Final.pdf (accessed 16 March 2020).

Furtună, T.F. and Bărbulescu, A. (2011). Data mining models applied in customer relationship management system. *Economy Informatics* 11 (1): 178–191.

Glas, A.H. and Kleemann, F.C. (2016). The impact of industry 4.0 on procurement and supply management. *International Journal of Business and Management Invention* 5 (6): 55–66.

Glave, T., Joerss, M., and Saxon, S. (2014). The hidden opportunity in container shipping. https://www.mckinsey.com/business-functions/strategy-and-corporate-finance/our-insights/the-hidden-opportunity-in-container-shipping (accessed 07 Jan 2020).

Harris, R. (2000). Buying and selling in a digital world. *Strategy & Leadership*, pp. 5–10.

Hesseldahl, A. (2018). If artificial intelligence changes everything at work, then education must change, too|CIO. *Cio*.

Hextall, A. (2020). New-age salespeople.

Johnson, E. (2018). INTTRA's rate management foray unnerves some partners. *Journal of Commerce*. https://www.joc.com/technology/inttra-rate-management-foray-unnerves-some-partners_20180503.html.

Johnson, E. (2020). Acquisition expands FourKites' freight visibility to the yard. *Journal of Commerce* https://www.joc.com/technology/supply-chain-visibility/acquisition-expands-fourkites'-freight-visibility-yard_20200325.html.

Josephson, M. (2014). History of the integrity, ethics and compliance movement: a cautionary tale for CEOs and corporate directors. *ETHIKOS* 28 (1): 13–15.

Kaschek, B. (2014). *Vertrieb für Logistikdienstleister*. Wiesbaden: Springer Fachmedien Wiesbaden https://doi.org/10.1007/978-3-658-04358-2.

Kersten, W., Seiter, M., von See, B., Hackius, N., and Maurer, T. (2017). *Trends and Strategies in Logistics and Supply Chain Management*. Hamburg: BVL International.

Kodwani, A.D. and Prashar, S. (2019). Assessing the influencers of sales training effectiveness before and after training. *Benchmarking: An International Journal* 26 (4): 1233–1254. https://doi.org/10.1108/BIJ-05-2018-0126.

KONTAINERS (2019). Who are kontainers?. https://kontainers.com/videos/kontainers-video-who-are-kontainers.html (accessed 27 February 2020).

Leuschner, R., Carter, C.R., Goldsby, T.J., and Rogers, Z.S. (2014). Third-party logistics: a meta-analytic review and investigation of its impact on performance. *Journal of Supply Chain Management* 50 (1): 21–43. https://doi.org/10.1111/jscm.12046.

Luo, C. (2017). Alibaba onetouch: facilitating cooperation between logistics firms and shipping companies happy to jump onboard Whatever you do, I do? *aCommerce*

McKinnon, A. Flöthmann, C., Hoberg, K., and Busch, C. (2017). *Logistics Competencies, Skills, and Training: A Global Overview*. Washington, D.C.: World Bank Group http://documents.worldbank.org/curated/en/551141502878541373/Logistics-competencies-skills-and-training-a-global-overview.

McKinsey (2016). Digital in industry: From buzzword to value creation. *McKinsey Digital*, pp. 1–9.

Naughton, B.J. (2007). *The Chinese Economy: Transitions and Growth*. Cambridge, MA: The MIT Press.

Van Nuys, A. (2019). New linkedIn research: upskill your employees with the skills companies need most in 2020. *LinkedIn – The learning blog*.

Porter, M.E. (2008). That shape strategy. *Harvard Business Review*.

Rackham, N. (2020). *SPIN®-Selling*. Taylor & Francis.

Riedl, B.J. and Chan, T. (2019). The digital imperative in freight forwarding. *The traditional offline quotation and booking process is lengthy and cumbersome, often necessitating several interactions to reach a final price*, pp. 1–13. https://www.bcg.com/publications/2018/digital-imperative-freight-forwarding (accessed 18 December 2019).

Shang, K.C. and Lu, C.S. (2012). Customer relationship management and firm performance: an empirical study of freight forwarder services. *Journal of Marine Science and Technology* 20 (1): 64–72.

Staff, J. (2016). Amazon gets OK to operate as NVOCC from China to US. *Journal of Commerce*: 1–8.

Tegmark, M. (2017). *Life 3.0: Being Human in the Age of Artificial Intelligence*. Knopf Doubleday Publishing Group.

Thiel, P.A. and Masters, B. (2014). Zero to one: notes on startups, or how to build the future. *Crown Business*.

US International Trade Administration (2019). China – eCommerce. https://www.export.gov/article?id=China-ecommerce (accessed 24 January 2020).

Villar, F. (2020). Evolution of digitalization in logistic sales.

Wei, H.-L. and Wang, E.T.G. (2010). The strategic value of supply chain visibility: increasing the ability to reconfigure. *European Journal of Information Systems* 19 (2): 238–249. https://doi.org/10.1057/ejis.2010.10.

Wood, J.C. and Kates, S. (2000). *Jean-Baptiste Say: Critical Assessments of Leading Economists, Critical Assessments Series*. Routledge.

Xu, M. and Walton, J. (2005). Gaining customer knowledge through analytical CRM. *Industrial Management & Data Systems* 105 (7): 955–971. https://doi.org/10.1108/02635570510616139.

Yang, P. (2016). The history of the Shanghai containerized freight index (SCFI). https://www.flexport.com/blog/shanghai-containerized-freight-index-scfi-history/ (accessed 28 November 2020).

Young, J. (2019). Global ecommerce sales to reach nearly $3.46 trillion in 2019. https://www.digitalcommerce360.com/article/global-ecommerce-sales/ (accessed 8 March 2020).

Zaveri, P. (2020). Microsoft teams now has 75 million daily active users, adding 31 million in just over a month. *Business I*, pp. 2–9.

24

Managing and Selecting Logistics Service Suppliers

Colin Cobb[1] and Dyci Sfregola[2]

[1] *Loloi, Inc., Dallas, TX, USA*
[2] *New Gen Architects, Atlanta, GA, USA*

Introduction

The transition from the vertical supply chain to a horizontal model that relies on collaborations with vendors and partners has led to an increased need to efficiently manage multiple suppliers. Whereas companies have drifted away from managing all aspects of the supply chain in-house, for instance, Ford Motor Company circa 1920, in favor of focusing on core competencies, internal supply chain functional areas have experienced a drastic increase in the need for supplier selection efforts and forging mutually beneficial partnerships (Risher 2016). This is especially true for logistics as companies look to outsource various operations like warehousing and transportation. While there are currently a plethora of technological tools available that professionals can leverage to help manage logistics, these technologies should not undermine the importance of creating strategic partnerships and developing and implementing processes that can be supported (and not replaced) by technology. The value of long-term strategic partnerships and a general thought process around how to develop them should not be overlooked.

Best-in-class supply chains have proven that clear, open communication and active collaborative relationships with strategic partners that go beyond transaction management have a positive effect on customer satisfaction. Moreover, according to the Oxford College of Procurement and Supply (2014), these long-term collaborative relationships can help lower both costs and risks. Perceptions and interpretations from parties straddling on both sides of the fence (the party proposing the logistical solution or the supplier and the team procuring the solution) must be managed early on as this is key to building trust, a cornerstone to a successful outsourcing relationship.

When it comes to selecting a logistics service provider (LSP), logistics professionals face the challenge of navigating the waters of designing a supplier relationship and process that aligns with their company's needs. However, needs and requirements are not always

The Digital Transformation of Logistics: Demystifying Impacts of the Fourth Industrial Revolution,
First Edition. Edited by Mac Sullivan and Johannes Kern.
© 2021 by The Institute of Electrical and Electronics Engineers, Inc.
Published 2021 by John Wiley & Sons, Inc.

completely defined, nor do they remain the same throughout the procurement and service partner management journey. To accomplish the ultimate goal of finding the optimal supplier and solution, logistics professionals can use supplier management processes and strategies that provide a set of clearly outlined checks and balances for everyone to follow. When embarking on a supplier selection journey, organizations sometimes choose a solution based on the wrong factors and end up with a solution they do not have the infrastructure to support in order to realize any true business value. This can be due to a variety of factors including, but not limited to, cross-collateralized supplier challenges, labor issues, IT infrastructure hindrances like legacy system constraints, financial boundaries, or the lack of internal alignment.

Identifying Business Needs, Capacity, and Capabilities

When it comes to defining or improving a logistics management strategy, teams often have exuberant aspirations of creating a best-in-class solution, with world-class partners. There is nothing wrong with these aspirations – everyone should have them – but an ideal state and logistics are not always compatible, especially considering financial and IT infrastructure constraints. The logistic ecosystem is one of the most complex and unpredictable environments in business. To achieve an ideal future state, a company should first identify the needs, capacity, and capabilities of its current organization. Conducting a supply chain-focused strengths, weaknesses, opportunities, and threats (SWOT) analysis is highly recommended. A focus on the entire supply chain as opposed to solely analyzing the logistics functions will reveal all relevant contingencies. Once this is done, company leadership can decide what can and should be outsourced and develop a supplier strategy that aligns with their strategic goals.

To prepare for solution development and to best utilize the discoveries of a SWOT analysis, a more technical mapping of logistical variables could add value to the initiative. These considerations reflect the characteristics that directly correlate to the logistical decision-making process. The variants can be grouped into a quadrant, which can be further subdivided into more specific value points required by a logistics procurement team. These four considerations are demand, freight compliance, transport mode, and service level requirements. Notice that mention pricing is not a part of the technical mapping, as it can create bias and falsely influence the decision-making process. These four variants further breakdown into specific activities that facilitate product movement and help to optimize the logistical process. The technical considerations implicated in Figure 24.1 will also help streamline the path to determining the cost-to-serve impact on the budget.

Logistical characteristic mapping does not attribute value to the impact of the variables, but rather provides identification to determine said values. Demand is associated with the volume level determinants and variables. Compliance identifies specific requirements that are regulated by customers, carriers, and governmental agencies that further affect how freight is identified, classified, transported, and consolidated under a variety of guidelines. The transport mode is determined and can be an answer to the service type required to fulfill requirements based on geography, lead time, and cargo type. Within the fourth quadrant, a team can identify specific customer and/or partner requirements necessary to meet expectations and further analyze the internal capability, service offerings, and potential LSP partner development. To conclude, logistical technical mapping will be specific to an organization and its partners.

Figure 24.1 Capability Matrix

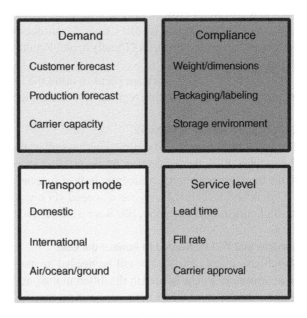

Motivational Example

Background

In order to illustrate how a company may go about outsourcing its logistics and supply chain activities, we will look at Benny's Burger Joint (BBJ), a food service company struggling to meet their logistics performance goals, which is inhibiting their ability to operate efficiently. After a strategic planning meeting, the senior vice president of supply chain sits down with the director of logistics, Sarah, for BBJ to discuss some strategic initiatives that have been handed from executive leadership to be executed primarily by the supply chain team. Sarah's main takeaway is that she needs to find a way to increase BBJ's on-time delivery of inventory from its warehouses to its restaurants, improve order fulfillment rates, and reduce expedited freight spend. She has been given no additional budget and no additional corporate resources to achieve these goals.

Sarah must now come up with a solution to bring these goals to fruition. She starts by defining the business metrics on which the logistics department will be judged and then identifies where and how she could leverage a new supplier relationship to meet her targets. Since the value chain is a system of complex, connected systems, the logistics department should not make any decisions or plans in a silo. Instead of sitting down with solely the procurement team, all relevant stakeholders and functional departments should be engaged in a collaborative effort to choose the supplier, or suppliers, that will best help the company achieve its strategic goals. Cross et al. (2006) found that while 80% of senior executives said that coordination across product, functional, and geographic lines was crucial for growth, only 25% thought their companies were effectively collaborating across boundaries. In BBJ's case, Sarah decides to reach out to Jake, the VP of restaurant relations; Alicia, an IT analyst; and Roberto, a senior inventory analyst to help define the business needs.

At their first meeting, the team keeps in mind the following considerations to help with alignment on the forthcoming requirements for the logistics service provider:

Current State Enterprise Network

BBJ currently has negotiated contracts with suppliers that allow them to keep all inventory at the supplier. BBJ's fleet of delivery trucks picks up inventory from the supplier and delivers to the franchise restaurants, aside from its beef suppliers who ship directly to restaurants.

Orders and inventory levels are managed via e-mail and Excel spreadsheets. It is also worth noting that at this point, BBJ has no carrier or 3PL relationships.

Services that Will Be Needed to Achieve the Goal

Warehousing and distribution will be needed as more franchisees open restaurants and more inventory will be needed in the different regions. BBJ may need to consider outsourcing transportation as this is not a core competency for the company and is expensive and time-consuming to manage. Data integration and real-time tracking of deliveries will be a necessity for accomplishing their goals as well as a system that will calculate relevant KPI percentages.

Industry Vertical Experience

BBJ's expertise lies in individual restaurant operations and franchisee training.

Technology Solutions

BBJ has an on-premise ERP solution that manages transactional financial data. The company's IT department created a proprietary TMS that tracks pick-up date, expected delivery date, and actual delivery date. All other data is managed via Google Sheets and e-mail, and there is also data in a data warehouse that can be pulled via a request to IT. These requests can take about three days to be filled.

Forecasted and Desired Demand

BBJ is expecting to open at least 90 new stores every year for the next five years across the country, driving a 45% increase in demand year over year. The bulk of the new stores is expected to be in the southwest.

Previous Wins and Challenges

Previous efforts to improve logistics KPIs were hindered by access to a single source of truth for data shared by internal departments and suppliers. The IT team at BBJ is confident this can be avoided if the right supplier is chosen. Another challenge is the time-consuming nature of aligning current suppliers with restaurant delivery needs.

The team also designs a communication strategy and meeting cadence to ensure continued internal alignment.

Deconstructing the Issues

Part of understanding the business needs is understanding customer needs. In his book, *Supply Chain Transformation: Building and Executing an Integrated Supply Chain Strategy*, Dr. J. Paul Dittmann (2012) describes two approaches to strategy development that can be applied here. The first is a supplier-forward approach where companies assess their supplier's capabilities and determine how well they will be able to meet their customer's needs. The second is a customer forward approach where companies choose the supplier that is best able to meet their customer's needs.

Through his research, Dittmann found that most organizations use the first approach, but not making customer needs a priority is risky, as organizations either fail to meet requirements altogether or prioritize supposed best practices so highly that they overengineer processes and technology, wasting resources on non-value-added activities. In this case, BBJ restaurants are the customers of BBJ corporate, and Jake will be able to communicate the voice of the customer to the team so that customer requirements can be considered as supplier requirements are defined.

Revisiting the final list of needs, lead times required to scale and maintain will vary by SKU. Following a customer-focused approach, Jake can work with restaurant operators to determine delivery time expectations that BBJ can use to narrow down the list of potential LSPs. Any LSPs that cannot manage lead times from food supplier, to warehouse, to restaurant (now or in the future) would not make the shortlist. With both needs and requirements of the LSP settled, the team can move to decide what type of service will be able to meet their specific needs. For companies unfamiliar with the services that LSPs provide or unfamiliar with logistics operations in general, a step before moving to solution design could be to brush up on the offerings of different providers so that time isn't wasted engaging service providers who will not be able to provide what is needed.

Defining a Solution

Once the team identifies the business needs, they can turn their attention to capacity and capabilities. Organizational capacity and capabilities refer to the internal load that the organization can support, which is relevant to what is being asked of the supplier partner. If an organization is to lead strategic procurement efforts, it must understand its limitations and governance measures to achieve a successful outcome. Dr. Robert Handfield of NC State University found that contributing factors to failures is poor or inadequate communication, planning, coordination, and synchronization of logistic activities related to requirements as well as a lack of understanding of constraints. Recognizing constraints and limitations after the supplier relationship has started can cause delays, inefficient use of resources, and a soured relationship with the supplier. Taking the time to clearly outline and detail the organization's realistic capacities and capabilities, focusing on core competencies, will provide a critical framework in supplier decisions and, in a future state, supplier scorecards.

Revisiting BBJ, the team can align on internal capacities and capabilities and determines that their ideal state of managing all logistics and distribution in-house is not feasible right now due mostly to human resources and labor constraints and budget. IT found a cloud-based point solution that could improve data sharing and planning speed; however, BBJ does not have anyone that has experience with any of the software platforms discussed, nor the resources to learn them. The customer liaison was able to share the customer's required delivery lead times but determined that their current fleet would not be able to meet them. Based on projected restaurant openings, an analysis of the inventory planning team showed they would need three additional full-time inventory analysts to effectively manage inventory in-house. Since distribution is not a core competency of the company, it made financial sense to invest in additional trainers and restaurant operations resources rather than hiring more truck drivers and inventory analysts.

Choosing a Solution

As a result of their findings, the team decides the best first step in meeting the new logistics goals is to seek out a third-party logistics (3PL) company that will manage their warehouse and distribution operations from BBJ's food suppliers to restaurants in regions where it is not feasible for the company to use their fleet for delivery and in situations where the supplier cannot store inventory. The team decided against using a fourth-party logistics (4PL) system because they do want to continue to maintain some control, so they do not want to completely outsource the management of logistics. More specifically, BBJ will benefit from a relationship with a 3PL because a 3PL will have access to technology and knowledge of mathematical equations to properly determine warehouse locations to best support restaurant operators based on today's network and forecasted growth. The 3PL can also handle the optimization of order picking and putting away and manage day-to-day warehouse operations using warehouse management system (WMS) software as well as provide SKU performance data to BBJ corporate inventory analysts to help determine inventory replenishment logic. They will be able to advise on how to leverage a transportation management system to optimize load and lane planning, fleet service management, and other transportation-specific functions.

Sourcing and Managing Suppliers in the Continuum

With needs, capacities, and capabilities identified, the team must now decide what type of supplier relationship they are seeking in a 3PL. Generally speaking, supplier relationships fall into one of five categories as outlined in *The Practice of Supply Chain Management* (Johnson and Pyke 2003) that can be used as a way to visualize and understand high-level differences between different sourcing relationships and can be considered the sourcing continuum:

Transactional	Ongoing relationship	Partnership	Strategic alliance	Backward integration
Arm's length	Medium-term contracts	Longer-term contracts	Long-term relationship	Ownership of the supplier
Clear part specifications	Some sharing of information	Extensive sharing of information	Full sharing of information and plans	Full sharing of information and plans
Computerized interaction	Some business with competitors	Increased trust	Limited or no business with competitors	One culture
Significant business with competitors	Good management relationship	Limited business with competitors	Extensive trust and merging of cultures	

Detailed supplier relationship management (SRM) processes and strategy are critical to relationship success with any logistics service provider. According to the APICS dictionary (2017), SRM is defined as, "a comprehensive approach to managing an enterprise's interactions with the organizations that supply the goods and services the enterprise uses. The goal of SRM is to streamline and make more effective the processes between an enterprise and its suppliers. SRM is often associated with automating procure-to-pay business processes, evaluating supplier performance, and exchanging information with suppliers." All supplier relationships are not created equal, and effective SRM is driven by segmentation. Failing to segment suppliers across the sourcing continuum can lead to missed value-realization opportunities by leveraging joint investments with strategic suppliers. Or, on the other hand, it can be detrimental to place too much emphasis on relationships with transactional suppliers.

Setting a good foundation for ongoing supplier relationship management means placing a supplier in the right segment before the relationship begins to allow for the alignment of expectations. Keep in mind that a successful supplier relationship is mutually beneficial for all parties involved and suppliers should know where they fall on the sourcing continuum to manage future requests, evaluations, and development expectations, which will be discussed later. As a reminder, BBJ is expecting considerable growth over the next few years, so they will be required to put a high level of trust in their chosen logistics service provider as they will rely heavily on their partner for distribution. Alicia also reminds the team that there are several planned internal initiatives over the next few years intended to transform the current IT infrastructure that includes data warehouse changes and implementation of data analytics tools, so data sharing will be crucial to relationship success. With these factors in mind, the BBJ team decides to seek out a strategic alliance relationship with a logistics service provider.

Strategies for Developing Strategic Alliance

The Vested Outsourcing Model

The vested outsourcing (VO) model, developed by the University of Tennessee, is rooted in a base of transparency and relationship scaling and is highly recommended Vested

Outsourcing Inc. (2020) as a framework for sourcing a supplier and entering into a strategic alliance relationship with them. VO researchers found that "while conventional contracts work well for simple relationships, they are not conducive to collaboration, innovation and sharing value, especially for complex, multi-dimensional business relationships." The VO concept is materialized by a simple approach of first classifying the sourcing relationship and developing a plan to scale into what could be a true partnership of transparent rewards for everyone involved if those goals are achieved. The execution of the strategy revolves around five ideas that Kate Vitasek, a highly experienced supply chain consultant, presented extremely well in her "five rules of vested outsourcing." The rules are as follows:

1) Focus on outcomes, not transactions.
2) Focus on the *what*, not the *how*.
3) Agree and use clearly defined and measurable outcomes.
4) Create pricing model incentives for cost/service trade-offs that optimize the business.
5) Organize and adapt insight, not oversight, governance structure (Vitasek and Ledyard 2016).

The first rule emphasizes that relationships with partners should be built around outcomes instead of transactions. For many service providers who are compensated per transaction, their focus is around completing transactions so that they can earn revenue with little focus on the outcomes of the transaction for the company that they are servicing (Vitasek and Ledyard 2016). Therefore, it is important for a company looking to outsource to work with the service provider and develop a shared vision of how the service will lead to positive outcomes and hold the provider to their promise to generate the desired outcome.

Next, a focus on the *what*, not the *how*, may seem obvious, but outsourcing companies often forget the important detail that you are utilizing a service provider because they can do something better than you are able to do it internally. Companies that fail to keep this in mind will often focus too much attention on telling the service provider how to go about delivering results (Vitasek and Ledyard 2016). Instead, the focus should be on communicating with the service provider clearly and precisely what it is that you hope to get out of the partnership. Let the service provider focus on how to get it done since that is the whole reason that they were contracted.

Once the desired outcomes of a partnership have been settled, the outsourcing company must come up with specific and measurable metrics that will determine whether the outcomes were met. Establishing these metrics in a contract is a way to ensure that the service provider is only paid when they have proven to have delivered the results that were promised as opposed to being paid for completing the transaction without delivering tangible results. Keep in mind that things cannot be improved if they cannot be measured, and try to think of the best metrics that can define a successful outcome.

Before the partnership can be established, a pricing model must be agreed upon by the service provider and the company they are providing service to. The model must have incentives that will push all parties to focus on business optimization. The pricing model should balance risk and reward for both parties, and the agreement again should specify that solutions, not just actions, will be delivered (Vitasek and Ledyard 2016).

Lastly, it is important for outsourcing companies to let the provider do the work that they were brought on to do. A governance structure will have to be set up to manage and oversee the partnership, but many companies go down the path of micromanaging their providers to the point that they can no longer do what they need to do in a timely and efficient manner. There is a balance though, and too little oversight can cause problems as well. Projects should not be blindly handed off to an outsourced partner to take care of on their own. This model favors having a small team handle oversight with a focus on gathering insight from the service provider, not playing the role of "Big Brother" (Vitasek and Ledyard 2016). This way the service provider can meet their required goals, and the outsourcing company can gain insight from partnership to help boost their own internal capabilities.

This model provides an excellent framework for planning and managing relationships with outsourced partners. It urges companies to think first about what they hope to get out of a partnership, then to enable the service provider to actually work toward those outcomes, and finally to measure success.

Pricing Considerations for Outsourcing

Once needs have been identified and the type of service needed is decided, a part of the service selection process will inevitably involve negotiating and comparing prices. Defining the supplier relationship for a logistics partner and discussing relevant topics like the costs associated with RFPs, supplier onboarding, and financial risk mitigation require input that goes beyond a transactional thought process. Low performance, high costs, and enterprise misalignment can be sensitive topics for cross-functional groups, especially if there are little transparency and understanding of priorities. The "right price" to pay for logistics services is often a point of contention, but the answer can usually be found at the decoupling point for scale, performance, and variable contribution margin (VCM) requirements as it relates to the financial expectations of the organization. To further understand this concept, we can assign specific values to performance metrics that directly correlate to a target variable contribution margin (VCM), based on the metrics assigned to that logistics purchaser. This will allow us to see the intersection of service provider cost and value added from the partnership to make assigning a reasonable price to the supplier more quantitative. An example of mathematically mapping and attributing value to impact variables would be:

1) Net earnings monthly target = $100 000 (reflective of a growth goal of x)
2) VCM target of sales required to achieve net earnings target= 40%
3) Performance target (fill rate) to achieve VCM target = 95%
4) Carrier on-time performance required to achieve estimated order count = 93%
5) Cost of logistics services required to achieve net earnings target = $60 000

The decoupling point becomes the point at which any combination of these numbers can influence the others by attributing a higher impact value while still achieving the NET goal.

Finding this point may be difficult for organizations whose approach to procurement and sourcing is more tactical than strategic. However, defining these inputs and coming up with the baseline of the supplier development strategy are paramount to developing a pathway to creating a program that can be presented to potential logistics suppliers.

The Roadmap for Success: Onboarding, Measurements, and Service Level Agreements

A Delphi study released in 2000 (Gray and Makukah 2004) found that most partnerships between logistics service providers and their customers remain operational in nature. Although logistics service providers claim to be true strategic partners, most companies do not think their LSP is able to provide the level of service required to be considered such. The failure to integrate on a strategic level suggests a lack of strategic management knowledge by relevant managers and stakeholders. In order to truly integrate an LSP and develop a strategic alliance, a company must work on the tactical elements of the partnership. This includes creating an execution plan that will maximize the chance of a successful strategic alliance partnership, ensuring smooth onboarding of any new partners, alignment on a visible system of measurements with the logistics service provider (LSP), and the ability for the ecosystem to work through strategic alignment and agreed upon continuous improvement efforts.

There is no "out-of-the-box" supplier onboarding process, but the process must have guidelines and a framework to ensure compliance and completion. Onboarding processes should be clearly defined for suppliers during the RFP process but should be flexible if the original process is deemed no longer feasible. The onboarding process is often the stage where challenges will start to appear that might prevent the team from following through with the original execution plan, so an agile approach that allows for iterations and changes is recommended. If a truly trust-based strategic alliance is to be realized, the team must work together to eliminate pitfalls caused by needlessly upholding a prerequisite when it is clear that a need for change in process has emerged. Keeping an open mind and being open to change will create a runway for continuous improvement, low bias, and agility, which are critical to success in an ever-changing, dynamic global logistics environment. There are a plethora of factors that neither partner will be able to control, from labor issues to tariffs to changing technologies and changing consumer behavior. The best way to weather the storm and manage risks is to work together in a mutually beneficial relationship that allows for growth for everyone involved.

A critical part of onboarding a strategic supplier is creating and visualizing all milestones and metrics that the client and supplier organizations deem necessary to accomplish their goals as identified by the alliance. To reiterate, these goals have been cross-functionally aligned and approved by all key stakeholders in the business and should, therefore, not be changed without full alignment across the alliance. This system of measurements will help crystallize the transparency efforts across the ecosystem. When looking to understand the key determinants and impact the logistics organization will have on the business, there are a few areas that can be useful in benchmarking. The supply chain operations reference model (SCOR), designed and maintained by the Association of Supply Chain Management (formerly APICS), provides a framework for three levels of metrics that are associated with five performance attributes. Of the five performance attributes (reliability, responsiveness, cost, agility, and asset management efficiency), reliability and responsiveness are most relevant to the logistics functions of the supply chain.

Without the proper data in the correct format, metrics and KPIs cannot be calculated. If we revisit the case of BBJ, there will be some new data to consider as the new strategic alliance with the 3PL will have resulted in new lead times from the warehouse to the restaurant as well as new P&L values due to shifted costs. The team can use any customer requirements, like delivery time expectations and sales forecasts, from the restaurant relations team. It can be noted that much of this was defined in the initial planning and discovery phase but must also be reflected in the metrics and KPIs that will measure the success of the alliance.

Key Performance Metrics

There are a variety of topline metrics that should be measured with supporting KPIs and milestones that will help facilitate success among partners in the ecosystem. Some generally accepted metrics that can be used are detailed below, and it is recommended to leverage the SCOR model and identify all relevant metrics and KPIs to be reviewed:

On-Time Delivery (Inbound)

Complete and on-time delivery (COTD) is defined by the APICS dictionary as a metric defining customer service. To be considered as complete and on time Atwater and Pittman (2019), all items in the order – in the correct quantity and with the correct line items – must arrive on time. As it relates to the logistics service provider, on-time delivery is a measurement of whether a service provider has achieved the prerequisite delivery requirement as outlined in the service level agreement. This measurement can have a direct impact on the entire supply chain and cause bottlenecks for additional processes required throughout the ecosystem. A team must work together to understand these impacts, as well as plan exception management processes to help maintain a preventative and proactive culture.

Order Accuracy

This is the percentage of all aspects of an order being transmitted and shipped by a service partner compared to the transmission request submitted by the upstream partner. This is a measurement of efficient processing and translations of data, goods, and standard operating procedures. Depending on the supply chain strategy and network design, this metric can be traced as a "cradle-to-grave" measurement potentially touching all partners within an ecosystem and therefore impacting all aspects of the business down to the end consumer.

Cost per Shipment

This is the direct expenditure to ship goods at a specific size, density, and weight that can be measured as an average of all routes or trade lanes, as well as by individual shipments based on dimensional requirements. This measurement is a direct impact to the P&L and gross-to-net calculation for an organization. It can be used in COGS calculations, as well as budgeting. Also, the *cost per shipment* can influence consumer behavior and make a difference in sales conversion.

Customer Order Cycle Time

This is sometimes called order fulfillment cycle time or order fulfillment time. This is the amount of time elapsed from when an order is placed to the time an order is fulfilled and delivered with all appropriate order milestones and acknowledgments delivered according to the service level agreement. This metric can help partners improve in efficiently getting a product off the dock, as well as fostering practices to help better plan warehouse space and labor allocations.

Carrier Scan Rate

This is the time a carrier scans and acknowledges the freight status according to location, time, and cargo count. This is a key metric in the eCommerce and omnichannel market. Because consumers are consistently monitoring where their freight is located, it is imperative that logistics service providers are transmitting freight status accurately and on time.

Inbound (Purchased) Order Cycle Time

Typically, this is the total time from when freight is delivered to (or picked up by) a carrier until the freight arrives at its destination, which is generally used to measure inbound transit lead times from a downstream stakeholder. This metric is key to planning and is an important part of the total product life cycle. Material resource planning, outbound transportation planning, capacity planning, and retail merchandise planning depend on this metric. It is always a good idea to perform time management studies to help develop this measurement, if the resources are available.

Note: It may be challenging to calculate this. While most tier 1–3 ERP systems manage this, smaller companies may find it difficult to manually calculate this metric in Excel. Arguably, best practice is to use an ERP module or a third-party system that can leverage EDI to extract purchase order and delivery data.

Inventory Accuracy

This is the percentage of variance or disparity between the product owner's financial accounting of product at a specific time and place compared to actual product levels at a specific location. This is one of the most cross-functionally scrutinized metrics within the ecosystem. Because there are direct impacts on the balance sheet and P&L, this measurement is not to be taken lightly. Improper measurements can negatively affect tax filings, financial strategies, materials and product planning, and sales. There are many technologies and software platforms in the logistics space that support real-time inventory visibility and continuous cycle counts to ensure accuracy.

Fill Rate

The fill rate is also called the customer service ratio and is defined by APICS as a measure of delivery performance of finished goods or other cargo, usually expressed as a percentage (Dittmann, 2012). How this is calculated will vary based on the company's production strategy (e.g. make-to-stock vs. make-to-order). More generally, the fill rate is the average percentage of total product fulfilled downstream calculated as total product fulfilled divided by 100. This metric provides visibility into efficiency in inventory planning, as well as fulfillment operations. This metric can also provide insight into the potential financial health of an organization or even effectiveness in supplier relationship management. For organizations working with retailers, this is a key metric and heavily scrutinized.

Data or Document Transfer Error Rate

This is the percentage of data transferred that is either inaccurate or is not received at the required time. This measurement is key to maintaining and improving the systemic flow of information throughout the ecosystem. Data inconsistencies and bottlenecks can cause partners to make the wrong decisions and compound unnecessary challenges to processes. This metric is key to ensuring informational transparency and accuracy.

Logistics Billing Accuracy

The percentage of correct billing for invoiced services by a provider compared to operational and financial records held by the organization paying for the services. This measurement has a direct impact on the P&L statement and GTN, and it also reveals inefficiencies in service provider processes. Strategically working with service provider partners on the chart of accounts, pricing tables and codes, and a system of checks and balances to ensure accuracy will be beneficial in helping to maintain a healthy score with this metric. Some well-known third-party services that can be used to calculate this metric are nVision Global and IPS Worldwide.

Many of these metrics can have consequential downstream effects and are not always the result of a direct touch associated with one partner, but rather the culmination of a collaborative effort. Transparency, capability, and capacity are key to ensuring that an accurate metric system directly correlates to the firm's growth strategy and expectations of all ecosystem partners.

Supplier Development and Managing the Supplier Relationship

One of the pillars of a sustainable partnership is a continuous review program that is jointly developed between stakeholders in the ecosystem. Understanding what is important to all businesses will streamline the continuous improvement process, optimize growth, and mitigate risk across the board. This is executed through a strategic communication and business review cadence strategy, where reviews of the current state are discussed, along with performance analyses and continuous improvement planning. This process can be facilitated by supplier scorecards and SRM software. While financials are certainly an important factor to consider in any business, partnerships with strategic suppliers should focus primarily on the total value added to the supply chain network and customer satisfaction, with the cost being only one point of many evaluation metrics on the overall scorecard. When developing an SRM strategy and related continuous improvement efforts, it is important to remember that the ultimate goal is serving the customer, not cost reduction and minimization.

Overall, supplier development is instituted through developing and employing a strategic communications strategy, implementing a reward system for all parties, and developing a continuous improvement strategy that is effective and efficient. Although tough to accomplish in a supply chain ecosystem, logistical simplicity that easily translates is always desirable and helps to promote the transparent culture required for success. When suppliers begin to experience successes and rewards that can be attributed to relationship management, the ecosystem flourishes, and inspiration begins to fuel the growth scale initiatives.

Conclusion

The Amazon effect has created a world where convenience, speed, cost value, and variety are king. With rapidly changing consumer demands and behavior, supply chain logistics is constantly evolving. People, processes, and technology are must remain dynamic to maintain or achieve a competitive advantage and meet customer needs. So, what does this mean for supplier relationships? The need for digital tools to support logistics ecosystems is a necessity, but they must be intuitive, easily integrated, and user-friendly, and they cannot be implemented with the notion of replacing skilled people and effective processes.

An organization must outline its business strategy to include a clear direction and insight into its own customers' needs in order for a supplier to be able to reach a partnership level. This pertinent information can be shared with strategic suppliers for continuous improvement efforts and can be used internally to drive selection for new suppliers. Organizations should choose strategic suppliers that can be flexible and adaptable as customer needs change while also balancing the need to achieve desired levels of cost and working capital. It may sound like an unrealistic goal, but best-in-class enterprises have demonstrated that a blended customer focus, a supplier-forward approach, is not only possible but also mutually beneficial to all parties involved.

As this chapter has discussed, creating mutually beneficial relationships and customizing the partnership framework have opened the floodgates to develop world-class systems and operations. Sourcing and managing suppliers can be accomplished quickly, but sourcing and co-managing an inclusive, scalable, and completely streamlined logistics ecosystem that exceeds expectations is the true definition of success.

Key Takeaways

- Technology can be leveraged to support sourcing and managing suppliers but is obsolete without both internal and external alignment, collaboration, and defined processes.
- The customer-based approach – using customer needs to define business needs and supplier criteria – should be used when selecting supplier partners.
- Using SRM technology and strategies can help to segment suppliers and strategically manage supplier relationships.
- Mutually beneficial supplier relationships are best achieved using the vested outsourcing model.
- The supplier partnership journey does not stop once a supplier is selected and onboarded – supplier development and continuous improvement efforts are the responsibility of all parties involved.

References

APICS (2017). Quick reference guide: supply chain operations reference model. *Version 12.0.*

Atwater, J. and Pittman, P. (2019). *APICS Dictionary: The Essential Supply Chain Reference, 16*e. Association for Supply Chain Management.

Cross, R., Martin, R., and Weiss, L. (2006). *Mapping the value of employee collaboration.* *McKinsey Quarterly* viewed 26 March 2020. https://www.mckinsey.com/business-functions/ organization/our-insights/mapping-the-value-of-employee-collaboration#.

Dittmann, J. (2012). *Supply Chain Transformation: Building and Executing an Integrated Supply Chain Strategy.* New York, NY: McGraw-Hill Education. https://www.amazon.com/ Supply-Chain-Transformation-Executing-Integrated/dp/0071798307

Gray, R. and Makukha, K. (2004). *Logistics partnerships between shippers and logistics service providers: the relevance of strategy. International Journal of Logistics Research and Applications* 7 (4): 361–377.

Handfield, R. (2017). The supply chain team's impact on capital projects. https://scm.ncsu. edu/scm-articles/article/the-supply-chain-teams-impact-on-capital-projects/ (accessed 28 March 2020).

Johnson, M. and Pyke, D. (2003). *The Practice of Supply Chain Management.* Germany: Kluwer Publishers.

Oxford College for Procurement and Supply (2014). The advantages of a well managed supplier relationship. https://www.oxfordcollegeofprocurementandsupply.com/the-advantages-of-a-well-managed-supplier-relationship/ (accessed 31 March 2020).

Risher, J. (2016). *From Offshoring to Reshoring: A Conceptual Framework for Manufacturing Location Decisions in a Slow-Steam World.* Georgia: Kennesaw State University.

Vested Outsourcing Inc (2020). https://www.vestedway.com/ (accessed 28 March 2020).

Vitasek, K. and Ledyard, M. (2016). Five game-changing rules for vested outsourcing – opinion – outsourcing – sourcingfocus.com. *Supply Chain Visions.* http:// sourcingfocus.com/site/opinionscomments/five_game-changing_rules_for_vested_ outsourcing/ (accessed 3 April 2020).

Section V

Conclusion

25

The Digital Transformation of Logistics

A Review About Technologies and Their Implementation Status

Johannes Kern

Tongji University, Shanghai, China

Introduction

In the past years, digital transformation, the profound changes taking place in society and industries through the use of digital technologies, has emerged as an important phenomenon on the agenda of researchers and practitioners (Vial 2019). For instance, a Google trend analysis shows that interest in the topic "digital transformation" increased 20-fold from 2014 to 2018 (Google 2019). The Fourth Industrial Revolution is predicted to holistically transform logistics and supply chain management with sensors placed in everything, networks created everywhere, anything becoming automated, and everything getting analyzed to significantly improve performance and customer satisfaction (Alicke et al. 2016).

A glimpse at this scenario can be caught already at "Asia No. 1," the fully automated fulfillment center of JD Logistics, which is the logistics division of Chinese e-commerce company JD.com, located in the northwest of Shanghai. Buzzing AGVs[1] cart merchandise to loading docks, and industrial robots place parcels on conveyor belts. Only five workers – whose job is just to service machines – operate the $40\,000\,m^2$ warehouse instead of the usually required 500 (Hornyak 2018). JD Logistics' vision goes further. "Smart delivery stations" in Changsha are using autonomous vehicles to perform last-mile delivery, and drones[2] are used to deliver consumer goods and medical supplies to remote areas. The Vice President of the firm explains that they have spent "the last decade building up advanced technology, logistics, supply chain and other capabilities [which helps to] achieve unparalleled operational efficiency for our online and brick and mortar operations and deliver a level of customer service that is unmatched globally (DC Velocity 2019)." Also, Amazon, the US e-commerce giant, invests heavily into the digital supply chain. Over $100\,000$ AGVs carry shelves of

---

1 Automated guided vehicles.
2 Also known as unmanned aerial vehicles (UAV).

The Digital Transformation of Logistics: Demystifying Impacts of the Fourth Industrial Revolution, First Edition. Edited by Mac Sullivan and Johannes Kern.
© 2021 by The Institute of Electrical and Electronics Engineers, Inc.
Published 2021 by John Wiley & Sons, Inc.

merchandise to workers who are filling orders instead of letting workers walk to fixed shelves. Compact robots that can lift about 700 kg assist warehouse workers and guide themselves to workers' stations via floor-mounted QR codes. Other large industrial robots with lifting arms are used to raise pallets of goods to upper floors (Rolfsen 2019). The company revealed that its robotics products have improved sorting accuracy company-wide by 50%. "We are always testing and trialing new solutions and robotics that enhance the safety, quality, delivery speed and overall efficiency of our operations," a spokesperson stated. "We believe that adding robotics and new technologies to our operations network will continue improve the associate and customer experience" (Amazon Improves Sorting Accuracy by 50 Percent 2019). Amazon is also leading in artificial intelligence (AI) technology. Its software can offer personalized recommendations for customers, forecast demands, or analyze images, voice, text, and videos (Machine Learning on AWS 2019).

However, while some companies can show such successful use cases, the reality in the logistics rank and file appears rather grim. A senior supply chain manager at a leading automotive original equipment manufacturer (OEM) (car makers) shared that when he oversaw truck deliveries for production facilities in Europe 10 years ago, tracking parts in emergency cases meant calling forwarders that then tried to somehow reach the truck driver to know the location and status. Although production line stops cost of more than 1000 USD per minute in the automotive industry, today the same approach is used in the case of risks of late part arrivals. Having a system that would provide a warning about delayed items so that the production could be adjusted sufficiently early was and still is just a dream for him. This case seems to be by no means an exception. In 2016, GT Nexus, a supply chain platform, and Capgemini, a consulting firm, conducted a study among executives of some of the largest global manufacturing and retail organizations from over 20 different countries around the world. Only 15% of the respondents said that the majority of data from the extended supply chain was accessible to their organization, and out of those only 23% stated that the majority of data was analyzed and used for decision making (GT Nexus Capgemini Consulting 2016). A similar picture can be seen for the implementation of some hyped technologies. At the 2018 Monitoring, Evaluation, Research, and Learning (MERL) Tech Conference, researchers wanted to present best practices from blockchain projects that were supposed to bring transparency to processes and operations in low-trust environments. However, even after studying 43 use cases, the researchers could not find any evidence about tangible result that blockchain was expected to achieve and had to conclude that all projects underdelivered on their claims (Mulkers 2018).

So how do these worlds fit together? Is there a digitalization chasm between born digital behemoths and traditional industry players who have not heard the wake-up call yet? Or do some technologies only add value in specific business models along the supply chain? Are *digital transformation, Industry 4.0, Fourth Industrial Revolution (4IR),* etc. just the latest buzzwords generated by consultants, authors, speakers, and academics to impress laymen and earn some quick bucks (Ettorre 1997)? To address these questions, this chapter will summarize current studies about the state of digitalization in logistics and supply chain management, evaluate the discussed key technologies, and provide an outlook about changes that are expected soon.

State of Digitalization in Logistics and Supply Chain Management

The logistics industry can be broken down in (i) *infrastructure* including seaports, airports, and warehousing; (ii) *execution* including road transport, sea transport, air transport, and courier, express, and parcel (CEP) delivery;[3] and (iii) *services and advisory*, especially forwarding. In 2014, this industry had a volume of 4.3 trillion USD, split as shown in Figure 25.1.

This structure will serve to guide the following analysis.

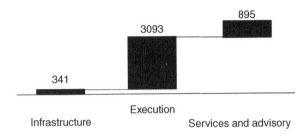

Figure 25.1 Logistics services costs worldwide in USD billion. *Source*: Based on Riedl et al. (2016). © John Wiley & Sons.

Logistics Infrastructure

Seaports

Seaports are the most common types of ports around the world, and they serve as the backbone of commercial shipping activities. Implementing Industry 4.0 for ports typically requires *automation* first. This can involve a variety of parts, ranging from *quay cranes* to horizontal transportation with *AGVs*, yard equipment, or the terminal gates. According to Drewry, a maritime research consultancy, automation could lead to a reduction of operational expenditures (especially for labor costs), increased safety, reduced risk for errors, less downtimes, the possibility to stack containers denser and hence utilize yard space better, and greener operations when electric equipment is used (Davidson 2016). However, there are also disadvantages connected such as high initial investments (especially as often automation equipment has to be fully installed), inflexibility, the necessity to have highly skilled labor, and remaining (or even lower) productivity (Silveira 2018). Some experts estimate that automation could cut operating expenses by 25–55% and raise productivity by 10–35%, while others predict a reduction of operating expenses at automated ports by 15–35% with productivity decreasing by 7–15% (King 2019). Asides from automation, ports could benefit from the implementation of AI. Klaus Lysdal, Vice President at the online freight forwarder iContainers, explains that ports should offer, for example, *real-time and precise container tracking*, especially since computers are already logging every single movement of a container, "when it's gated, when it's loaded or unloaded, when it's discharged, etc. These data

---

3 The focus of this work lies on international freight, where rail transport has a small share. Therefore, although rail freight is considered a part of logistics execution, it will not be discussed here.

are generally available now but often with a delay while it's being confirmed (iContainers 2018)." Berths, ship's allotted place at a wharf or dock, could be *planned in real time* to better utilize berthing slots and employed labor, predictive maintenance employed for key assets such as cranes and vehicles to minimize downtime and *yards automatically planned* to swap assets, reroute containers and make adjustments in real-time (Chu et al. 2018).

The digital transformation for ports is happening slower than in other logistics areas. Drewry estimates that only 1% of container terminals worldwide are fully automated and 2% semiautomated (Davidson 2018). It seems difficult to justify the high upfront capital expenditures considering the disadvantages stated above, especially when incorporating the manifold operational challenges such as shortage of capabilities, poor data quality, siloed operations, and difficulties to handle exceptions (Chu et al. 2018). However, the Port Equipment Manufacturers Association is optimistic about the future implementation of Industry 4.0 at seaports. "Despite a slow start, robotization of container terminal handling and transport systems is now taking off. More than 1100 driverless cranes are in operation worldwide today in container yards and are fast becoming a standard product. Yet horizontal transport between the quayside and yard storage blocks has not yet reached the same level of automation maturity as yard operations. In many cases, automated yards are served by manned horizontal transfer vehicles (. . .). AGVs however, have been deployed and proven for horizontal transport at a number of facilities and Automated Shuttle Carrier technology is also now being adopted (PEMA 2017)." Ports in China are leading the way, for example, the Qingdao New Qianwan Container Terminal (QQCTN) or the Shanghai Yangshan Port Phase 4. According to the General Manager of QQCTN "labor costs have been reduced by 70 percent because of this automatic terminal, while efficiency increased by 30 percent, because we can work at night." Despite of project costs of 468 million USD, he claimed that the automated terminal became profitable 10 months after opening (iContainers 2018). Shanghai Yangshan port is the world's busiest port, accounting for 3.2% of world's total container throughput. Its latest terminal is fully automated, and "no person is required for loading/unloading containers," which led to labor cost reductions of 70% and handling efficiency increases of 50% (New China TV 2019). Globally, a study by the B2B research group MarketsandMarkets predicts that the semiautomated and fully automated container terminal market is currently worth 9.09 billion USD and will increase by 20% to 10.89 billion USD by 2023 (MarketsandMarkets 2018). Figure 25.2 illustrates a digitally transformed seaport.

Airports

Airports are vital economic generators that provide gateways to their cities, regions, and countries, handle goods transported by air, and offer connections between aircraft and other modes of transportation (Airports Council International 2017). Thanks to the automation of bag drop, passport checks, and other flow processing tasks, airports today often implemented some "Industry 2.0"-level technology such as self-service. With the technologies of Industry 4.0, a whole set of new solutions is conceivable. At an "Airport 4.0," full connectivity with all airport stakeholders and superior proactive/reactive adaptation to real-time operational needs and customer requests is possible. For air cargo handling, various technologies could be adopted. For instance, *processes could be automated* for taxiing or docking. A *Total Airport Management Center* with real-time operation management could facilitate collaborative decision making of air and ground ops. Buildings could be

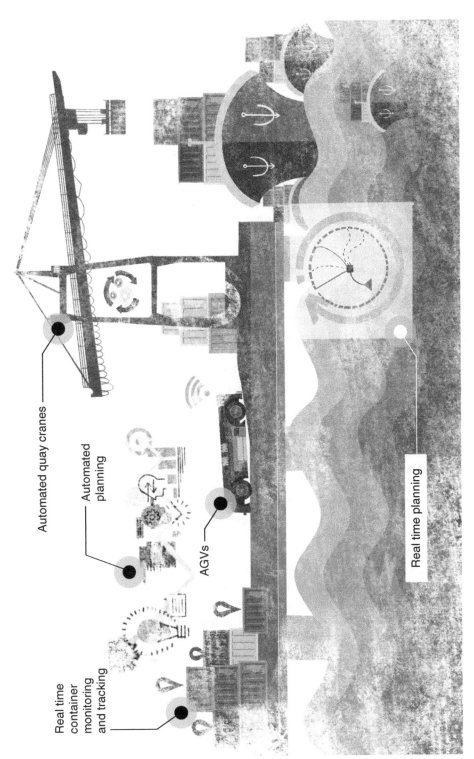

Figure 25.2 Digital Transformation of Seaports.

intelligently managed with real-time energy and utilities management, *intelligent preventive maintenance*, and data-based asset management. Finally, *predictive technology for automated planning management* could be implemented for key airport resources (Blondel et al. 2015; Streichfuss 2016).

Employing such technologies would lead to several benefits. The consulting firm Arthur D. Little assumes a 10–15% capacity increase airside and up to 30% capacity gain landside (for cargo and passenger handling combined). Operational expenditures related to airport revenues could be reduced by 14%, especially by saving staff required for administrative processes and through reduced energy consumption (Blondel et al. 2015).

Despite of plenty digital initiatives that are – often publicly appealing – launched, airport digitalization is happening slower than expected. Operators remain attached to their traditional business models, lack investments for digital initiatives, and are challenged to define relevant business cases (Bréchemier et al. 2017). Based on a 2018 study, the consulting firm Sia Partners confirms that the air cargo industry has been slow to embrace digitization. "The industry still relies on paper-based processes to exchange shipment information along the complex supply chain. (. . .) The limited offer pricing transparency, the lack of visibility over the localization and the status of goods transported is not up to speed with current customer expectations. (. . .) IoT[4] and Big data remain of marginal use and struggle to develop," they concluded (Berland 2018). Although it is acknowledged that business contexts and strategies are different, some key challenges are common for most digital airport transformation initiatives. These include positioning airports in the broader ecosystem (joining forces with the right partners and stakeholders in the value chain), recognizing data as an enterprise asset (including leveraging its value, managing its quality, promoting its ethical usage, and protecting it), influencing relevant regulation, developing a digital culture across the enterprise, acquiring and retaining talent, and dealing with data privacy laws and regulations (Airports Council International 2017). Nevertheless, the biggest risk to any airport is to ignore the digital transformation (Airports Council International 2017). It will likely bring new disruptors entering the industry, fast-changing business ecosystems and hypercompetition. The market research firm Profound Market Intelligence forecasted that airport expenses into digital based solutions will grow by around 40% from 9.5 billion USD in 2014 to 13.2 billion USD in 2020 (Blondel et al. 2015). Figure 25.3 illustrates a digitally transformed airport.

Warehousing

Warehouses are nodes in logistics networks where goods are stored, handled, or distributed. Various technologies are discussed in the context of Industry 4.0 inside a warehouse, also depending on its main function (Tjahjono et al. 2017; Pfohl 2018).[5] *IoT* objects, such as products, equipment, or assets, can communicate their physical context information, e.g. location, status, or history, to the Internet (Schlick et al. 2014). In an IoT-enabled warehouse, the data these objects send can create transparency and enable a synchronization of processes and an informed decision making. For instance, a sensor on a product could report critically increasing temperatures, so that the warehouse storage temperature could automatically be adjusted

---

4 Internet of Things.

5 For a study about conventional automation equipment with a focus on transshipment hubs, cf. (Pfohl et al. 2019).

Figure 25.3 Digital Transformation of Airports.

(Aleksandrova 2019). *Big data analytics* lets firms better understand their business and market and make timely business decisions (Chen et al. 2012). By forecasting future demands, it can help to make better inventory management decisions to avoid out-of-stock situations as well as overstocking. Also, warehouse operations, such as capacity or shift planning, can be optimized based on consumer behavior, which translates into better customer service (Sunol 2018). *Augmented reality* (AR) devices such as smartphones, wearable devices, or smart glasses combine the physical and digital worlds in real time. In a warehouse, they could among others facilitate picking, storing, or shipping by showing the item's storage location, picking route, or check orders (Stoltz et al. 2017). The automotive company Volkswagen uses smart glasses already for order picking in its warehouses to improve process security and efficiency. A camera inside the smart glasses is used as hands-free operable barcode reader to identify correctly and incorrectly picked parts (Kern and Wolff 2019). *Robots* can automatically carry out a complex series of movements, increasingly with humanlike capabilities (Oxford University Press 2018). Especially in order preparation – often accounting for 50% of storage and handling costs in warehouses – pick-and-place robots can increase production rates by speeding up the processes of picking parts up and placing them in new locations (van Marwyk 2016; RobotWorx 2018). Also in the depalletizing processes, robots enable fully automated solutions that are faster and more ergonomic compared with manual ones (Mayer 2016). *AGVs* can be used to tow and/or carry materials inside warehouses, where they navigate either along physical or magnetic guidelines, by laser, or with satellite navigation (Ullrich 2015). Operations can, for example, be improved through this technology by transporting shelves directly to warehouse workers instead of letting the workers spend time and energy to walk through picking bays and to pick materials (Audi 2018; Kern and Wolff 2019). *Drones* can conduct infrastructure monitoring and inventory management by using bar codes, QR codes, or RFID in combination with IoT technologies. They may also provide auditability details such as geolocational and other sensor data and reinforce the auditability of other inputs – particularly in areas difficult to access (Companik et al. 2018). Through drones, costs for inventory auditing can be cut half compared to a warehouse worker, and the need for humans to climb warehouse racks and perform other dangerous work be eliminated (Appelbaum and Nehmer 2017).

Although these technologies have the potential to transform the way warehouses are operating, there are still technical and management hurdles to overcome.[6] Issues regarding IoT devices involve technical heterogeneity (a lack of standards and especially the need for interoperability – infrastructures that allow connecting and integrating a diverse set of technologies), privacy concerns, and security risks where poorly designed objects, absence of policies, or lack of control over disclosure mechanisms enable untrusted entities to obtain the location and further critical data of other entities (Haller et al. 2008; Elkhodr et al. 2013; Ben-Daya et al. 2017). Big data analytics is also still facing a variety of challenges. Difficulties include data capturing, storing, searching, sharing, analysis, and visualization. Therefore, on the hardware and software sides, more advanced storage and input/output techniques, more favorable computer architectures, more efficient data-intensive techniques such as cloud computing, and more progressive technologies (big data platforms with sound architecture, infrastructure, approach, and properties) will be needed (Chen et al. 2012). Battery life or processor overheating of AR devices are technical barriers that cause headaches when used

---

6 Benefits and challenges of employing these technologies depend significantly on the actually used software/hardware solutions.

over an extended period. In addition, total costs of ownership are still rather high – especially if wearable devices are, for example, due to hygiene issues considered as personal equipment and if compared with alternative solutions (Stoltz et al. 2017). AGVs and drones also suffer from some technical drawbacks such as limited carrying capacity, lack of flexibility regarding the size of the carrying load, and limited travel distance (which translates into considerable time spent at charging stations). Often, they also require a modification of the warehouse layout and high investment costs (Companik et al. 2018). MHI, the material handling and logistics trade association, conducted a global survey among a wide range of company types and industries about emerging disruptive technologies and innovations. Regarding technology adoption, 43% of respondents in this study reported that they have sensors and automatic identification in use, 43% inventory and network optimization, 32% robotics and automation, 30% predictive analytics, 26% Internet of Things, and 14% autonomous vehicles and drones (MHI and Deloitte 2019). However, the respondents also intend to increase their investment in warehouse Industry 4.0 technologies. Fifty-seven percent plan to spend more than 1 million USD over the next two years (a 10% increase compared with last years' study), 34% plan more than 5 million USD, and 22% more than 10 million USD. Investments in the next three years are planned in software systems for warehousing, distribution, and logistics (37% of survey participants), AGVS and robotics (33%), IoT (30%), predictive analytics (28%), mobile and wearable technologies (28%), AI technologies (24%), and AGVs and drones (16%) (MHI and Deloitte 2019). Figure 25.4 illustrates a digitally transformed warehouse.

Logistics Execution

Road Transport

Around 3.4 million trucks are sold globally every year, with heavy commercial vehicles accounting for more than 60%. Through technologies of the 4IR, the global truck industry is now on the brink of major changes (Nürk and Maier 2014). *IoT sensors* are the enabler of truck connectivity. They are the base to communicate the vehicle and cargo status, location and loading information, or traffic conditions to the Internet (Ben-Daya et al. 2017). Through remote diagnostics functions, vehicle conditions can be monitored, repair shops nearby recommended, and spare parts ordered, long before a vehicle actually breaks down. This could reduce maintenance costs by up to 5% and increase a truck's service life (Kelp and Krenz 2012). Cargo quality control will be possible over the Internet with sensors that indicate temperature and humidity fluctuations or vibrations (Ben-Daya et al. 2017). With drivers' wages and benefits accounting for almost 40% of marginal costs per kilometer driven, it is easily understandable why *autonomous trucks* are considered the "next big thing in trucking" (Choe et al. 2017). Benefits of them also include an increase in possible truck hours of service (allowing for drivers to rest while productivity occurs simultaneously), a reduced number of driving violations, and tackling of driver shortage. Driver's health and wellness could be improved and drivers easier retained (American Transportation Research Institute 2016). Through *fleet management systems* (FMS), carriers can analyze operational data collected from trucks, e.g. through IoT technology. FMS services include vehicle management solutions that, for example, enable preventive maintenance, driver management solutions that monitor among others driving performance, and network solutions that support the integration of enterprise resource planning (ERP) systems (Riedl et al. 2018). Employing FMS can enhance service quality and efficiency as improved

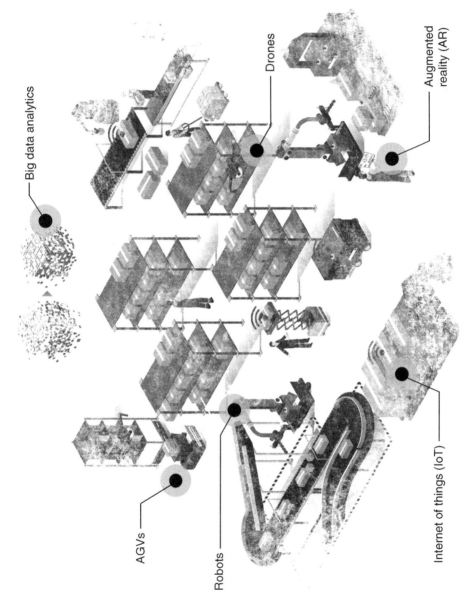

Big data analytics

Drones

Augmented
reality (AR)

AGVs

Robots

Internet of things (IoT)

Figure 25.4 Digital Transformation of Warehouses.

dispatching allows for additional jobs, driver behavior analysis and optimization can lead to higher vehicle safety plus fuel efficiency, and overall reporting can be automated (Leasing Associates 2018). *Transportation management systems* (TMS) enable the digital integration of all transportation-related activities and participants in the logistics process and can also act as "onboard digital operating system" that hosts various in-truck applications (Riedl et al. 2018). TMS typically offer a wide range of functions to move freight such as route planning and optimization, load building, operations execution, freight audit and payment, yard management, order visibility, and carrier management. This leads to increased shipment efficiency, reduced costs, real-time supply chain visibility, and enhanced customer service (Rouse 2015). In combination with autonomous driving[7] and IoT technology, TMS could plan routes dynamically in real-time, based on weather and traffic conditions and if certain parking/resting locations are available (American Transportation Research Institute 2016). Freight *marketplaces* combine the traditional freight brokerage model with the e-commerce model to match shippers and carriers quickly and efficiently (Jain et al. 2017). They act as platforms where freight capacity can be purchased but are not involved in negotiating shipper–carrier contracts or providing freight capacity (Riedl et al. 2018). Shippers can benefit through a plethora of product choices with high price transparency and trust through reviews/ratings that will lead to superior service and more competitive pricing (Lock 2018). Marketplaces might help to improve the underutilization of assets with current empty truck running rates typically estimated around 30–40% for China and 20–25% for the United States (Agenbroad et al. 2016). They also have considerable advantages for carriers who can access a network of potential customers readily available and serve them at low transaction costs (Lock 2018).

Although these technologies can bring large benefits, they also lead to drawbacks. Cybersecurity becomes a key challenge for data-driven life cycles and business models. Seamless digital integration significantly increases the vulnerability compared with traditional, less networked systems (Armengaud et al. 2017). As Jason Krajewski, Director of Connectivity at commercial vehicle manufacturer Daimler Trucks explains, "the instant you put a wireless connectivity point on your vehicles, you are introducing attack vectors." An emerging security risk scenario is a ransomware attack where trucks are deactivated until a payment is received. "Most of the technologies that are used now for telematics would not readily lend themselves to that type of attack, but the CAN bus[8] is vulnerable. With the wrong tie-in, you could in theory flood the brake system with a denial of service attack and have the brakes lock up. Without resetting everything on the vehicle, you may not be able to get the truck started again," Jason explained, although he also highlighted that "I have not heard of that happening (Menzies 2018)." An increasing number of advanced technologies inside trucks will also lead to increased investment costs. Although there are no systems commercially available yet, figures of 30 000 USD per automated truck system in the future are discussed (American Transportation Research Institute 2016). It can be assumed that

---

7 Due to the risks of distractions (e.g. smartphone usage regulations) while driving, it is difficult to implement real-time route planning in today's standard vehicles (American Transportation Research Institute 2016).

8 The Controller Area Network (CAN bus) is the "nervous system" of a car, enabling communication between all parts of the body (CSS Electronics 2019).

costs will go down as these technologies become more widely adopted; however we must remember that the price for the Google autonomous car system in 2016 was around 150 000 USD (Technology And Costs 2016). Also for FMS or TMS systems, the main disadvantage are the costs for hardware, software licenses, maintenance, and required expertise to manage the software (although cloud-based solutions at low monthly subscription fees become increasingly common) (McCrea 2018). Sensor quality is another challenge. IoT sensors ideally are robust to minimize maintenance and diagnose and fix themselves (Newark 2018). Today, maintenance companies are already spending a majority of their time on identifying and maintaining faulty sensors, which might become even worse if the number of sensors increases without dramatic quality improvements (Menzies 2018). Marketplaces might suffer from too low service quality that further compounds the traditional lack of trust in the industry. Also, quotations offered by carriers are typically standardized that might not meet shippers' demands. Generally, marketplaces largely meet the basic requirements of shippers and carriers but, however, often fall short to offer value-added services and to create a customer-centric experience (Jain et al. 2017). Nonetheless, technology adoption is steady. A survey by the International Road Transport Union found that a majority of 65% of transport operators in Europe plan to invest in digital solutions in the coming six months. The focus are investments in technology to optimize their day-to-day operations (65% of respondents), new fleet management solutions (38%), and electronic/digitized documentation (e-CMR, 25%). Two in five respondents believe that autonomous vehicles will be the most important changes in their businesses, while one in four believes that digital marketplaces and transport providers will be the most important ones, over the next 5–10 years (International Road Transport Union 2018). Many insiders predict that the industry is at least a decade away from technologies that would allow a truck to operate without a driver in the cab (Cannon 2017). Still, trials already took place across different countries around the world, including the United States, Sweden, Germany, Australia, and China, and enthusiasm for autonomous trucks did not cool off ever since (O'Byrne 2018). In 2019, industry pioneer Daimler launched a truck that reached Level 2 automated driving[9] with an advanced driver assistance system supporting forward and lateral directions (Daimler 2019). However, the gradual shift to autonomous trucking might drastically change the nature of a driver's role to become more of a conductor or attendant who cares for that truck while it is being driven (Cannon 2017). Gartner, a market research firm, assumes that up to now, only about 35–40% of companies adopted a TMS. This is especially divided among companies with 100 million USD freight spend, where the TMS adoption rate is already 50%, companies with 25–100 million USD freight spend with an adoption rate of 25%, and companies with under 25 million USD freight spend who have an adoption rate of roughly 10% (Young 2019). Freight marketplaces are also expected to grow sharply over next years. The consulting firm Frost & Sullivan expects that digital freight matching revenue will grow from around 11.2 billion USD[10] today to 79.4 billion USD by 2025 (Hampstead 2019). Figure 25.5 illustrates the digitally transformed road transport.

---

9 For a taxonomy on levels of driving automation, cf. SAE Recommended Practice J3016 (SAE International 2018).

10 Including Trucking as a Service (TaaS), a business model where shippers can gain access to trucks on demand (Hampstead 2019).

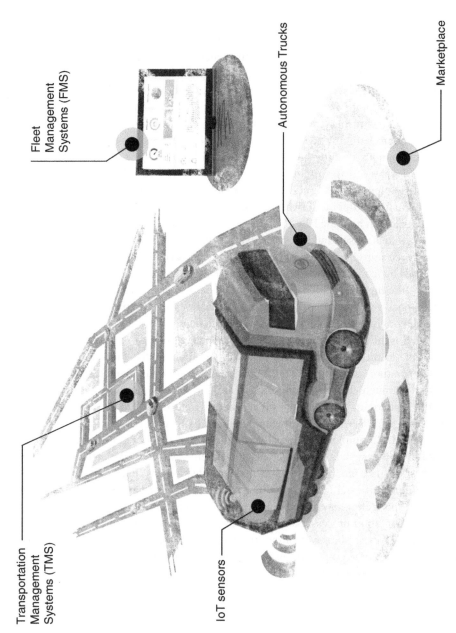

Fleet Management Systems (FMS)

Autonomous Trucks

Marketplace

Transportation Management Systems (TMS)

IoT sensors

Figure 25.5 Digital Transformation of Road Transport.

Sea Transport

Shipping is the life blood of the global economy that enables intercontinental trade, bulk transport of raw materials, and import/export of affordable food and manufactured goods. Today, vessels carry around 90% of all world trade (International Chamber of Shipping 2019). The shipping industry is cautiously embracing a variety of technologies arising from digitalization, and more and more carriers are taking measures to digitalize internal processes, develop integrated information technology infrastructures, and offer real-time transparency on shipments (UNCTAD 2018). Similar to the discussion about autonomous trucks, *autonomous ships* are foreseen as a game changer that will increase efficiency, improve safety, and relieve humans from unsafe and repetitive tasks (Marr 2019). Technological advances in modular control systems and communication will enable wireless monitoring and control functions both on and off board, which will allow for ships to operate remotely under semiautonomous or fully autonomous control (Fraunhofer Center for Maritime Logistics and Services 2016). Automation is expected to enter vessels gradually, starting from automated shipping decision support systems and remotely controlled ships with crew ashore up to fully autonomous vessels (Fraunhofer Center for Maritime Logistics and Services 2016). Research funded by the European Union showed that around 7 million USD per autonomous vessel could be saved over a typical ship lifetime through a reduced crew and improved ship efficiencies. Also, safety and security could be drastically enhanced. Incidents caused by "collision and foundering," with a share of 50% of all total losses in the past decades – the highest incident category, could be decreased around ten times compared with manned vessels (MUNIN 2016). Transformations are also predicted for the container, in which original design has been largely unchained for more than 50 years. *Smart containers*, such as the pilot "Container 42," are equipped with sensors and communication technology, able to record and transmit position, vibrations, pitch, noise, air pollution, humidity, temperature, and more (We are 42 2019). These data can be collected in the cloud and analyzed through IoT platforms, such as the Port of Rotterdam Authority's (Havenbedrijf Rotterdam 2019). The United Nations Centre for Trade Facilitation and Electronic Business (UN/CEFACT) already published its first set of standards within the frame of the "Smart Containers Business Requirements Specifications" initiative to ensure interoperability and easy data exchange within different systems in the container supply chain (van Marle 2019). *Blockchains* are open, distributed ledgers that can record transactions between two parties efficiently and in a verifiable and permanent way (Iansiti and Lakhani 2017). This technology provides an opportunity for the industry to streamline and secure its documentation process (Sanders and Egloff 2019). Various blockchain technology initiatives and partnerships were announced for cargo tracking and end-to-end supply chain visibility, recording vessel information (e.g. for insurance purposes), and digitalizing and automating paper filings and documents (UNCTAD 2018). For instance, the blockchain-based bill of lading created by the container shipping company Maersk and the information technology company IBM showed that administrative costs could be reduced by as much as 15% of the value of shipped goods by tracking shipping containers and eliminating paper documents. It could also help to reduce expenditures caused by data entry errors, procedural delays, and discrepancies (IBM 2017). With *big data analytics*, data collected by the myriad of sensors on modern vessels can leveraged effectively for decision support. This can result in among others improved vessel performance, hull and machinery maintenance, and operations (Schwarz 2016). The marine engine developer Winterthur

Gas & Diesel (WinGD) partnered up with Propulsion Analytics to develop an Engine Diagnostic System that analyzes data on the performance of engines/subcomponents in real time and provides live advice. Thermophysical simulation models, machine learning techniques, and expert/knowledge-based algorithms are also combined to diagnose and troubleshoot abnormalities, allowing to help prevent breakdowns before they happen (Safety4sea 2019). This results in reduced maintenance costs and improved asset efficiency through automated data and reporting (Backas and Yeoman 2018). Also, information technology equipment and services company Fujitsu and technology enterprise Kongsberg teamed up to develop a vessel fuel optimization service that uses AI to learn ship captains' strategies and ships' performances, combines it with meteorological and hydrographic forecasts (wind, waves, and ocean currents), and recommends optimal routes that maximize energy efficiency, safety, and profitability (Magnussen 2019). FMS in sea transport are used to optimize and facilitate the management of a fleet. Functions include often crewing (e.g. crew planning, work and rest hour registration), maintenance management, (spare part) inventory management, or fuel management. Employing FMS increases business operation efficiency and lead to overall cost reductions (Mastex 2015). FMS can be part of control towers that allow continuous routing and vessel performance optimization by leveraging data and onboard sensors (Sanders and Egloff 2019). *Marketplaces* are also discussed in sea transportation. They and other digital platforms can facilitate operations, trade, and exchange of data. This results in reduced costs and easier interactions between different actors along the shipping supply chain (UNCTAD 2018).

While these technologies promise transparency and more integration, there are also certain disadvantages and risks connected. With the increase in digitalization and interconnectedness of all ship systems, e.g. communications, navigation, and onboard automation, the risk of cyberattacks and hence the need for cybersecurity for vessels and their offices increase rapidly. Vulnerabilities can occur through external threats, such as operating system and application update patches, or internal attacks whether unwittingly through the use of corrupted flash drives or through malicious attacks by discontented employees (Cyprus Shipping Chamber 2018). Cyberattacks can have critical business impacts. For instance, when Maersk was affected by the malware NotPetya in 2017, 50 000 endpoints and thousands of applications and servers across 600 sites in 130 countries got infected. Lewis Woodcock, the firm's head of cybersecurity compliance, remembered that "the severity for me was really taken in when walking through the offices and seeing banks and banks of screens, all black. There was a moment of disbelief, initially, at the sheer ferocity and the speed and scale of the attack and the impact it had." It is estimated that Maersk lost up to 300 million USD through this serious business interruption (Palmer 2019). Another challenge is data ownership and interoperability. While a myriad of new opportunities for applications and business models are conceivable, implementation challenges remain to be solved. An increasing number of companies compete in the ship data services realm, some from adjacent industries, with overlapping offerings. Communications satellite operators try to differentiate themselves by offering IoT platforms, bridge electronics companies announce AI-driven routing optimization tools, or marine electronics vendors are developing platforms for vessel data collection, storage, and analysis. Therefore, it becomes unclear who will own the data (especially if an OEM is paying for its collection) and if platforms will be interoperable with each other or data be locked in into one service (Benecki 2019). This is also seen as a key issue by the United Nations Conference on Trade and Development, which points out that establishing interoperability, while ensuring cybersecurity and the protection of commercially sensitive or private data, is vital (UNCTAD 2018). While initially hyped, blockchain technology could often not deliver on the promises and might have a lack of technical benefits. In 2018, a research team from the US Agency for International Development reported that "we found no documentation or evidence of the results blockchain was purported to have achieved in these claims. We also did not find lessons learned or practical insights, as are available for other technologies in development" (Mulkers 2018). IT security expert Felix von Leitner summarized that blockchain is a solution to a problem that companies often do not have. Manipulation-evident accounting can be better done without blockchain, and blockchain is not suitable for time-critical transactions or high transaction numbers (von Leitner 2019). Despite these challenges, companies start to cautiously embrace the various technologies (UNCTAD 2018). Leading firms in the industry are optimistic about the future of autonomous vessels and see the challenge in refining it, increasing reliability, and making it economically viable and safe enough to get required approvals. Oskar Levander, Vice President at Kongsberg, predicts that autonomous ships will start as small vessels that operate in confined areas along a certain route. Indeed, the company could already demonstrate the remote operation of a commercial vessel in the harbor of Copenhagen. Industry insiders predict that cargo ships will travel in the open ocean without crews already between 2020 and 2025 (Walker 2019). Also, the blockchain platform of Maersk and IBM is in the process of being rolled out. After some startup issues where Maersk's rival carriers hesitated to join, in 2019, the container shipping companies MSC and CMA CGM signed up as members that will also operate blockchain nodes (Allison 2019). The uptake of big data analytics by the shipping industry has been slow, and only a handful of marine companies currently leverage its potential despite of studies assuming a performance increase potential of 5–10%. Constantine Komodromos, CEO of a digital chartering marketplace VesselBot, points out that "usually a big majority of the shipping industry adopts new technology after it's been tried and tested elsewhere. So it's not a matter of not wanting to change, but we are late to adjust in comparison to other markets, which has delayed the advancements in technology." However, there is room for hope. A survey among shipping industry leaders showed that 94% of respondents believe that it is time for the industry to move toward smart shipping (Trelleborg Marine Systems 2018). To understand the adoption of marketplaces, the international freight marketplace Freightos conducted a study about carrier connectivity. They concluded that "at this stage the industry is still very analog with only a few carriers having modern APIs and web sites." In its digital carrier maturity assessment, carriers achieved a rating of 38% across categories, broken down into a score of 26% in customer connectivity (categories are related to connections with customer systems through application programming interface [API] and as fallback electronic data interchange [EDI]), 45% in online experience (one-stop-shop, B2C-like, e-commerce experiences for customers), and 45% in transformation (top-down focus on changing systems, process, and culture – internally and externally) (Freightos 2019). Figure 25.6 illustrates the digitally transformed sea transport.

Air Transport

Air transport is essential to many facets of modern life, moving perishable goods, pharmaceuticals, or electronic devices from one side of the world to the other. Although air cargo represents less than 1% of world trade by volume, it accounts for over 35% by value

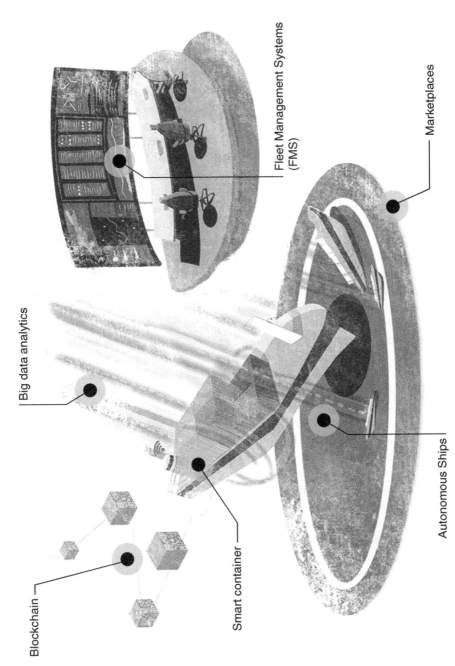

Blockchain

Big data analytics

Smart container

Fleet Management Systems
(FMS)

Marketplaces

Autonomous Ships

Figure 25.6 Digital Transformation of Sea Transport.

(IATA 2018). The next challenge will be to modernize and digitalize air cargo to increase the value of air freight for shippers (IATA 2018). Base for integrating the air cargo value chain is an end-to-end paperless transportation process with underlying regulatory framework, modern electronic messages, and high quality of data. The industry attempts to achieve these *electronic freight documents* through the International Air Transport Association (IATA)'s "IATA e-Freight initiative" where customs documents, transport documents, and commercial and special cargo documents should be digitized. This should lead to higher operational efficiency through reduced processing time, increased cost-effectiveness, higher data quality and accuracy, compliance with international and local regulations, and the development of innovative services and solutions (Sauv 2018). Another technology are *smart unit load devices (ULDs)*. The IATA estimates that around 900 000 ULDs, pallets/containers used to load luggage, freight, and mail on wide-body aircraft and specific narrow-body aircrafts, are in operation today (IATA 2019). Companies are now working on equipping these ULDs with tracking devices to increase transparency in the air cargo industry. For instance, the ULD firm Unilode Aviation Solutions developed a system that can track Bluetooth-equipped ULDs over an aircraft's WiFi system. In a try run, data about location, temperature, humidity, and light of the aircraft belly could be successfully transmitted and was accessible via a smartphone app. According to Unilode, this "will open up new avenues in transportation safety by allowing live in-flight temperature monitoring and automated load sequencing control before take-off." (Brett 2019). *Marketplaces* are also set to increase transparency and reduce the speed to make bookings in the air cargo market. Platforms such as cargo.one allow their customers to receive multiple quotes from multiple airlines, make bookings 24/7 that are immediately confirmed, and track the progress of shipments (Cargo One 2019a). Airlines benefit through quick access to an increasing number of customers, reduced complexity and costs for the booking process, and access to real-time market data. David Kerr, CEO of the cargo airline CargoLogicAir, explains that "being capable now to better reach small and medium sized forwarders (. . .) will also help us increase our short-term capacity sales and thus our load factors, to grow even more sustainably (Cargo One 2019b)." Predicting available cargo capacity, which equals saleable space, is a highly complex endeavor. Demand is volatile; aircraft belly capacity is not fixed but depending on various factors such as weather, passenger cabin occupancy, or fuel carried; various routing options are feasible for the same destination; and the booking period is very short (Berland 2018). To tackle this complexity, *big data analytics* solutions could be used that derive the available capacity and forecast demands based on historic information and latest trends. This could be the base for optimized market price segmentation and minimum acceptable prices for given shipment densities (Berland 2018).

While the digital transformation could bring substantial benefits in the form of new value-added services, operational efficiencies, revenue gains, productivity, and improved partner relations, there are some hurdles to overcome (Berland 2018). The airline industry is slow in adopting new technologies, lacks transparency and communication between stakeholders, and is short of young talents (IATA 2018). Sami Tähtinen, CTO of the cloud-based Integration Platform-as-a-Service provider Youredi, clarifies that embracing new technology can be a problem for air cargo as it is slowed by high levels of regulation and that making such investment would lead to high financial risks. Especially as most industry players are still running their operations on legacy infrastructures, adopting new technology becomes very difficult and time-consuming (Muir 2017). A survey conducted by Air Cargo Community Frankfurt in 2017 showed that while most respondents are convinced that market transparency will increase in the future, for instance, through marketplaces, most companies see this rather as a threat than as a chance. IT projects are seen as highly unpredictable, data confidentiality is considered as only restrictedly ensured, and EDI links lead to dependence on a limited number of service providers or become overly complex (Bierwirth and Schocke 2017). Hence, it is understandable that in 2017, more than 50% of the global air trade relied on paper-based processes and shipments could generate up to 30 paper documents. The global electronic shipping document penetration rate in the form of e-AWB[11] adoption reached only around 68% by the end of 2018, and according to Boris Hueske, Head of Digital Transformation at the cargo airline Lufthansa Cargo, even that took "decades of discussion (Klare 2018; Sauv 2018)." The challenge is that although most parties are aware of the benefits in the long run, maintaining two processes during the transition is time-consuming and expensive (Branson 2018). The Air Cargo Community Frankfurt survey also revealed that while 85% of respondents have IT connections to their partners, the information that is actually transmitted is insufficient along several dimensions. Study participants see high improvement potential with respect to data timeliness, completeness, and accuracy, with 26% of companies admitting that the electronically transmitted data have to be manually corrected (Bierwirth and Schocke 2017). Freightos also conducted a study about airline carrier connectivity. With a cross-category score of 22%, the digital maturity of air carriers lags behind ocean carriers (38%). Manel Galindo, CEO of digital freight solutions provider WebCargo, stresses that this is likely due to the low budgets of airline cargo businesses compared with passenger businesses. The customer connectivity score is 11% (API real-time connection with customer systems), the online experience score is 40% (one-stop-shop, B2C-like, e-commerce experiences for customers), and the transformation score is 10% (top-down focus on changing systems, process, and culture – internally and externally) (Freightos 2019). Some industry players, such as Amsterdam airport's information provider Cargonaut, realized that changes and more collaboration are urgently needed. Cargonaut's CEO Nanne Onland, who wants to drive forward the integration and exchange of data in the logistics industry, states that "we have to do better. Together (Klare 2018)." Investments into digitalization are increasing. For instance, Lufthansa Cargo ramped up its foothold in the tech sphere through cooperation with technology startups to meld a "comprehensive air cargo experience with a technology-driven, fresh view of global logistics processes (Chamlou 2018)." The American Airlines took a major change to update its systems and processes to be better prepared for the future, which included the transformation of 91 legacy systems into less than 10 (Branson 2018). Also, the cargo airline Etihad Cargo saw the introduction of several transformative programs across its commercial and operational processes as well as its physical and digital infrastructure. This culminated in the migration to a booking portal in October 2018 that accounted already for 16% of the carrier's bookings by March 2019 (Alghoul 2019). Figure 25.7 illustrates the digitally transformed air transport.

Courier, Express, and Parcel Delivery (CEP)

With *courier delivery* services, documents, small samples, patterns, or important spare parts up to 5 kg are personally accompanied during all stages of transportation from sender to addressee, without rerouting. *Express delivery* services refer to delivery services that carry goods from the

---

11 Electronic Air Waybills, the online version of a printed Master Air Waybill.

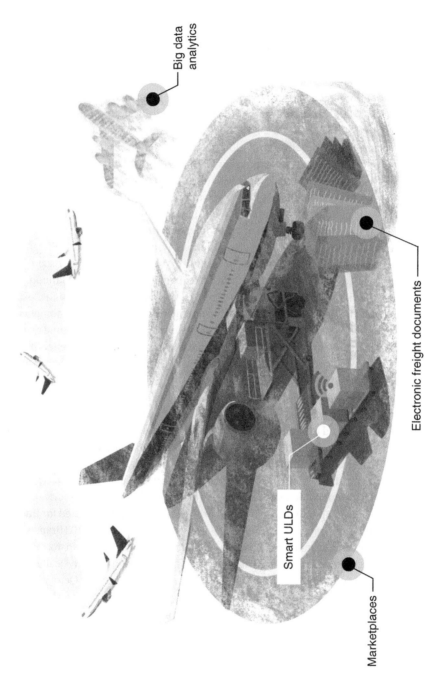

Big data analytics

Electronic freight documents

Smart ULDs

Marketplaces

Figure 25.7 Digital Transformation of Air Transport.

sender to the receiver by combining large numbers of units and distributing them internally in accordance with a flexible transportation program with guaranteed delivery times. And with *parcel delivery* services, standardized lightweight parcels (including documents) are conveyed in accordance to a fixed, defined transportation program using logistic networks with fixed running times for their specific goods. These services are today often used interchangeably or with different meanings, and service providers offer all or a mix of these concepts, collectively referred to CEP (Sage 2008). Particularly e-commerce is driving growth in the market but also leads to increasing customer expectations toward lower prices, greater convenience, and seamless experience (Bateman et al. 2015). Aside from most of the technologies already discussed in the transportation and warehouse sections of this chapter, additional tools related to the digital transformation are specifically mentioned in the CEP area.

Crowdshipping, also called crowd logistics, crowdsourced delivery, or collaborative logistics, makes use of the excess capacity of available transport modes by using premeditated trips to perform such deliveries (Serafini et al. 2018). Customers who need to send a package are matched and connected with drivers who have unused space in their vehicle and are willing to deliver the package. This would be especially helpful for "last-mile" deliveries in urban areas, where it could become a flexible, eco-friendly, low-cost solution that can also provide a "sense of community" (Punel et al. 2018). For example, the retailer Walmart, the logistics provider DHL, or Amazon conducted trials where they engaged individuals on their way to provide last-mile delivery services by using an ad hoc mobile application (Serafini et al. 2018). Peter Hesslin, CEO of DHL Freight Sweden, explained regarding the trials that were conducted in Stockholm that it "is not only a service for those requesting flexible deliveries; it is also a service for those who would consider delivering a package and earning a little extra money (. . .) As soon as the package arrives at one of DHL's collection locations, the recipient and the deliverer confirm the fee and delivery details, all within the mobile app (DHL 2013)." Shippers also looked back positively on the trials where deliveries were handled mainly by students. One shipper explained that his customers felt excited about the concept and moreover perceived it as beneficial for the environment, considering that packages were delivered by private individuals along their daily routes is was also seen as beneficial to the environment (DHL 2013). Despite of the apparent advantages of crowdshipping, some concerns must be addressed before the concept can reach maturity. Trust in the service must be established as couriers are typically occasional drivers. Delivery conditions might be affected as typically the probability of loss, theft, or damage of transported packages increases with crowdshipping. Users might also have privacy concerns about their personal information with such services (Punel et al. 2018). *Anticipatory logistics* aims to ship goods that will likely be ordered by a customer near the place of future delivery. Based on forecasts, inventory is transported in the direction of the recipient and possibly rerouted. Once an order is actually placed, it can be almost immediately fulfilled. This enables quick order response times and low total inventory in the network (Stölzle et al. 2018). For instance, Amazon that filed a patent for "anticipatory package shipping" as early as 2013 delivers goods to delivery warehouses that customers in this region have not yet ordered, but likely will. Inputs for this forecast are data points based on purchase and search history, wish lists, or time a mouse pointer stays on a certain offer. "It's not about sending the article off before the customer orders it. Rather, we want to predict the demand and distribute the goods in the network in such a way that we can bring them to the customer quickly and economically when ordering,"

Norbert Brandau, a location head at Amazon, stated (Berthold 2019). However, there are some challenges associated with anticipatory logistics. Collecting massive amounts of personal data might lead to data security and privacy issues. In addition it requires open data exchange between a logistics provider and its customers and costly and complex analyses of massive amounts of data (Kückelhaus and Chung 2016). Delivery *drones* could transform goods delivery for rural areas and cities. They could increase delivery efficiency (particularly in areas with difficult accessibility), reduce costs, and increase the speed of delivery (Bamburry 2015). Researchers from the Mayo Clinic in Rochester, USA, concluded that using delivery drones could be particularly a feasible mode for the transport of medical products in times of critical shortage. They highlighted that "in the resource-intensive environment of a disaster, speed is valued, and the special capabilities of UAVs, including the capacity to travel over closed roads and terrain without risk to a flight crew, are particularly valuable (Thiels et al. 2015)." CEP companies also see the technical and commercial feasibilities for delivery drones in the healthcare sector first. In 2019, the package delivery and supply chain management company UPS announced that it received the approval[12] from the US Federal Aviation Administration to operate a full "drone airline." The company's plan is to start by setting up UAV solutions specific to hospital campuses nationwide in the United States and to expand later to other industries (Etherington 2019). However, before drones can be used by mainstream consumers, they will have to overcome technical and regulatory challenges. Drones allow for little security to prevent theft of packages once delivered to their destination – especially as it is very visible for bystanders that packages get delivered. Drones might also be shot down as a means of entertainment and target practice. Concerns revolve further around possibilities of hacking (e.g. to get access to consumer data) and transformation into weapons (if a bird can bring down a plane, what might a drone be capable of?) (Bamburry 2015). Considering what drones have to be able to operate in a wide range of unstructured and changing environments, advanced low-cost sensors and positioning systems must be integrated and extend localization and navigation via GPS-enabled platforms (D'Andrea 2014). Today, drone technology that has been mostly developed for the consumer market is several orders of magnitude below the critical system failure tolerance of commercial aviation of once in one billion hours of flight (DiNota et al. 2019). *Parcel lockers* are unattended delivery machines or systems of reception boxes that enable to both receive and send parcels. They are intended to solve the last-mile deliver problem. When couriers arrive at their destination, they often do not meet anybody at home to receive the parcel. With parcel lockers, customers get informed of deliveries via electronic messages once parcels arrive and have the possibility to access their packages 24/7. For CEP providers, this can lead to efficiency gains and reduce delivery costs (Iwan et al. 2016). With 200 000 installations, China has more parcel lockers than the rest of the world combined[13] (Sotolongo

---

12 The permission authorizes UPS to fly an unlimited number of drones with an unlimited number of remote operators in command so that it can scale its operations according to customer demand. The "Part 135 Standard" certification it obtained is the highest level of certification that has no limits on the size (traditionally drone and cargo must not exceed 25 kg) or scope of operations (e.g. also flying at night is permitted) (Etherington 2019).

13 For instance, the European countries with the most developed locker networks, Germany, Poland, and Spain, have about 15 000 lockers combined (Sotolongo et al. 2019).

et al. 2019). A local customer explains that "it makes me feel safer to pick up the parcels from the lockers rather than open the door to a strange person, especially at night." Deliverymen also appreciate the convenience. "I don't have to climb up and down [and] it also solves the problem that many customers are unwilling to open their doors to us," one claims. Another one adds that "We don't have to take liability for the theft of the parcel left on the doorstep when the customers are not at home (Yang 2017)." A disadvantage of parcel lockers is that essentially the final leg of a parcel's journey must be made by the customers themselves. Also, CEP companies have to be willing to participate and allow to deliver goods to parcel locker, even with systems set up by competitors. Furthermore, implementing and utilizing such lockers requires the support of local residents and owners of an intended location (Iwan et al. 2016).

While crowd logistics or anticipatory logistics are still at an early stage, drones and especially parcel lockers are already rather mature technologies (Stölzle et al. 2018). However, Martin Sarch, a Partner at the postal advisory firm Leapfrog Business Consulting, is not optimistic about a large rollout of the latter. "Smart parcel lockers are innovative and need to be tested for utility, customer acceptance and a true evaluation of their cost, capital and ongoing. They may definitely have a place in office blocks and shopping areas, but before posts spend millions of dollars on the parcel systems and renting thousands of locations, a cost benefit [analysis] must be done first, as someone has to pay (Sarch 2018)." Although this also holds true for delivery drones, a study by the Association for Unmanned Vehicle Systems International concluded that the drone industry could add about 82 billion USD in economic activity over the next 10 years (Bamburry 2015). In a 2016 McKinsey study about the future of parcel delivery, the authors believe that most prerequisites for new business models and new technologies in last-mile delivery are already prevalent (Joerss et al. 2016). Leading CEP companies such as DPD, DHL, FedEx, Australia Post, or UPS are already making billion USD investments to digitalize and expand their network in the parcel delivery business (Bateman et al. 2015). In its "Strategy 2025," DHL announced that it will be spending roughly 2 billion EUR on initiatives to enhance customer and employee experience and improve operational excellence. The group states that systematic digitalization throughout its businesses is a lever for achieving significant progress and that the pace of the digital transformation that has already been triggered within the group will be stepped up. "We are convinced that future growth will come from a consistent focus on our profitable core logistics businesses – and digitalization will become the greatest lever," Frank Appel, CEO of Deutsche Post DHL Group, explained (DHL 2019). Similarly, the Chinese CEP provider SF Express stated that "with the gradual evolution from mobile Internet to IoT technology and the rapid arrival of industry 4.0 and the omni-channel new retail era, the original traditional supply chain of enterprises must respond to the advent of digital economy through changes, transformation, innovation and optimization (S.F. Holding 2019)." JD Logistics that focused historically on drone deliveries, logistics automation, and smart vehicles raised around 200 million USD to further invest in logistics-related companies and technology (Shu 2019). A disruption of the last-mile delivery might become reality in the coming years, with the speed at which different countries will adapt depending on opportunity costs, regulation, and public acceptance (Joerss et al. 2016). Figure 25.8 illustrates the digitally transformed Courier, Express, and Parcel Delivery.

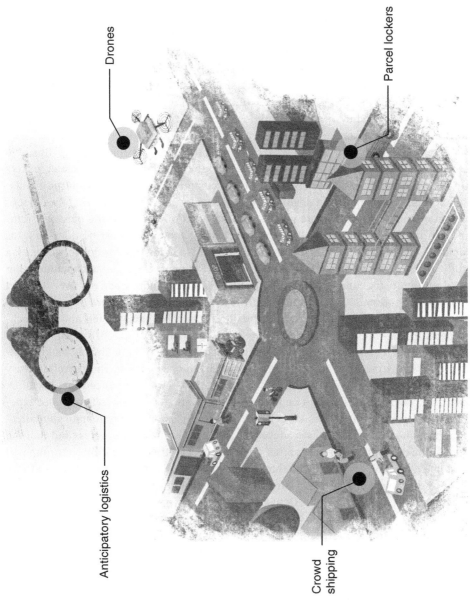

Drones

Parcel lockers

Anticipatory logistics

Crowd shipping

Figure 25.8 Digital Transformation of Courier, Express, and Parcel Delivery.

Logistics Services and Advisory

With revenues of around 870 billion USD, third-party logistics (3PL) providers account for the majority of the logistics services and advisory market (Armstrong and Associates 2019). 3PLs are activities carried out by a logistics service provider on behalf of a shipper, leading to economies of scale and scope (Berglund et al. 1999). Domestic and international transportations (including freight forwarding) and warehousing account for the largest share of 3PLs revenues (Langley 2019). Aside from these areas, where impact and status of the digital transformation was discussed in previous sections of this chapter already, services that 3PLs conduct can be categorized into three areas. *Logistics management services* include consulting services (i.e. the development of customized logistics concepts), transportation management (planning, monitoring, and conducting transports; handling and storage of goods), and IT services (services such as track and trace, API, and EDI). *Customized value-added services* include packaging and labeling (packing, commissioning, and labeling of goods on behalf of shippers), manufacturing services (product assembly and contract manufacturing on behalf of shippers), and special transports and handling (transportation and handling of special goods). *Customer relationship management (CRM)* includes order processing (acting as interface between shippers and customers), administrative support (support for shippers and customers with tasks such as billing and claims handling), and on-demand services (arrangement of service times according to customer needs) (Hofmann and Osterwalder 2017). Digitalization has brought forth new technologies that force 3PLs to adopt to changing environments with differing customer expectations.[14] *Robotic process automation (RPA)* tools perform "if, then, else" statements on structured data by using a combination of user interface interactions or by connecting to APIs to drive client servers, mainframes, or code. With RPA, instead of redesigning systems, human operators are replaced by digital agents. It is therefore seen as a way to quickly achieve a high return on investment by reducing costs and quickly linking legacy applications (van der Aalst et al. 2018). 3PL organizations typically employ a large labor force that is occupied with automation prone activities such as quoting, scheduling, and tracking shipments, securing proof of delivery, or generating and collecting invoices. Using RPA might free up lots of staff to, for example, focus on harder-to-automate tasks such as exception management. Estimations for potential cost savings through RPA range from 25 to 50% (Gould 2019). *Chatbots* are computer programs that can hold a conversation with humans using natural language speech. Such programs can assist in human computer interaction and examine and influence the behavior of the user by asking questions and responding to the user's questions. Chatbots mimic intelligent conversations where they give answers that are the best intelligent response to a certain input sentence (Abdul-Kade and John 2015). When integrated with a content source (like ERP or CRM applications) to assume what an end user wants to do, they can be particularly powerful in logistics and supply chain management. End-to-end visibility through shipment tracking, real-time status updates, product stock/

---

14 It is important to note the other technological developments in the areas of domestic and international transportations (including freight forwarding) and warehousing that are discussed previously, considering that they account for the largest share of 3PL revenues (Langley 2019).

order requests, or inventory management are just some areas where chatbots can possibly be utilized (Patel 2018). Chatbots provide 24-hour service, ensure that users can get instant responses, answer consistently, and are always "friendly and approachable with endless patience." For companies this can lead to improved customer satisfaction, cost savings, increased customer interaction and sales, and gaining of deeper understanding of customers' needs and greatest pain points (AIMultiple 2019). *Three-dimensional (3D) printing*, or additive manufacturing, is a method of manufacturing in which materials, such as plastic or metal, are deposited onto one another in layers to produce a 3D object.[15] Originally mainly used to create prototypes, technical advances have enabled 3D printers to make objects that are comparable with traditionally manufactured items (Schubert et al. 2014). Technical Improvements make additive manufacturing a feasible alternative in higher value-added production and higher-order designs. For instance, the speed it takes to produce parts via 3D printing is rapidly accelerating; the assortment of materials available is expanding into a broad assortment of metals such as stainless, gold, bronze, steel, nickel, or even carbon fiber; and possible tolerances can reach beyond +/− 0.2 mm. This potentially could create a world of massive single lot production close to end customers. Decisions where to produce items would then be about proximity to customers and shipping times instead of, e.g. labor costs, as the defining economic inputs for 3D printing cost roughly the same amount regardless of the country they are located in (Roberto and Wilcox 2018). Through 3D printing, the number of shipments – especially emergency freight – could drastically be reduced, inventory holding patterns shift, and the expectation toward 3PL's manufacturing services radically change (Tipping and Kauschke 2016). Therefore, freight firms such as DSV set up labs where they create 3D models for customers. The company believes that the value of additive manufacturing will increase from improved logistics, cost reduction by substitution, and cost-effective product development to whole value chain repositioning and forward/backward integration (Botman 2018). *Supply chain simulation software* possesses a variety of advantages compared with experience-based or trial-and-error approaches for supply chain design. In simulations, complex systems with their dynamic processes are emulated in an experimental model in order to arrive at insights that are transferable to reality (VDI-Fachbereich Fabrikplanung und -betrieb 2014). With simulation software, it is possible to visualize the variability and interconnectivity of supply chains and predict the expected performance for alternative designs. This allows for cost-effective and objective decision making without disrupting day-to-day operations, especially when employed in a Software-as-a-Service (SaaS) model. Also, *platforms* can support 3PLs to easier serve their customers as they allow for easy arrangement and management of, e.g. transportation services. Paired with data mining and transportation planning algorithms, platforms such as marketplaces can be a significant upgrade to the IT solutions currently offered by 3PLs, allowing for more efficient and transparent service. Knowledge from such platforms can also be integrated into 3PLs' IT solutions or lead to the development of new transportation management concepts and consulting services (Hofmann and Osterwalder 2017). Digital freight sales platforms, such as DHL's "myDHLi Quote &

---

15 For a case study about the additive manufacturing company *Materialise NV*, cf. (Wolff and Kern 2019).

Book," can also support instant quoting, automated execution, and improved process handling together with dynamic pricing (Toczauer 2019). Platforms further facilitate trade as they allow easier cross-border transportation and the offering of digitally enabled logistics services. In addition, they might allow 3PLs to easier service customers' demand for same-day delivery, particularly in current and future megacities (World Economic Forum 2016). Robots can be integrated with label printing and, if required gluing, stations to build *Robotics Labelling Systems* that allow for label placement by robots (Motion Controls Robotics 2019). Systems optimized for functionality, structure, actuation, and sensor configuration and control can efficiently and accurately apply labels on even randomly oriented objects (Lin et al. 2016). This can reduce labor costs and human errors and improve productivity. For instance, the IT solution service provider Million Tech Development's automatic label applicator labels 10 800 products in eight hours, which is a threefold productivity increase compared with a manual solution (Million Tech Development Ltd 2019).

While these solutions bring doubtlessly advantages, they also lead to new challenges that have to be overcome. The "smartness" level of current RPA and chatbot solutions is still below the technical maturity level that would lead to widespread adoption. It will still take time until, e.g. due to AI technology, more complex and less defined tasks can be supported. The goal is to have IT tools learn in a similar way as humans, for example, by observing human problem resolving capabilities, so that they can adapt and handle nonstandard cases (van der Aalst et al. 2018). Another task is to identify how to control computer agents, such as RPA and chatbots, to avoid security, compliance, and economic risks. Who is responsible if an agent makes incorrect decisions due to contextual changes or because it "learned" incorrect behavior from humans? What happens if it does not behave as intended and insults customers? Chatbots can also irritate customers as they are still not capable to improvise and thus must rely on preprogrammed algorithms that impossibly can forecast all potential situations. Also, while researchers are working on "humanizing" digital agents, so far chatbots are not able to "read between the lines" of a conversation and to recognize sarcasm, irony, and humor (Abdul-Kade and John 2015; InternetDevels 2018; van der Aalst et al. 2018). Supply chain simulation software and platforms can drastically improve operations and empower users to proactively manage supply chains. However, as they are typically available for any potential user, shippers could easily cut out the middleman in the form of a 3PL and substitute them. James Cooke, Principal Analyst at Nucleus Research explains that "until recently, only very large companies could afford to make investments in supply chain design tools and regularly evaluate the makeup of their network. SaaS offerings are changing the game and empowering even smaller retailers to punch above their weight class and keep their network agile (Nucleus Research 2015)." Logistics marketplaces affect almost the entire 3PL value chain, especially transport management as one of 3PLs' main activities. Platforms are likely to replace the classical transportation management and CRM of 3PLs and erode the value of IT solutions that 3PLs offer today. Also vehicle manufacturers can extend their offerings to platforms and add service capabilities in the area of order processing and administration so that they potentially replace management-related activities of 3PLs (Hofmann and Osterwalder 2017). Another crucial challenge for 3PLs will be workforce shifts as most staff at 3PLs is employed in jobs that are at high risk of automation.

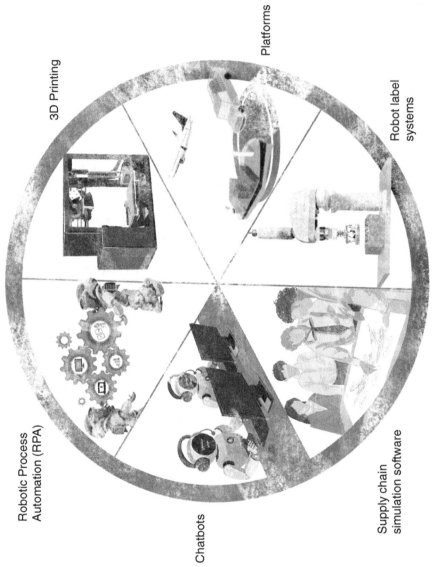

Figure 25.9 Digital Transformation of Logistics Services and Advisory.

Studies show that in developed countries like the United States, Germany, or Japan, a workforce displacement of around 15–35% by 2030 in such functions is likely[16] (Manyika et al. 2017). 3PLs must therefore incorporate a different skillset into their training and hiring activities, focusing more on creativity, originality, and innovation and data- and software-related skills such as data management, data science, and human–machine interface-related know-how (Lorenz et al. 2016; World Economic Forum 2018). In a 2019 3PL study, 93% of 3PL customers agreed that IT capabilities were a necessary element of the 3PL expertise, but only 55% agreed that they are satisfied with 3PLs' IT capabilities (Langley 2019). This gap between expectations and capabilities remained consistent over years, indicating that shippers' greater expectations toward 3PLs' data reporting and analysis capabilities were only partly met by the increases in 3PLs' IT capabilities. However, with shippers' stronger focus on IT skills, 3PLs are improving and promoting their IT capabilities. The study revealed that within the next two years, 72% plan to invest in supply chain visibility/control towers, 67% in TMS, 67% in warehouse management systems (WMS), 58% in predictive analytics, 52% in package receiving/tracking software solutions, and 42% in ERPs (Langley 2019). While several 3PLs recognize 3D printing as a means to improve customer satisfaction and cut delivery times, only 5% of them were making investments into 3D printing. This might increase, considering that the global 3D printing market grew with over 20% per year since 2013 and is expected to reach a volume of 180–490 billion USD by 2025 (Keeney 2016; Langley 2019). Derrick Johnson, Vice President at UPS, states that "we are now seeing that 3D printing has reached an inflection point as lower costs and technological advances have put it within reach of more people (UPS 2016). Figure 25.9 illustrates the digital transformation of Logistics Services and Advisory.

Conclusion

This chapter set out with the tale of "two worlds of digitalization" in the logistics industry. While some companies can show use cases of the digital transformation with fully automated warehouses, AGVs, and robots, lots of companies are still struggling to gain visibility in their supply chain or have a hard time making sense of the data they collect. To answer the questions how these two worlds fit together and whether there is a digitalization chasm between born digital behemoths and traditional industry players, this chapter summarized current studies about the state of digitalization in logistics and supply chain management. The situation for *logistics infrastructure* is ambivalent. On the one hand, the digital transformation of seaports and airports is happening slower than in other logistics areas. For instance, only 3% of container terminals are considered (semi)automated. Major barriers here are the high upfront investments, operational challenges (for instance, which

---

16 It should be noted that massive technological unemployment is improbable as new, different jobs will replace those that are destroyed. For instance, the Organisation for Economic Co-operation and Development (OECD) concludes that historically, the net effects of major technological revolutions on employment have been positive and that there are few signs that this trend will radically change in the next years (OECD 2019).

business case to define and poor data quality), and a mindset that is still focused on traditional business models. On the other hand, warehouses are already halfway through. Approximately 30–40% of warehouses adopted some type of Industry 4.0 technology, such as sensors, robotics and automation, or predictive analytics.

Similar figures are also observable in some areas of *logistics execution*. In road freight, 35–40% of companies adopted a TMS, and most transport operators plan to invest in FMS and other technology to optimize their day-to-day operations. For sea freight, shipping companies are cautiously embracing the digital transformation and prefer to "wait and see." The potential of big data remains largely untapped; few carriers use modern ways to connect with their customers, e.g. via marketplaces; and blockchain adoption remains low. In air freight, carrier connectivity lags even behind ocean carriers. Suffering from low budgets, companies are struggling to implement digital solutions that do not immediately lead to cost benefits. And in CEP delivery, most technologies discussed are largely mature but, however, are still in early rollout phases. Opportunity costs, regulatory hurdles, and public acceptance require case-by-case evaluations per technology, region, and particular application.

In *logistics services and advisory*, only around half of the customers are satisfied with their 3PLs' IT capabilities, and industry players have not been investing enough to increase this share. They focus their investments on supply chain visibility/control towers, TMS, or WMS, largely disregarding some innovative technologies such as 3D printing. This might be due to insufficient maturity levels of some new technologies; but it could also be related to the disruptive impact that they potentially have on a 3PL's business model.

Overall, the state of digitalization for traditional industry players is on a low to medium level. Universal barriers such as high costs, lack of standards, and unclear benefits might have cautioned them in the past to embrace digitalization and instead keep relying on their existing approaches (Sandberg and Hemilä 2018). Nevertheless, an increasing number of companies understood the impact that the digital transformation will have on them and intend to follow the call of the times. Their heavy investments into the digital transformation might not immediately materialize, but within the next years, they will have gained a leading edge that others likely will not be able to catch up anymore. Once they can offer outstanding performance, excellent customer satisfaction, and extraordinarily competitive prices, different worlds of digitalization will become more evident.

Companies in logistics and supply chain management must set the course today to be able to participate in a world where sensors are placed in everything, networks are created everywhere, anything becomes automated, and everything gets analyzed. Likely "no regret" investments are improvements in the data collection across the value chain as well as into the necessary capabilities to analyze big data streams (World Economic Forum 2016). At the same time, the biggest risk for either company is to ignore the digital transformation. Zeljko Ivkovic, CIO Greater China at DB Schenker summarizes it: "Digital disruption has redefined the need to exponentially leverage and adapt to an increasingly changing environment. Companies and even industries that do not pro-actively adapt their digital strategy will not only be left behind but vanish faster and more surprisingly than this was the case in the past."

Key Takeaways

- There is a digital divide among digital behemoths and traditional industry players: While some companies have fully automated warehouses, AGVs, and robots, others are struggling to gain visibility in their supply chain or analyze collected data.
- Automation at *seaports* can reduce operational expenditures, risk for errors, and downtimes. However, it requires high investments, is inflexible, and needs skilled labor. Only 3% of container terminals worldwide are fully automated or semiautomated.
- Implementing a fully connected *Airport* 4.0 concept would increase capacity by over 10% while reducing costs for administrative staff and energy. However, airports are low at adopting innovative solutions, often not knowing how to position themselves in a broader ecosystem, influence relevant regulation, or recognize data as enterprise asset.
- A variety of technologies, such as Internet of Things devices, big data analytics, augmented reality, robots, or drones, are discussed in the context of the digital transformation in *warehouses*. They are recognized to potentially transforming the way how warehouses are operating, creating transparency, and improving and speeding up processes. Although technical and management hurdles are yet to overcome, some type of Industry 4.0 technology is already implemented in 30–40% of warehouses.
- In *road transportation*, transport management and fleet management systems already optimize day-to-day operations for about 40% of companies. Meanwhile, the "next big thing" in the form of autonomous trucks is predicted not to arrive within the next decade.
- The *shipping* industry is also cautiously embracing a variety of technologies like smart containers, big data analytics, or marketplaces. They and other digital solutions can facilitate operations, trade, and transparency, although increasing the risk of cyberattacks. Industry experts foresee that already in the next five years, cargo ships will travel in the open ocean without crews.
- While the digital transformation for the *airline* industry could bring substantial benefits in the form of operational efficiencies, revenue gains, productivity, and improved partner relations, hurdles such as a lack of transparency and the industry's tendency to slowly adopt new technologies have to be overcome. Relevant technologies include electronic freight documents, marketplaces, smart loading devices, and big data analytics to better forecast available capacity demands.
- Aside from most of the mentioned technologies, additional tools are mentioned in the *courier, express, and parcel delivery part*. Delivery drones could transform goods delivery for rural areas and cities; parcel lockers solve the last-mile deliver problem. Industry players announced to invest billions to systematically digitalize.
- In the *Logistics Services and Advisory* part, third-party logistics explore technologies that improve the logistics management and value-added and customer relationship management services that they offer. Beyond addressed technologies, advanced solutions include robotic process automation, chatbots, platforms, or supply chain simulation software. 3PLs receive low customer satisfaction levels regarding their IT capabilities, so most companies plan to further invest in technology, mainly related to supply chain visibility, TMS, warehouse management systems, and predictive analytics.
- The digitalization maturity of most industry players is currently on low to medium level. Most now understood the impact of the digital transformation and intend to heavily investment into new technology, ready to overcome barriers such as high costs, lack of standards, and unclear benefit.

Acknowledgments

I would like to thank Prof. Dr. Tariq Samad, Mac Sullivan, and Dr. Axel Neher for their helpful comments on an earlier draft of this article. Special thanks also go to Wang Wei (汪 巍) who created the illustrations for this chapter.

References

van der Aalst, W.M.P., Bichler, M., and Heinzl, A. (2018). Robotic process automation. *Business & Information Systems Engineering* 60 (4): 269–272. https://doi.org/10.1007/s12599-018-0542-4.

Abdul-Kade, S.A. and John, D. (2015). Survey on Chatbot design techniques in speech conversation systems. *International Journal of Advanced Computer Science and Applications* 6 (7) https://doi.org/10.14569/IJACSA.2015.060712.

Agenbroad, J., Creyts, J., Mullaney, D. et al. (2016). Improving efficiency in Chinese trucking and logistics. *Proceeding and Insights from the Design Charrette on Chinese Logistics and Trucking Efficiency*. Shenzhen: Rocky Mountain Institute. https://www.rmi-china.com/static/upfile/news/nfiles/201803221629074970.pdf (accessed 25 January 2020).

AIMultiple (2019). Top 12 benefits of Chatbots: comprehensive guide [2019 update]. *appliedAI Blog*. https://blog.aimultiple.com/chatbot-benefits (accessed 25 January 2020).

Airports Council International (2017). *Airport Digital Transformation*, p. 32. https://store.aci.aero/product/airport-digital-transformation-best-practice (accessed 25 January 2020).

Aleksandrova, M. (2019). streamlining your warehouse management with digitalization. https://easternpeak.com/blog/streamlining-your-warehouse-management-with-digitalization (accessed 25 January 2020).

Aleph, V.C. (2018). Michael Eisenberg on FreighTech. https://www.youtube.com/watch?v=3BBzpWiSn3U (accessed 25 January 2020).

Alghoul, R. (2019). Etihad Cargo's digital transformation ushers in New Era of Cargo Management. http://wam.ae/en/details/1395302757911 (accessed 25 January 2020).

Alicke, K., Rachor, J., and Seyfert, A. (2016). *Supply Chain 4.0 – The Next-Generation Digital Supply Chain*. McKinsey Supply Chain Management Practice https://www.mckinsey.com/~/media/McKinsey/Business%20Functions/Operations/Our%20Insights/Supply%20Chain%2040%20%20the%20next%20generation%20digital%20supply%20chain/08b1ba29ff4595ebea03e9987344dcbc.pdf (accessed 25 January 2020).

Allison, I. (2019). IBM, Maersk finally sign up 2 big carriers for shipping blockchain. *Coindesk*.

Amazon Improves Sorting Accuracy by 50 Percent (2019). Supply & demand chain executive. https://www.sdcexec.com/warehousing/news/21072140/amazon-improves-sorting-accuracy-by-50-percent (accessed 25 January 2020).

American Transportation Research Institute (2016). Identifying autonomous vehicle technology impacts on the trucking industry. *Arlington*. https://atri-online.org/wp-content/uploads/2016/11/ATRI-Autonomous-Vehicle-Impacts-11-2016.pdf (accessed 25 January 2020).

Appelbaum, D. and Nehmer, R.A. (2017). Using drones in internal and external audits: an exploratory framework. *Journal of Emerging Technologies in Accounting*. 14 (1): 99–113. https://doi.org/10.2308/jeta-51704.

Armengaud, E., Sams, C., von Falck, G. et al. (2017). Industry 4.0 as digitalization over the entire product lifecycle: opportunities in the automotive domain. In: *EuroSPI* (eds. J. Stolfa, S. Stolfa, R. V. O'Connor, and R. Messnarz.), 334–351. Springer https://doi. org/10.1007/978-3-319-64218-5_28.

Armstrong & Associates (2019). Global 3PL market size estimates. https://www.3plogistics. com/3pl-market-info-resources/3pl-market-information/global-3pl-market-size-estimates (accessed 25 January 2020).

Audi, A.G. (2018). Driverless transport vehicles in use for the Audi A8 | Audi Blog. https://blog. audi.de/driverless-transport-vehicles-in-use-for-the-audi-a8/?lang=en (accessed 9 October 2018).

Backas, J. and Yeoman, A. (2018). The digitalization of the marine industry and the future of risk. https://www.eniram.fi/digitalization-marine-industry-future-risk (accessed 25 January 2020).

Bamburry, D. (2015). Drones: designed for product delivery. *Design Management Review* 26 (1): 40–48.

Bateman, T., Buhler, B., and Pharand, A. (2015). Adding value to parcel delivery. https://www. accenture.com/_acnmedia/accenture/conversion-assets/dotcom/documents/global/pdf/ dualpub_23/accenture-adding-value-to-parcel-delivery.pdf (accessed 25 January 2020).

Ben-Daya, M., Hassini, E., and Bahroun, Z. (2017). Internet of things and supply chain management: a literature review. *International Journal of Production Research*: 1–24. https:// doi.org/10.1080/00207543.2017.1402140.

Benecki, P. (2019). Competing for data, maritime executive magazine. https://www.maritime-executive.com/editorials/competing-for-data (accessed 25 January 2020).

Berglund, M., van Laarhoven, P., Sharman, G. et al. (1999). Third-party logistics: is there a future? *The International Journal of Logistics Management* 10 (1): 59–70. https://doi. org/10.1108/09574099910805932.

Berland, P. (2018). How can digital transform the air cargo industry? Panorama of air cargo digital initiatives. http://transport.sia-partners.com/sites/default/files/air_cargo_digital_ opportunities_vf_no_logo.pdf (accessed 25 January 2020).

Berthold, G. (2019). Anticipatory logistics: why delivery will take place before the order in the future, lead innovation management. https://www.lead-innovation.com/english-blog/ anticipatory-logistics (accessed 25 January 2020).

Bierwirth, B. and Schocke, K.-O. (2017). Lead-time optimization potential of digitization in air cargo. In: *Digitalization in Supply Chain Management and Logistics: Smart and Digital Solutions for an Industry 4.0 Environment. Proceedings of the Hamburg International Conference of Logistics (HICL)*, vol. 23, 75–98. Berlin: epubli GmbH.

Blondel, M., Zintel, M., and Suzuki, H. (2015). *Airports 4.0: Impact of Digital Transformation on Airport Economics*, 19. Arthur D. Little https://www.adlittle.com/sites/default/files/ viewpoints/2015-05-Arthur_D_Little_T_T-Impact_of_Digital_on_Airport_Business_Model. pdf (accessed 25 January 2020).

Botman, M. (2018). "DSV start up": innovation nurturing in the world of logistics service providers. https://www.supplychainmagazine.nl/wp-content/uploads/2018/12/Presentatie-DSV-Partnerbijeenkomst-SCM.pdf (accessed 25 January 2020).

Branson, J.R. (2018). Adapt or die. https://www.aacargo.com/about/digital-disruption-air-cargo.html (accessed 25 January 2020).

Bréchemier, D., Hanke, M., and Streichfuss, M. (2017). *Rise to the Challenge – The Risks and Opportunities of Digitalization for Airports*. Munich: Roland Berger https://www.rolandberger. com/publications/publication_pdf/think_act_digital_airport.pdf (accessed 25 January 2020).

Brett, D. (2019). Unilode and onasset complete bluetooth ULD trial. *AirCargoNews*. https://www.aircargonews.net/airlines/unilode-and-onasset-complete-bluetooth-uld-trial (accessed 25 January 2020).

Cannon, J. (2017). Digitizing the trade transaction: the 'next big thing' in trucking. *Commercial Carrier Journal*. https://www.ccjdigital.com/digitizing-the-trade-transaction-the-next-big-thing-in-trucking (accessed 25 January 2020).

Cargo One (2019a). book air cargo capacities across multiple airlines within seconds. https://www.cargo.one/#Features (accessed 25 January 2020).

Cargo One (2019b). Leading airlines distribute their short-term cargo capacities with Cargo. one. https://www.cargo.one/airlines (accessed 25 January 2020).

Chamlou, N. (2018). Lufthansa cargo broadens digitalization efforts with cargo.one investment. *AirCargoWorld*. https://aircargoworld.com/allposts/lufthansa-cargo-broadens-digitalization-efforts-with-cargo-one-investment (accessed 25 January 2020).

Chen, H., Chiang, R.H.L., and Storey, V.C. (2012). Business intelligence and analytics: from big data to big impact. *MIS Quarterly. JSTOR* 36 (4): 1165–1188. https://doi.org/10.2307/41703503.

Choe, T., Garza, M., Rosenberger, S.A. et al. (2017). The future of freight how new technology and new thinking can transform how goods are moved. https://www2.deloitte.com/content/dam/insights/us/articles/3556_FoM_Future-of-freight/DUP_FoM_the-future-of-freight.pdf (accessed 25 January 2020).

Chu, F., Gailus, S., and Lisa Liu, L.N. (2018). *The Future of Automated Ports*. McKinsey & Company. https://www.mckinsey.com/industries/travel-transport-and-logistics/our-insights/the-future-of-automated-ports (accessed 25 January 2020).

Companik, E., Gravier, M.J., and Farris, M.T.I. (2018). Feasibility of warehouse drone adoption and implementation. *Journal of Transportation Management* 28 (2): 33–50.

CSS Electronics (2019). CAN bus explained - a simple intro. https://www.csselectronics.com/screen/page/simple-intro-to-can-bus/language/en (accessed 25 January 2020).

Cyprus Shipping Chamber (2018). *Vulnerability Management Case Study*. Limassol. https://maritimecyprus.files.wordpress.com/2018/06/cyprus-shipping-chamber-vulnerability-management-case-study.pdf (accessed 25 January 2020).

Daimler, A.G. (2019). Daimler trucks invests half a billion Euros in highly automated trucks. https://media.daimler.com/marsMediaSite/ko/en/42188247 (accessed 25 January 2020).

D'Andrea, R. (2014). Guest editorial can drones deliver? *IEEE Transactions on Automation Science and Engineering* 11 (3): 647–648. https://doi.org/10.1109/TASE.2014.2326952.

Davidson, N. (2016). Container terminal automation: pros, cons & misconceptions. *Port Technology* 70: 30–31.

Davidson, N. (2018). *Retrofit Terminal Automation: Measuring the Market*. London: Port Technology https://globalmaritimehub.com/wp-content/uploads/2018/03/Neil_Davidson__presentation_2018.pdf (accessed 25 January 2020).

DHL (2013). DHL crowd sources deliveries in stockholm with my ways. https://www.dhl.com/en/press/releases/releases_2013/logistics/dhl_crowd_sources_deliveries_in_stockholm_with_myways.html#.XZ7xh3hR270 (accessed 25 January 2020).

DHL (2019). Strategy 2025: Deutsche Post DHL Group Accelerates Growth in Core Businesses and Invests EUR 2 Billion in Digital Transformation. https://www.dpdhl.com/en/media-relations/press-releases/2019/strategy-2025-deutsche-post-dhl-group-accelerates-growth-in-core-businesses-and-invests-eur-2-billion-in-digital-transformation.html (accessed 25 January 2020).

DiNota, A., Douglas, S., and Marcontell, D. (2019). *Why the Skies Aren't Filled With Delivery Drones . . . Yet*. Forbes. https://www.forbes.com/sites/oliverwyman/2019/10/07/why-the-skies-arent-filled-with-delivery-drones-yet/#7b8c654a3bc9 (accessed 25 January 2020).

Economist (2018). The global logistics business is going to be transformed by digitisation. *The Economist* 427 (9089): 20–22. https://www.evernote.com/shard/s218/nl/49618278/a979c191-c6aa-4be4-858d-e3cc31a2df25 (accessed 25 January 2020).

Elkhodr, M., Shahrestani, S., and Cheung, H. (2013). The Internet of Things: vision & challenges. In: *IEEE 2013 Tencon-Spring*. IEEE, pp. 218–222 (accessed 25 January 2020).

Etherington, D. (2019). UPS gets FAA approval to operate an entire drone delivery airline. https://techcrunch.com/2019/10/01/ups-gets-faa-approval-to-operate-an-entire-drone-delivery-airline (accessed 25 January 2020).

Ettorre, B. (1997). What's the next business buzzword? *Management Review* 86 (8): 33–35.

Fraunhofer Center for Maritime Logistics and Services (2016). *The Autonomous Ship, Maritime Unmanned Navigation through Intelligence in Networks Project*.

Freightos (2019). *Digital Carrier Connectivity 2019 Benchmarking the Digital Maturity of Top International Freight Carriers*.

Google (2019). Google trends. https://trends.google.com/trends/explore?date=today 5-y&q=%2Fm%2F0g5r88p,industry 4.0 (accessed 25 January 2020).

Gould, R. (2019). Transforming 3PL, transportation and logistics international. http://www.tlimagazine.com/sections/columns/2538-transforming-3pl (accessed 25 January 2020).

GT Nexus Capgemini Consulting (2016). The current and future state of digital supply chain transformation. http://mktforms.gtnexus.com/rs/979-MCL-531/images/GTNexus-Digital-Transformation-Report-US-FINAL.pdf (accessed 25 January 2020).

Haller, S., Karnouskos, S., and Schroth, C. (2008). The Internet of Things in an enterprise context. In: *Future Internet Symposium*, 14–28. https://www.alexandria.unisg.ch/46642/1/fis2008-haller-final.pdf (accessed 25 January 2020).

Hampstead, J.P. (2019). *Digital Freight Brokerage Growth to Accelerate Sharply Over Next Five Years*. Freightwaves. https://www.freightwaves.com/news/technology/digital-freight-brokerage-growth-to-accelerate-sharply-over-next-five-years (accessed 25 January 2020).

Havenbedrijf Rotterdam (2019). Rotterdam sends hyper-smart container on trip around the world. https://www.portofrotterdam.com/en/news-and-press-releases/rotterdam-sends-hyper-smart-container-on-trip-around-the-world (accessed 25 January 2020).

Hofmann, E. and Osterwalder, F. (2017). Third-party logistics providers in the digital age: towards a new competitive arena? *The Log* 1 (2): 9. https://doi.org/10.3390/logistics1020009.

Hornyak, T. (2018). *The World's First Humanless Warehouse Is Run Only by Robots and Is a Model for the Future*. CNBC. https://www.cnbc.com/2018/10/30/the-worlds-first-humanless-warehouse-is-run-only-by-robots.html (accessed 25 January 2020).

Iansiti, M. and Lakhani, K.R. (2017). The truth about blockchain. *Harvard Business Review* 95 (1): 118–127.

IATA (2018). IATA cargo strategy. https://www.iata.org/whatwedo/cargo/Documents/cargo-strategy.pdf (accessed 25 January 2020).

IATA (2019). Unit load devices (ULD). https://www.iata.org/whatwedo/cargo/unit-load-devices/Pages/index.aspx (accessed 25 January 2020).

IBM (2017). *The Paper Trail of a Shipping Container*. Armonk. https://www.ibm.com/downloads/cas/VOAPQGWX (accessed 25 January 2020).

iContainers (2018). The future of automation at terminals and ports. https://www.icontainers.com/us/2018/10/09/the-future-of-automation-at-terminals-and-ports (accessed 25 January 2020).

International Chamber of Shipping (2019). Shipping and world trade. http://www.ics-shipping.org/shipping-facts/shipping-and-world-trade (accessed 25 January 2020).

International Road Transport Union (2018). European road transport industry readies for tech transformation. https://www.iru.org/resources/newsroom/european-road-transport-industry-readies-tech-transformation (accessed 25 January 2020).

InternetDevels (2018). *The Advantages and Disadvantages of Using Chatbots for Business*. Corporate Blog. https://internetdevels.com/blog/pros-and-cons-of-using-chatbots-for-business (accessed 25 January 2020).

Iwan, S., Kijewska, K., and Lemke, J. (2016). 'Analysis of parcel lockers' efficiency as the last mile delivery solution – the results of the research in Poland. *Transportation Research Procedia* 12: 644–655. https://doi.org/10.1016/j.trpro.2016.02.018.

Jain, S., Chaudhary, P.K., and Patchala, S. (2017). Technology-led disruptive freight marketplaces – the future of logistics industry. https://www.tcs.com/content/dam/tcs/pdf/Industries/travel-and-hospitality/Building-Disruptive-Freight-Marketplaces.pdf (accessed 25 January 2020).

Joerss, M., Schröder, J., Neuhaus, F. et al. (2016). Parcel delivery: the future of last mile. https://www.mckinsey.com/~/media/mckinsey/industries/travel transport and logistics/our insights/how customer demands are reshaping last mile delivery/parcel_delivery_the_future_of_last_mile.ashx (accessed 25 January 2020).

Keeney, T. (2016). 3D printing market: analysts are underestimating the future. https://ark-invest.com/research/3d-printing-market (accessed 25 January 2020).

Kelp, R. and Krenz, W. (2012). Connected trucks: the time has come. https://www.oliverwyman.com/content/dam/oliver-wyman/global/en/files/archive/2012/The_time_has_come_Perspectives_2012_2.pdf (accessed 25 January 2020).

Kern, J. and Wolff, P. (2019). The digital transformation of the automotive supply Chain–an empirical analysis with evidence from Germany and China. https://www.innovationpolicyplatform.org/system/files/imce/AutomotiveSupplyChain_GermanyChina_TIPDigitalCaseStudy2019_1.pdf (accessed 25 January 2020).

King, M. (2019). Ports are failing at automation, Mckinsey study concludes, Lloyd's loading list. https://www.lloydsloadinglist.com/freight-directory/news/Ports-are-failing-at-automation-McKinsey-study-concludes/73563.htm#.XSWJI-1R1PY (accessed 25 January 2020).

Klare, B. (2018). *Digitization of the Air Cargo Industry Is Slowly Taking Off*. innoFRAtor. https://www.innofrator.com/en/digitalisierung-der-luftfrachtbranche-nimmt-langsam-fahrt-auf (accessed 25 January 2020).

Kückelhaus, M. and Chung, G. (2016). Logistics trend radar. https://www.logistics.dhl/content/dam/dhl/global/core/documents/pdf/dhl-logistics-trend-radar-2016.pdf (accessed 25 January 2020).

Langley, C.J. (2019). 2019 Third-Party Logistics Study - The State of Logistics Outsourcing. In: *19th Annual Study*. https://www.kornferry.com/content/dam/kornferry/docs/article-migration/2019-3PL-Study.pdf (accessed 25 January 2020).

Leasing Associates (2018). *Benefits of a Fleet Management System*. https://www. theleasingcompany.com/benefits-fleet-management-system (accessed 25 January 2020).

von Leitner, F. (2019). Blockchain. https://ptrace.fefe.de/Blockchain (accessed 25 January 2020).

Lin, C.-Y., Wagshum, A.T., and Le, T.-S. (2016). Robotic label applicator: design, development and visual servoing based control. In: *MATEC Web of Conferences* (eds. M. Kavakli, M.J.E. Salami, A. Amini, M.A.B.M. Basri, A.B. Masli, S.C.H. Li, and M. Pal.), 56, p. 06001. doi: 10.1051/ matecconf/20165606001.

Lock, M. (2018). *The Marketplace Model: A Proven Model for Success & Why the Shipping Industry Should Adopt It*. Cerasis. https://cerasis.com/marketplace-model (accessed 25 January 2020).

Lorenz, M., Küpper, D., Rüßmann, M. et al. (2016). Time to accelerate in the race toward industry 4.0. https://www.bcg.com/publications/2016/lean-manufacturing-operations-time-accelerate-race-toward-industry-4.aspx (accessed 8 October 2018).

Machine Learning on AWS (2019). Amazon. https://aws.amazon.com/machine-learning (accessed 25 January 2020).

Magnussen, M.V. (2019). *Fujitsu and Kongsberg Digital Tackle Greenhouse Gas Emissions with AI-powered Fuel Optimization Service*. https://www.fujitsu.com/emeia/about/resources/ news/press-releases/2019/emeai-20190605-fujitsu-and-kongsberg-digital-tackle.html (accessed 25 January 2020).

MarketsandMarkets (2018). Automated container terminal market by degree of automation (semi-automated and fully automated), project type (brownfield and greenfield), offering (equipment, software, and services), and geography - global forecast to 2023. https://www. marketsandmarkets.com/Market-Reports/automated-container-terminal-market-109170291. html (accessed 25 January 2020).

van Marle, G. (2019). *New Standards Published to Advance the Use of Data from Smart Containers*. The Loadstar. https://theloadstar.com/new-standards-published-to-advance-the-use-of-data-from-smart-containers (accessed 25 January 2020).

Marr, B. (2019). *The Incredible Autonomous Ships of the Future: Run by Artificial Intelligence Rather than a Crew*. Forbes. https://www.forbes.com/sites/bernardmarr/2019/06/05/ the-incredible-autonomous-ships-of-the-future-run-by-artificial-intelligence-rather-than-a-crew/#47afec6c6fbf (accessed 25 January 2020).

van Marwyk, K. (2016). Of robots and men – in logistics. Towards a confident vision of logistics in 2025. In: *Den Wandel Gestalten Driving Change* (eds. T. Wimmer and C. Grotemeier), 195–212. BVL.

Mastex (2015). *Fleet Management Software*. Alblasserdam. https://mastexsoftware.com/ uploads/Downloads/MXSuite_Brochure.pdf (accessed 25 January 2020).

Mayer, M. (2016). SSI SCHÄFER sets new standards as general contractor for a new logistics concept at brose - YouTube. https://www.youtube.com/watch?time_continue=156&v=AJtXtIFz8-A (accessed 27 September 2018).

McCrea, B. (2018). *Transportation Management Systems Market 2018*. SupplyChain247. https:// www.supplychain247.com/article/transportation_management_systems_market_2018 (accessed 25 January 2020).

Manyika, J., Lund, S., Chui, M. et al. (2017). *Jobs lost, jobs gained: workforce transitions in a time of automation*. McKinsey Global Institute https://www.mckinsey.com/~/media/ McKinsey/Industries/Public%20and%20Social%20Sector/Our%20Insights/What%20the%20 future%20of%20work%20will%20mean%20for%20jobs%20skills%20and%20wages/MGI-Jobs-Lost-Jobs-Gained-Report-December-6-2017.pdf (accessed 25 January 2020).

Menzies, J. (2018). The benefits, and risks, of connected trucks. *TruckNews*. https://www.trucknews.com/features/benefits-risks-connected-trucks (accessed 25 January 2020).

MHI and Deloitte (2019). *2019 MHI Annual Industry Report*. https://www.mhi.org/publications/report (accessed 25 January 2020).

Million Tech Development Ltd. (2019). Intelligent Robotics Labelling System (IRLS). https://milliontech.com/solutions/label-printing/irls-intelligent-robotics-labeling-system (accessed 25 January 2020).

Motion Controls Robotics (2019). Robotic label placement. https://motioncontrolsrobotics.com/robotic-label-placement (accessed 25 January 2020).

Muir, J. (2017). Air cargo industry must undergo digital transformation. *Air Cargo Week*. https://www.aircargoweek.com/air-cargo-industry-undergo-digital-transformation (accessed 25 January 2020).

Mulkers, Y. (2018). Blockchain: what's it good for? Absolutely nothing, report finds. *Medium*. https://medium.com/the-data-intelligence-connection/blockchain-whats-it-good-for-absolutely-nothing-report-finds-16c0d7c01a40 (accessed 25 January 2020).

MUNIN (2016). Research in maritime autonomous systems project results and technology potentials. http://www.unmanned-ship.org/munin/wp-content/uploads/2016/02/MUNIN-final-brochure.pdf (accessed 25 January 2020).

New China TV (2019). Exploring world's biggest automated container terminal in Shanghai. *New China TV*. https://www.youtube.com/watch?v=p55N5WeYdcI (accessed 25 January 2020).

Newark (2018). Smart sensor technology for the IoT. *Techbriefs*. https://www.techbriefs.com/component/content/article/tb/features/articles/33212 (accessed 25 January 2020).

Nucleus Research (2015). Leading supply chain design vendors enabling retailers to compete with e-commerce giants, nucleus research value matrix finds. *Marketwatch*. https://www.marketwatch.com/press-release/leading-supply-chain-design-vendors-enabling-retailers-to-compete-with-e-commerce-giants-nucleus-research-value-matrix-finds-2015-02-17 (accessed 25 January 2020).

Nürk, C. and Maier, M.A. (2014). Truck market 2024 - sustainable growth in global markets. www.google.com.hk/url?sa=t&rct=j&q=&esrc=s&source=web&cd=3&cad=rja&uact=8&ved=2ahUKEwiL8ufGsIHkAhVTeXAKHRzmDL0QFjACegQIARAC&url=https%3A%2F%2Fwww2.deloitte.com%2Fcontent%2Fdam%2FDeloitte%2Ftr%2FDocuments%2Fprocess-and-operations%2Ftruck-studie-201 (accessed 25 January 2020).

O'Byrne, R. (2018). 6 Supply chain trends from 2017 and their ongoing progress. *Logistics Bureau*. https://www.logisticsbureau.com/6-key-supply-chain-and-logistics-trends-to-watch-in-2017 (accessed 25 January 2020).

OECD (2019). *OECD Employment Outlook 2019*. OECD (OECD Employment Outlook). doi: https://doi.org/10.1787/9ee00155-en.

Oxford University Press (2018). *robot, n.2 : Oxford English Dictionary, Oxford English Dictionary*. http://www.oed.com/view/Entry/166641?rskey=XozBiF&result=2&isAdvanced=false#eid (accessed 24 September 2018).

Palmer, D. (2019). Ransomware: The key lesson Maersk learned from battling the NotPetya attack. *ZDnet*. https://www.zdnet.com/article/ransomware-the-key-lesson-maersk-learned-from-battling-the-notpetya-attack (accessed 25 January 2020).

Patel, U. (2018). How chatbots give your supply chain a competitive advantage. https://www.softwebsolutions.com/resources/chatbots-in-supply-chain.html (accessed 25 January 2020).

PEMA. (2017). *Container Terminal Automation*, p. 17. https://www.pema.org/wp-content/uploads/downloads/2016/06/PEMA-IP12-Container-Terminal-Automation.pdf (accessed 25 January 2020).

Pfohl, H.-C. (2018). *Logistiksysteme*, 9e. Springer.

Pfohl, H.-C., Wolff, P., and Kern, J. (2019). Transshipment hub automation in China's CEP sector. In: *Urban Freight Transportation Systems*, 1e (eds. R. Elbert, C. Friedrich, M. Boltze, and H.-C. Pfohl). Elsevier.

Punel, A., Ermagun, A., and Stathopoulos, A. (2018). Studying determinants of crowd-shipping use. *Travel Behaviour and Society* 12: 30–40. https://doi.org/10.1016/j.tbs.2018.03.005.

Riedl, J., Farag, H., and Korenkiewicz, D. (2016). *Transportation and Logistics in a Changing World: The Journey Back to Profitable Growth*. The Boston Consulting Group. https://www.bcgperspectives.com/content/articles/transportation-travel-tourism-transportation-logistics-changing-world (accessed 25 January 2020).

Riedl, J., Jentzsch, A., Melcher, N.C., Gildemeister, J. et al. (2018). Why road freight needs to go digital - fast. https://www.bcg.com/publications/2018/why-road-freight-needs-go-digital-fast.aspx (accessed 25 January 2020).

Roberto, B. and Wilcox, S. (2018). 3-D printing: a game-changer for manufacturing. https://www.infosysconsultinginsights.com/wp-content/uploads/2018/02/infosys_3d-printing-pt1-game-change-manufacturing_ebook.pdf (accessed 25 January 2020).

RobotWorx (2018). Pick and place robots. https://www.robots.com/applications/pick-and-place (accessed 24 September 2018).

Rolfsen, B. (2019). Amazon's growing robot army keeps warehouses humming, bloomberg environment. https://news.bloombergenvironment.com/safety/amazons-growing-robot-army-keeps-warehouses-humming (accessed 25 January 2020).

Rouse, M. (2015). *Transportation Management System (TMS)*. TechTarget. https://searcherp.techtarget.com/definition/transportation-management-system-TMS (accessed 25 January 2020).

S.F. Holding (2019). 2018 S.F. Holding Co., Ltd. Annual Report. https://www.sf-express.com/cn/sc/download/20190318-IR-1-2018.PDF (accessed 25 January 2020).

SAE International (2018). Taxonomy and definitions for terms related to driving automation systems for on-road motor vehicles. https://www.sae.org/standards/content/j3016_201806 (accessed 25 January 2020).

Safety4sea (2019). Technologies changing the future of shipping at the centre of SMART4SEA Conference. https://safety4sea.com/technologies-changing-the-future-of-shipping-at-the-centre-of-smart4sea-conference (accessed 25 January 2020).

Sage, D. (2008). Express delivery. In: *Handbook of Logistics and Supply-Chain Management*, 3e (eds. A.M. Brewer, K.J. Button and D.A. Hensher), 455–466. Emerald.

Sandberg, E. and Hemilä, J. (2018). 'Digitalization in industrial logistics and supply chains - The Contemporary Situation in Sweden and Finland Digitalization. In: *Proceedings of the 23rd International Symposium on Logistics* (eds. K. Pawar, A. Potter, C. Chan, and N. Pujawan). Bali: Centre for Concurrent Enterprise, Nottingham University Business School, pp. 222–229.

Sanders, U. and Egloff, C. (2019). Digital transformation in the shipping industry. https://www.bcg.com/industries/transportation-travel-tourism/center-digital-transportation/shipping.aspx (accessed 25 January 2020).

Sarch, M. (2018). Thinking smart parcel lockers? Don't do it!, Parcel and postal technology international. https://www.parcelandpostaltechnologyinternational.com/opinion/thinking-smart-parcel-lockers-dont-do-it.html (accessed 25 January 2020).

Sauv, D. (2018). e-AWB implementation playbook. https://www.iata.org/whatwedo/cargo/e/eawb/Documents/e-awb-implementation-playbook.pdf (accessed 25 January 2020).

Schlick, J., Stephan, P., Loskyll, M. et al. (2014). Industrie 4.0 in der praktischen Anwendung. In: *Industrie 4.0 in Produktion, Automatisierung und Logistik* (eds. T. Bauernhansl, M. ten Hompel, and B. Vogel-Heuser.), 57–84. Wiesbaden: Springer https://doi.org/10.1007/978-3-658-04682-8.

Schubert, C., van Langeveld, M.C., and Donoso, L.A. (2014). Innovations in 3D printing: a 3D overview from optics to organs. *British Journal of Ophthalmology* 98 (2): 159–161. https://doi.org/10.1136/bjophthalmol-2013-304446.

Schwarz, S. (2016). Meeting the challenges of compliance and data management. *Riviera.* https://www.rivieramm.com/opinion/meeting-the-challenges-of-compliance-and-data-management-33601 (accessed 25 January 2020).

Serafini, S., Nigro, M., Gatta, V. et al. (2018). Sustainable crowdshipping using public transport: a case study evaluation in Rome. *Transportation Research Procedia* 30: 101–110. https://doi.org/10.1016/j.trpro.2018.09.012.

Shu, C. (2019). JD.com's logistics arm raises a $218 million investment fund. *TechCrunch.* https://techcrunch.com/2019/06/24/jd-coms-logistics-arm-raises-a-218-million-investment-fund (accessed 25 January 2020).

Silveira, R.M.S. (2018). Container terminal automation in port of santos – a business case analysis. *Rotterdam.* https://thesis.eur.nl/pub/43890/Muricy-Rodrigo.pdf (accessed 25 January 2020).

Sotolongo, J., Rozycki, M., and Kerr, I. (2019). JD vs Alibaba in the last mile: what's happening behind the great wall. https://www.parcelandpostaltechnologyinternational.com/analysis/jd-vs-alibaba-in-the-last-mile-whats-happening-behind-the-great-wall.html (accessed 25 January 2020).

Stoltz, M-H., Giannikas, V., McFarlane, D. et al. (2017). Augmented reality in warehouse operations: opportunities and barriers. *IFAC-PapersOnLine* 50 (1): 12979–12984. https://doi.org/10.1016/j.ifacol.2017.08.1807.

Stölzle, W., Schmidt, T., Kille, C. et al. (2018). Digitalisierungswerkzeuge in der Logistik: Einsatzpotenziale, Reifegrad und Wertbeitrag. *Göttingen.* http://logistik-digitalisierung.de/wp-content/uploads/2018/10/Digitalisierungswerkzeuge-in-der-Logistik-Einsatzpotentiale-Reifegrad-Wertbeitrag_V4-1.pdf (accessed 25 January 2020).

Streichfuss, M. (2016). *Digitalization of Airports*, 16. Roland Berger: Shanghai https://www.rolandberger.com/publications/publication_pdf/2016_10_13_ais2016___digital_airports.pdf (accessed 25 January 2020).

Sunol, H. (2018). Should you adopt predictive analytics in warehouse management? https://articles.cyzerg.com/should-you-adopt-predictive-analytics-in-warehouse-management (accessed 25 January 2020).

Technology And Costs (2016). Googlesautonomousvehicle. http://googlesautonomousvehicle.weebly.com/technology-and-costs.html (accessed 25 January 2020).

Thiels, C.A., Aho, J.M., Zietlow, S.P. et al. (2015). Use of unmanned aerial vehicles for medical product transport. *Air Medical Journal* 34 (2): 104–108. https://doi.org/10.1016/j.amj.2014.10.011.

Tipping, A. and Kauschke, P. (2016). Shifting patterns - the future of the logistics industry. https://www.pwc.com/sg/en/publications/assets/future-of-the-logistics-industry.pdf (accessed 25 January 2020).

Tjahjono, B., Esplugues, C., Ares, E. et al. (2017). What does industry 4.0 mean to supply chain? *Procedia Manufacturing* 13: 1175–1182. https://doi.org/10.1016/J.PROMFG.2017.09.191.

Toczauer, C. (2019). Panalpina, DHL GF join growing list of digital freight sales platforms. https://aircargoworld.com/allposts/panalpina-dhl-gf-join-growing-list-of-digital-freight-sales-platforms (accessed 25 January 2020).

Trelleborg Marine Systems (2018). Use of big data in the maritime industry. https://www.patersonsimons.com/wp-content/uploads/2018/06/TMS_SmartPort_InsightBee_Report-to-GUIDE_01.02.18.pdf (accessed 25 January 2020).

Ullrich, G. (2015). *Automated Guided Vehicle Systems, Information Systems Frontiers*. Berlin, Heidelberg: Springer Berlin Heidelberg https://doi.org/10.1007/978-3-662-44814-4.

UNCTAD (2018). Review of maritime transport 2018. https://unctad.org/en/PublicationsLibrary/rmt2018_en.pdf (accessed 25 January 2020).

UPS (2016). 3D printing: the next revolution in industrial manufacturing. https://www.ups.com/assets/resources/media/en_US/3D_Printing_executive_summary.pdf (accessed 25 January 2020).

VDI-Fachbereich Fabrikplanung und –betrieb (2014). *Simulation of Systems in Materials Handling, Logistics and Production - Fundamentals*. VDI-Gesellschaft Produktion und Logistik. https://www.vdi.de/richtlinien/details/vdi-3633-blatt-1-simulation-von-logistik-materialfluss-und-produktionssystemen-grundlagen (accessed 25 January 2020).

DC Velocity (2019). JD.com says its fulfillment technology will support 'boundaryless' shopping for other retailers. https://www.dcvelocity.com/articles/20190107-jd-com-opens-its-fulfillment-network-to-other-retailers (accessed 25 January 2020).

Vial, G. (2019). Understanding digital transformation: a review and a research agenda. *The Journal of Strategic Information Systems* 28 (2): 118–144. https://doi.org/10.1016/j.jsis.2019.01.003.

Walker, J. (2019). Autonomous ships timeline – comparing Rolls-Royce, Kongsberg, Yara and more, Emerj artificial intelligence research. https://emerj.com/ai-adoption-timelines/autonomous-ships-timeline (accessed 25 January 2020).

We are 42 (2019). About; we are 42. https://weare42.io/about (accessed 25 January 2020).

Wolff, P. and Kern, J. (2019). Materialise: teaching along the silk road. In: *Management Practices in Asia*, 203–215. Cham: Springer International Publishing https://doi.org/10.1007/978-3-030-19662-2_15.

World Economic Forum (2016). Digital transformation of industries: logistics industry. http://reports.weforum.org/digital-transformation/wp-content/blogs.dir/94/mp/files/pages/files/wef-dti-logisticswhitepaper-final-january-2016.pdf (accessed 25 January 2020).

World Economic Forum (2018). The future of jobs. https://www.weforum.org/reports/the-future-of-jobs-report-2018 (accessed 25 January 2020).

Yang, B. (2017). Intelligent courier locker offers 'last mile' solution. *ShanghaiDaily*. https://archive.shine.cn/metro/society/Intelligent-courier-locker-offers-last-mile-solution/shdaily.shtml (accessed 25 January 2020).

Young, A. (2019). Expectations, TMS market size & growth. https://blog.intekfreight-logistics.com/tms-market-size-and-growth-expectations (accessed 25 January 2020).

Glossary

Vocabulary	Definition	Definition Source
3-D Printing	A technology that uses printer like devices to apply layers of metal or plastic in patterns that result in a finished product. 3D printing is similar to ink jet printing in that a processor controls how much material is applied, and where it is applied. 3D printing may offer the possibility of building spare parts on demand virtually anywhere in the world that it may be needed versus carrying an inventory of old parts. It can also be used to create items with very complex designs. The time required to 'print' an item makes this process not economical for large volume requirements.	Council of Supply Chain Management Professionals (CSCMP), 2020
3pl	See "Third Party Logistics"	
4PL	See "Fourth Party Logistics"	
Active and Passive Sensors	Active means that data are actively sent out from the Thing. This might be necessary if the condition of the Thing changes quickly and the knowledge about these changes are decisive for the next step. In a passive mode the Thing reveals its data when it is triggered from outside, e.g. a box labelled with a barcode passing a barcode reader.	Chapter 3
Active Learning	Active learning indicates that the trainer strives to create a learning environment in which the learner can learn to restructure the new information and their prior knowledge into new knowledge about the content and to practice using it. Active learning engages learners mainly in two aspects – doing things and reflecting on the things they are doing.	Chapter 21
Additive Manufacturing	See "3-D Printing"	

Vocabulary	Definition	Definition Source
Advanced Shipping Notice	Advanced Shipping Notice (ASN) refers to detailed shipment information transmitted by the shipper to a customer or consignee in advance of delivery, designating the contents (individual products and quantities of each) and nature of the shipment. In EDI data standards this is referred to as an 856 transaction. May also include carrier and shipment specifics including time of shipment and expected time of arrival. The ASN data can be valuable in providing digital knowledge about what is in a shipment in a way that it can be used to eliminate manual data entry of each shipment.	Council of Supply Chain Management Professionals (CSCMP), 2020
Agreement on the International Carriage of Perishable Foodstuffs and on the Special Equipment to be used for such Carriage	A 1970 United Nations treaty that establishes standards for the international transport of perishable food between the states that ratify the treaty.	Chapter 17
AGV	See "Automated guided vehicle"	
AI	See "Artificial Intelligence"	
AIS	See "Automatic Identification System"	
algorithm	A clearly specified mathematical process for computation; a set of rules, which, if followed, give a prescribed result.	Council of Supply Chain Management Professionals (CSCMP), 2020
API	See "Application Programming Interface"	
Application Programming Interface	An API, is a way for software to communicate with each other without a platform having direct access to a database. An example of this would be when a user comes onto Kayak's website, they fill in their information and instantly are able to see flights relevant to the user. Kayak can provide this information seamlessly to the user by leveraging API technology to integrate with the airlines. When Kayak receives the user's requirements, they make an API call to the airlines which returns the schedules, rates, and other amenities in real time. If this were EDI, the user would have to wait until the data was pushed to them by the data source. EDI also would not allow for specific, relevant, on demand data to be available.	Chapter 1
AR	See "Augmented Reality"	
Arbitrage	The simultaneous purchase and sale of the same item or service, such as transportation capacity on a vessel, in different markets to profit from unequal prices.	*arbitrage*, 2020

Vocabulary	Definition	Definition Source
Arbitrageur	A person who engages in arbitrage.	Chapter 11
Artifacts	They are phenomena that can be seen, heard, or felt which are representative of organizational culture. Artifacts can include visible products, such as architecture, language, technology, artistic creations, clothes, manners of address, emotional displays, myths and stories told about the organization, a published lists of values and observable rituals or ceremonies.	Chapter 20
Artificial Intelligence	Understanding and computerizing the human thought process in order to solve complex problems.	Council of Supply Chain Management Professionals (CSCMP), 2020
ASN	See "Advanced Shipping Notice"	
Asset Light Model	A business strategy which attempts to maintain the minimum amount of assets necessary to keep the business running.	Chapter 11
ATP	See "Agreement on the International Carriage of Perishable Foodstuffs and on the Special Equipment to be used for such Carriage"	
Audit	The inspection and examination of a process, financial results, or quality system to ensure compliance to requirements. An audit can apply to an entire organization or may be specific to a function, process or production step.	Council of Supply Chain Management Professionals (CSCMP), 2020
Augmented Reality	Augmented Reality Devices support decisions and processes by providing instructions via augmented and virtual reality technologies (e.g. Headsets or smart glasses showing the worker on a small display in the glasses what to do next).	Chapter 3
Automated Decision	Based on pre-defined checking and decision finding processes an application automatically executes a pre-defined solution process or gives a proposal to a human decider.	Chapter 3
Automated guided vehicle	Driverless, mobile vehicles programmed to transport materials in a facility. They are used in many warehouse applications across numerous industries to compensate employee shortages, reduce labor cost and increase warehouse efficiency.	Lodamaster Group, 2020
Automatic Identification System	A satellite based track and trace solution that gathers data available at the vessel level.	Chapter 16
B2B	See "Business to Business"	
B2C	business to consumer	Descartes Systems Group, 2020

Vocabulary	Definition	Definition Source
Batch Size 1	Batch production is a method of manufacturing where identical or similar items are produced together for different sized production runs. Batch size 1 refers to the economically efficient manufacturing of highly customized products.	Fraunhofer Institute for Production Technology, 2021
BCO	See "beneficial cargo owner"	
Belt and Road Initiative	The Belt and Road Initiative (previously named "One Belt One Road") is a major strategic project by the Chinese government, which explains that it is aimed at *"promoting orderly and free flow of economic factors, highly efficient allocation of resources and deep integration of markets; encouraging the countries along the Belt and Road to achieve economic policy coordination and carry out broader and more in-depth regional cooperation of higher standards; and jointly creating an open, inclusive and balanced regional economic cooperation architecture that benefits all."*	Lokhande, 2017
Benchmarking	The process of comparing performance against the practices of other leading companies for the purpose of improving performance. Companies also benchmark internally by tracking and comparing current performance with past performance. Benchmarking seeks to improve any given business process by exploiting "best practices" rather than merely measuring the best performance. Best practices are the cause of best performance. Studying best practices provides the greatest opportunity for gaining a strategic, operational, and financial advantage.	Council of Supply Chain Management Professionals (CSCMP), 2020
beneficial cargo owner	The company that is acting as the Importer of Record is also known as the BCO. They are responsible for the specific legal requirements for the country they are importing into. For example, the BCO has the legal responsibility to ensure their product follows all legal requirements as well as customs requirements like a bond and documentation requirements.	Chapter 16
Big Data	Refers to the accumulated data contained in multiple databases across an organization, or even extending to some supply chain partners. Separate databases may exist to support a variety of individual business processes, making analysis of that data quite difficult in any consolidated manner. The term Big Data is generally used when discussing how to do a consolidated analysis, or the difficulty of consolidating multiple databases.	Council of Supply Chain Management Professionals (CSCMP), 2020
Bill of Materials	A comprehensive list of parts, items, assemblies, and other materials required to create a product, as well as instructions required for gathering and using the required materials.	Arena Solutions, 2021

Vocabulary	Definition	Definition Source
Bitcoin	A cryptocurrency which isn't managed by a bank or agency but in which transactions are recorded in the blockchain that is public and contains records of each and every transaction that takes place.	Digital Currency Group, 2020
Blended Learning	In Blended Learning, traditional face-to-face training and digital formats are combined with the help of collaborative tools, such as learning communities, working groups, impulse talk sessions, etc.	Chapter 21
BOM	See "Bill of Materials"	
Bot	A "bot" or "robot" in the context of RPA is the same as a unique software instance (Lacity and Willcocks, 2016). RPA bots are not like AI and machine learning algorithms, they don't "learn" through iterations of large data sets in the same way that these algorithms do. Instead the bots must be given specific instructions in order to complete their task.	Chapter 5
Bottlenecks	A constraint, obstacle or planned control that limits throughput or the utilization of capacity.	Council of Supply Chain Management Professionals (CSCMP), 2020
BPM	See "Business Process Management"	
BPO	See "Business Process Outsourcing"	
BPS	Here: Bosch Production System, Bosch's Lean Management reference model.	Gnoni *et al.*, 2013
BRI	See "Belt and Road Initiative"	
Business Process Management	Business Process Management (BPM) includes methods, techniques, and tools to support the design, enactment, management, and analysis of operational business processes.	Van Der Aalst, Ter Hofstede and Weske, 2003
Business Process Outsourcing	A method of subcontracting various business-related operations to third-party vendors.	Dittrich and Braun, 2004
Business to Business	As opposed to business-to-consumer (B2C). Many companies are now focusing on this strategy, and their sites are aimed at businesses (think wholesale) and only other businesses can access or buy products on the site. Internet analysts predict this will be the biggest sector on the Web.	Council of Supply Chain Management Professionals (CSCMP), 2020
Business Transformation	The process of shifting from a current as-is state to a desired future target state and thus to the establishment of a new way of being. It is to reinvent the organization and discover a new or revised business model based on a vision for the future by executing a portfolio of initiatives, which are interdependent or intersecting.	Barrett, 2015
C2C	Consumer to consumer	Chapter 15
CA	See "Controlled Atmosphere"	

Vocabulary	Definition	Definition Source
CAD	See "Computer Aided Design"	
CAN Bus	The Controller Area Network (CAN bus) is the "nervous system" of a car, enabling communication between all parts of the body (CSS Electronics, 2019).	Chapter 25
CAPEX	A term used to describe the monetary requirements (CAPital Expenditure) of an initial investment in new machines or equipment.	Council of Supply Chain Management Professionals (CSCMP), 2020
Carrier	A firm which transports goods or people via land, sea or air.	Council of Supply Chain Management Professionals (CSCMP), 2020
Center of Excellence	A team tasked with concentrating existing expertise and resources in a discipline or capability to attain and sustain world-class performance and value.	Hou, 2020
Centralization	Handling a large area with multiple markets, e.g. Europe, via one central distribution center and billing from one central finance center, managing the website, sales and customer support from one central location	Chapter 11
Change management	Change Management subsumes tasks, measures and activities which aim at effecting a specific comprehensive change for implementing a new strategy, new structures, systems, processes or behaviours. It assesses the past, compares it to the present and determines the ideal future state from the current business state.	Doppler and Lauterburg, 2019
Channel	A method whereby a business dispenses its product, such as retail or distribution channel, call center or web based electronic storefront.	Council of Supply Chain Management Professionals (CSCMP), 2020
Chatbots	Computer programs that can hold a conversation with humans using natural language speech. Such programs can assist in human computer interaction and examine and influence the behavior of the user by asking questions and responding to the user's questions.	Chapter 25
Cold Chain	A temperature-controlled supply chain. An unbroken cold chain is an uninterrupted series of storage and distribution activities which maintain a given temperature range. It is used to help extend and ensure the shelf life of products such as fresh agricultural produce, seafood, frozen food, photographic film, chemicals and pharmaceutical drugs.	Descartes Systems Group, 2020

Vocabulary	Definition	Definition Source
Collaborative Work	Work requiring communication skills, for example, the skills that would be needed by an employee working for a call center.	Chapter 5
Company mission	The mission describes what business the organization is in and where it wants to be in the future.	Chapter 20
Company Purpose	Explains why this company exists and how it is making a difference for someone else.	Chapter 20
Company Vision	A vision, sometimes also referred to as "strategic intent", is the desired future state of a company, its aspiration.	Chapter 20
Competence Management	A set of practices that identify and optimize the skills and competencies required to deliver on the business strategy.	Chapter 21
Competence Model	A competence model is a narrative description of the competencies for a specific job category, occupational group, division, department or analysis units. It can provide identification of the competencies employees need to develop in order to be prepared or improve in their job.	Chapter 21
Computer Aided Design	Computer-aided design (CAD) is, the use of computers to aid in the creation, modification, analysis, or optimization of a design.	Chapter 4
Conditional Statements	Conditional statements set up conditions that could be true or false. These conditions lead to a result that may or may not be true.	Grupa, 2020
Container yard	A physical facility from which ocean carriers accept and deliver ocean containers, as well as issue and receive back empty containers.	Flexport, 2020
Contract Rates	Rates that are negotiated in advance and remain valid for a certain period of time depending on the contract.	Chapter 16
Control Tower	A single point of command over some business function.	Chapter 21
Controlled atmosphere	A strategy for slowing the aging of perishables which involves different stages: reduction of oxygen, increase of carbon dioxide, removal of carbon dioxide and removal of ethylene.	Chapter 17
cost of goods sold	The amount of direct materials, direct labor, and allocated overhead associated with products sold during a given period of time, determined in accordance with Generally Accepted Accounting Principles (GAAP).	Council of Supply Chain Management Professionals (CSCMP), 2020
CPS	See "cyber-physical system"	

Vocabulary	Definition	Definition Source
Cross-Border eCommerce	The online selling of goods to consumers in different countries. It can be between a business and a consumer (B2C), between two businesses, often brands or wholesalers (B2B) or between two private persons (C2C).	Chapter 15
Crowd Shipping	Also called crowd logistics, crowdsourced delivery or collaborative logistics, makes use of the excess capacity of available transport modes by using premeditated trips to perform such deliveries.	Chapter 25
Culture	A group phenomenon that cannot exist solely within a single person. It resides in joint behaviors, values, and assumptions, commonly experienced through "unwritten rules."	Chapter 20
Customer Service	Activities between the buyer and seller that enhance or facilitate the sale or use of the seller's products or services.	Council of Supply Chain Management Professionals (CSCMP), 2020
Customs Bond	A part of the customs clearance process where Customs will inspect the documents submitted as well as the physical goods to ascertain the value declared is in accordance with the international markets and approve the assessment based on the appropriate classification.	Chapter 15
Customs Broker	An individual or firm licensed to represent an importer or exporter in front of customs authorities. A customs broker must ordinarily be authorized to act as such by the local government.	Descartes Systems Group, 2020
cyber-physical systems	Cyber-physical systems (CPS) are systems with embedded software (as part of devices, buildings, means of transport, transport routes, production systems, medical processes, logistic processes, coordination processes, and management processes), which directly record physical data using sensors and affect physical processes using actuators; evaluate and save recorded data, and actively or reactively interact both with the physical and digital world; are connected with one another and in global networks via digital communication facilities (wireless and/or wired, local and/or global); use globally available data and services; and have a series of dedicated, multimodal human-machine interfaces.	Chapter 8
Data Integration	The process of combining data from different sources into a single, unified view.	Talend Inc., 2021
DC	See "Distribution Center"	

Vocabulary	Definition	Definition Source
dematerialization	An idea which says that closed systems inexorably become less structured, less organized, less able to accomplish interesting and useful outcomes, until they slide into an equilibrium of grey, tepid, homogenous monotony.	Chapter 17
Demurrage	A fee aid by the consignee when their cargo remains on terminal property after the last free day.	Chapter 14
Descriptive Analytics	Analytical method where selected information is gathered and the monitored situation is described (e.g. condition, environment, process).	Chapter 3
Diagnostic Analytics	Root causes of deviations between a set target value and the actual monitored value are analyzed.	Chapter 3
Differentiation	In the postponement supply chain model, this is the point where an end product assumes unique characteristics through final assembly configuration and/or packaging.	Council of Supply Chain Management Professionals (CSCMP), 2020
Digital Integration	Digital integration entails the introduction of a single platform for digital communication on the Silk Road that facilitates integration of all service providers.	Chapter 2
Digital Platforms	See "Marketplaces"	Chapter 1
Digital Silk Road	The aim of the Digital Silk Road is to bring advanced IT infrastructure to BRI (Belt and Road Initiative) countries, including broadband, e-commerce hubs and smart cities.	Chapter 19
Digital Twin	A virtual model of a real thing based on 'sensed' real life data.	Chapter 3
Digitalization	"Digitalization" describes the ongoing digital transformation that affects industries, politics, and society. It is changing business and operating models in industry and economics and transforms supply chains. Innovative technologies enable companies to develop digital supply chain models, which offer more agility and collaboration structures. Due to digitalization, business processes undergo an automation process, becoming more flexible and leading to digital management approaches.	Raab and Griffin-Cryan, 2011
Digitization	The process of making analog processes digital.	Chapter 19
Direct Costs	A cost that can be directly traced to a cost object since a direct or repeatable cause-and-effect relationship exists. A direct cost uses a direct assignment or cost causal relationship to transfer costs. Direct costs can consist of materials used and labor directly involved in production.	Council of Supply Chain Management Professionals (CSCMP), 2020

Vocabulary	Definition	Definition Source
Distribution Center	A Distribution Center (DC) is the warehouse facility which holds inventory from manufacturing pending distribution to the appropriate stores.	Council of Supply Chain Management Professionals (CSCMP), 2020
Downstream	Referring to the demand side of the supply chain. One or more companies or individuals who participate in the flow of goods and services moving from the manufacturer to the final user or consumer.	Council of Supply Chain Management Professionals (CSCMP), 2020
drayage	Transportation of materials and freight by a "side-less cart" on a local basis to or from a congested area such as a shipping port or exhibition area to a more open loading / unloading which allows for large trucks, etc. Intermodal freight carriage may also be referred to as drayage. A charge associated with the pick up or delivery of an ocean container.	Council of Supply Chain Management Professionals (CSCMP), 2020
DSR	See "Digital Silk Road"	
Dynamic routing	Dynamic Routing or Adaptive Routing, as name suggests changes the routing table once any changes to network occurs or network topology changes. During network change, dynamic routing sends a signal to router, recalculates the routes and send the updated routing information.	Parahar, 2020
Earning before interest and taxes	An indicator of a company's profitability. EBIT can be calculated as revenue minus expenses excluding tax and interest. EBIT is also referred to as operating earnings, operating profit, and profit before interest and taxes.	Murphy, 2020a
EBIT	See "Earnings before interest and taxes"	
eCommerce	A type of industry where the buying and selling of products or services is conducted over electronic systems such as the Internet and other computer networks. Electronic commerce draws on technologies such as mobile commerce, electronic funds transfer, supply chain management, Internet marketing, online transaction processing, electronic data interchange (EDI), inventory management systems, and automated data collection systems.	Descartes Systems Group, 2020
Edge Computing	Bringing computation and data storage closer to the devices where it's being gathered, rather than relying on a central location that can be thousands of miles away.	Shaw, 2019
EDI	See "Electronic Data Interchange"	
Efficient Market Hypothesis	A hypothesis that states that prices reflect all information, so that commodities or stocks always trade at their fair value on exchanges, making it impossible for investors to purchase undervalued stocks or sell stocks for inflated prices.	Downey, 2020

Vocabulary	Definition	Definition Source
Electronic Commerce	Also written as e-commerce. Conducting business electronically via traditional EDI technologies, or online via the Internet. In the traditional sense of selling goods, it is possible to do this electronically because of certain software programs that run the main functions of an e-commerce website, such as product display, online ordering, and inventory management. The definition of e-commerce includes business activity that is business-to-business (B2B), business-to-consumer (B2C).	Council of Supply Chain Management Professionals (CSCMP), 2020
Electronic Data Interchange	a) The transfer of structured data, by agreed message standards, from one computer application to another by electronic means and with a minimum of human intervention; b) The electronic exchange of documents between businesses and organizations, or between businesses and government agencies."	Descartes Systems Group, 2020
Electronic Freight Documents	Documents needed to move freight represented digitally for ease of storage and sharing.	Chapter 25
Electronic Product Code	The Electronic Product Code (EPC) is a universal identifier that provides a unique identity for every physical object anywhere in the world.	Chapter 8
Enterprise Resource Planning System (Whole entry is then "Enterprise Resource Planning System")	Enterprise resource planning systems are part of the transaction management foundation on which all IT operations and analytics of a company rest. ERP systems consist of components, for example for financials and human resources, that are the necessary base for other Supply Chain macro processes to function and communicate with one another. Examples of ERP systems are the application suites from SAP, Oracle, Microsoft and others.	Chopra, 2019
EPC	See "Electronic product code"	
ERP	See "Enterprise Resource Planning"	
ERP System	See "Enterprise Resource Planning System"	
Ethylene	A hydrocarbon gas, which is being produced as a function of the ripening process of fruits and vegetables.	Chapter 17
Ex-Factory Date	The date which production of a product is planned to finish.	Chapter 16
eXtensible Markup Language	A flexible markup language for structured electronic documents. XML is based on SGML (standard generalized markup language), an international standard for electronic documents. XML is commonly used by data-exchange services (like blog feeds) to send information between otherwise incompatible systems.	Descartes Systems Group, 2020

Vocabulary	Definition	Definition Source
FBA	See "Fulfillment by Amazon"	
Federal Maritime Commission	A regulatory agency that controls services, practices, and agreements of international water common carriers and noncontiguous domestic water carriers.	Chapter 16
Fill rate	The percentage of order items that the picking operation actually fills within a given period of time.	Council of Supply Chain Management Professionals (CSCMP), 2020
Fleet Management System	A system for managing and tracking fleets of commercial vehicles and ships.	Chapter 25
FMS	See "Fleet Management System"	
Forward Stock Location	Small warehouses that are geographically close to the end customers.	Flash Global, 2021
Fourth Party Logistics	Fourth Party Logistics (4PL) refers to an independent, singularly accountable, non asset based integrator of a clients supply and demand chains. The 4PL's role is to implement and manage a value creating business solution through control of time and place utilities and influence on form and possession utilities within the client organisation. Performance and success of the 4PL's intervention is measured as a function of value creation within the client organization	Win, 2008
Freight Forwarder	An organization which provides logistics services as an intermediary between the shipper and the carrier, typically on international shipments. Freight forwarders provide the ability to respond quickly and efficiently to changing customer and consumer demands and international shipping (import/export) requirements.	Council of Supply Chain Management Professionals (CSCMP), 2020
FSL	See "Forward Stock Location"	
Fulfilment by Amazon	A service provided by Amazon that provides storage, packaging, and shipping assistance to sellers. This takes the burden off of sellers and grants them more flexibility in their selling practices.	Feedvisor, 2020
Fused Deposition Modeling	Fused Deposition Modeling is an approach in Additive Manufacturing where a continuous filament of a thermoplastic material is used as base material. The filament is fed from a coil, through a moving heated printer extruder head. The molten material is forced out of the extruder's nozzle and deposited onto a platform. Once the first layer is completed, the extruder and the platform are parted away in one step, and the second layer can then be directly deposited onto the growing item.	Chapter 4

Vocabulary	Definition	Definition Source
Global Trade Item Number	A unique number that comprises up to 14 digits and is used to identify an item (product or service) upon which there is a need to retrieve pre-defined information that may be priced, ordered or invoiced at any point in the supply chain. The definition covers raw materials through end user products and includes services, all of which have pre-defined characteristics. GTIN is the globally-unique EAN. UCC System identification number, or key, used for trade items (products and services). It's used for uniquely identifying trade items (products and services) sold, delivered, warehoused, and billed throughout the retail and commercial distribution channels. Unlike a UPC number, which only provides information specific to a group of products, the GTIN gives each product its own specific identifying number, giving greater tracking accuracy.	Council of Supply Chain Management Professionals (CSCMP), 2020
globalization	The process of making something worldwide in scope or application.	Council of Supply Chain Management Professionals (CSCMP), 2020
Graphical User Interface	A type of user interface through which users interact with electronic devices via visual indicator representations.	OmniSci, 2020
GS1	The new name of EAN International. The GS1 US is the new name of the Uniform Code Council, Inc® (UCC®) the GS1 Member Organization for the U.S., the association that administrates UCS, WINS, and VICS and provides UCS identification codes and UPCs. GS1 subgroups also manage the standards for electronic product codes (EPCGlobal) and Rosettanet.	Council of Supply Chain Management Professionals (CSCMP), 2020
GTIN	See "Global Trade Item Number"	
GUI	See "Graphical User Interface"	
HACCP	See "Hazard Analysis and Critical Control Point"	
Hard Skills	Teachable abilities or skill sets that are easy to quantify, such as a degree, certificate or proficiency in a foreign language.	Doyle, 2020
hardware-software interoperability	The ability of software to recognize, act on, and control hardware independently of their supplier.	Chapter 8
Hazard Analysis and Critical Control Point	An internationally recognized system for reducing the risk of safety hazards in food.	Chapter 17
Heavyweight IT	A knowledge domain driven by IT professionals through digital technology and made possible by software engineering.	Chapter 5

Vocabulary	Definition	Definition Source
Hedging	Hedging is a risk management strategy employed to offset losses in investments.	Investopedia, 2020
HS Code	An internationally standardized system of names and numbers for classifying traded products developed and maintained by the World Customs Organization (WCO) (formerly the Customs Co-operation Council), an independent intergovernmental organization with over 170 member countries based in Brussels, Belgium."	Descartes Systems Group, 2020
HTTP	See "Hypertext Transfer Protocol"	
Hypertext Transfer Protocol	The Internet protocol that allows World Wide Web browsers to retrieve information from servers.	Council of Supply Chain Management Professionals (CSCMP), 2020
Immutable	When the logic of a program or data cannot be changed.	Chapter 12
Import Declaration	A statement made by the importer (owner of the goods), or their agent (licensed customs broker), at the destination country's customs boarder, providing information about the goods being imported. The Import Declaration collects details on the importer, how the goods are being transported, the tariff classification and customs value.	Australian Border Force, 2021
Incoterms	International terms of sale developed by the International Chamber of Commerce to define sellers' and buyers' responsibilities.	Council of Supply Chain Management Professionals (CSCMP), 2020
Independent Work	Work done by one person without interacting with others like generating accounting statements.	Chapter 5
Inductive coupling	Inductive coupling allows vehicles to transfer electrical power between two circuits (one on the vehicle, and one on the ground) through a shared magnetic field. The power system can be built on both continuous energy supply from the driving path to the AGV, and an on-board battery enabling the vehicle to drive safely even in case of interruptions of the ground's inductive power supply.	Chapter 8
Infrastructure as a Service	Rented server space to organizations that offers highly reliable infrastructure and Internet connectivity.	Chapter 9
Institutionalization	Institutionalization focuses on making change stick by addressing organizational patterns that hinder change.	Chapter 19
Intangible rewards	Intangible rewards, also known as intangible benefits, are rewards provided to an employee that don't have an inherent monetary value, and are often applied in response to a particular achievement. Some examples are praise, thanks, and public recognition.	HRZone, 2020

Vocabulary	Definition	Definition Source
Intermodal Transportation	Transporting freight by using two or more transportation modes such as by truck and rail or truck and oceangoing vessel.	Council of Supply Chain Management Professionals (CSCMP), 2020
Internet+	Internet+ (Chinese 互联网+) is a business model where traditional industries align themselves with technology and the Internet. In China in the future, this model is expected to promote the widespread use of information technology in industrialization and advance and exploit networking, digitalization and smart technologies. An example is "Internet+Manufacturing Industry" which is similar to the concept "Industry 4.0".	Brodie, 2015
IoT Sensors	A collection of interconnected physical devices that monitor, report, send and exchange data.	Chapter 7
Key Performance Indicator	A measure which is of strategic importance to a company or department. For example, a supply chain flexibility metric is Supplier On-time Delivery Performance which indicates the percentage of orders that are fulfilled on or before the original requested date.	Council of Supply Chain Management Professionals (CSCMP), 2020
KPI	See "Key Performance Indicator"	
landed cost	Cost of product plus relevant logistics costs such as transportation, warehousing, handling, etc. Synonym: Total Landed Cost, Net Landed Cost.	Council of Supply Chain Management Professionals (CSCMP), 2020
Last Mile	The last leg of the supply chain. It is often less efficient, comprising up to 28% of the total cost to move goods. This has become known as the last-mile problem. The last-mile problem can also include the challenge of making deliveries in urban areas where retail stores, restaurants, and other merchants in a central business district often contribute to congestion and safety problems.	Descartes Systems Group, 2020
LC	See "Letter of Credit"	
Lead Time	The total time that elapses between an order's placement and its receipt. It includes the time required for order transmittal, order processing, order preparation, and transit. Variants are supplier lead time, manufacturing / assembly lead time, and customer order lead time.	Council of Supply Chain Management Professionals (CSCMP), 2020
Lean	A business management philosophy that considers the expenditure of resources for any goal other than the creation of value for the end customer to be wasteful, and thus a target for elimination.	Council of Supply Chain Management Professionals (CSCMP), 2020

Vocabulary	Definition	Definition Source
Letter of Credit	An international business document that assures the seller that payment will be made by the bank issuing the letter of credit upon fulfilment of the sales agreement.	Council of Supply Chain Management Professionals (CSCMP), 2020
Lightweight IT	A socio-technical knowledge domain that is enabled by solution driven user demand, mass adoption of digital technology and innovative processes.	Chapter 5
Localization	Localization is the customization of all components of a product for a particular target market.	Andra, 2016
Logistics	The process of planning, implementing, and controlling procedures for the efficient and effective transportation and storage of goods including services, and related information from the point of origin to the point of consumption for the purpose of conforming to customer requirements. This definition includes inbound, outbound, internal, and external movements.	Council of Supply Chain Management Professionals (CSCMP), 2020
Machine Learning	Machine Learning is about getting computers to learn and act like humans do, and improve their learning over time in autonomous fashion, by feeding them data and information in the form of observations and real-world interactions.	Faggella, 2020
Manual Decision	Decisions made by a human rather than Software.	Chapter 3
MAP	See "Modified atmosphere packaging"	
marginal cost	The cost to produce one additional unit of output. The change in total variable cost resulting from a one-unit change in output.	Council of Supply Chain Management Professionals (CSCMP), 2020
Marketplaces	Marketplaces are a place where people physically met to exchange goods and services. Today, this is all being driven by platforms that leverage technology to match buyers and sellers in a more efficient and scalable fashion. Utilizing filters, a marketplace can ensure that only the most relevant information is provided to the user ensuring a more efficient and pleasing experience.	Chapter 16
mass customization	A marketing and manufacturing technique which combines the flexibility and personalization of custom-made products with the low unit costs associated with mass production.	Dollarhide, 2020
Master Production Schedule	The master level or top level schedule used to set the production plan in a manufacturing facility.	Council of Supply Chain Management Professionals (CSCMP), 2020

Vocabulary	Definition	Definition Source
Material Master Data	Data, typically stored in ERP systems, that describes a certain item, such as weight, dimension, Country of origin, packaging information, etc.	Chapter 3
Mental Work	Work or tasks that require cognitive ability.	Chapter 5
Micro-Learning	Micro-learning helps to transform learning contents into fragmented and bite-size learning nuggets, which are easier for people to learn anywhere and anytime. Micro-learning is a digital learning experience designed to be effective, engaging and enjoyable.	Chapter 21
Middleware	A type of software that is able to connect disparate software components or applications. Typically used to provided a level of integration between software components where were acquired from different developers.	Council of Supply Chain Management Professionals (CSCMP), 2020
Mobile Learning	Mobile learning refers to learning processes supported by the use of mobile and wireless information and communication technologies which have as a fundamental characteristic the learners' mobility, who may or may not be physically or geographically distant from one another and also from formal educational spaces, such as classrooms, instruction and training rooms or the work place.	Chapter 21
Modified atmosphere packaging	The use of special semi-permeable plastic film inside a load unit to maintain a delicate balance of different gases that will slow aging.	Chapter 17
Multilateral	Cooperation of several countries to solve political, social or technical problems which are cross-border.	Krause, 2016
Multitenant	This means that one instance of the software is serving multiple users or tenants. Underlying code is the same across all users and cannot usually be customized for an individual customer.	Chapter 9
Network Nodes	A fixed point in a firm's logistics system where goods come to rest; includes plants, warehouses, supply sources, and markets.	Council of Supply Chain Management Professionals (CSCMP), 2020
Networking Technology	Enables connections between things in a fast and reliable way so that they can interact (network) with each other.	Chapter 3
Non-Vessel-Operating Common Carrier	A Non-Vessel-Operating Common Carrier (NVOCC) is an ocean carrier that transports goods under its own House Bill of Lading without operating ocean transportation vessels. A NVOCC signs contracts directly with the Carriers that own the vessels e.g. Maersk, CMA, COSCO, and can then sell this space to Freight Forwarders or directly to a BCO.	Chapter 1

Vocabulary	Definition	Definition Source
Nonce	A nonce in bitcoin, or "a number only used once," is the number that bitcoin miners must find first before they can create a new block of transactions, and thus be handsomely rewarded with newly minted cryptocurrency.	Chapter 6
nonconformity	A quality management event that captures the failure to meet specified inspection or testing requirements.	Council of Supply Chain Management Professionals (CSCMP), 2020
NVOCC	See "Non Vessel Owning Common Carrier"	
OCC	See "Organizational Culture Change"	
Ocean Network Express	A Japanese container shipping company that is headquartered in Tokyo and Singapore. It was formed in 2016 as a joint venture of the Japanese shipping companies Nippon Yusen Kaisha, Mitsui O.S.K. Lines, and K Line, inheriting the container shipping operations of its parent companies.	Micro Shipping Cultural Communication, 2020
OCR	See "Optical Character Recognition"	
Odette Barcode Label	Barcode labels that are following the standards of the European Association of the Automotive Industry (Odette). Odette is the equivalent of the Automotive Industry Action Group (AIAG) in North America or the German Association of the Automotive Industry (VDA) in Germany.	Chapter 3
OEM	See "Original Equipment Manufacturer"	
Offshoring	The practice of moving domestic operations such as manufacturing to another country.	Council of Supply Chain Management Professionals (CSCMP), 2020
Omnichannel	Distribution strategy that includes multiple methods of capture and delivery of merchandise to the customer. Channels covered typically include distribution to a DC, Retail Store, and direct to Consumer with order coming via traditional Purchase Orders, EDI, Web and Phone. All forms may ship from a single DC, but it could also include consumer store pick-up such as Walmart's "Site to Store" program, and shipment direct to a consumer from a retail location vs. a distribution center.	Council of Supply Chain Management Professionals (CSCMP), 2020
ONE	See "Ocean Network Express"	
Operational Expenditures	Operating expenditures refer to day-to-day costs that are necessary to keep a business running.	Maverick, 2020
OPEX	See "Operational Expenditures"	

Vocabulary	Definition	Definition Source
Optical Character Recognition	The mechanical or electronic conversion of scanned images of handwritten, typewritten or printed text into machine-encoded text. It is widely used as a form of data entry from some sort of original paper data source, whether documents, sales receipts, mail, or any number of printed records. It is a common method of digitizing printed texts so that they can be electronically searched, stored more compactly, displayed on-line, and used in machine processes such as machine translation, text-to-speech and text mining.	Descartes Systems Group, 2020
Oracle	An American company selling database software and technology, cloud engineered systems, and enterprise software products.	Oracle, 2020
Organizational capacity and capabilities	Refers to the internal load that the organization can support, relevant to what is being asked of the supplier partner. If an organization is to lead strategical procurement efforts, it must understand its own limitations and governance measures to achieve a successful outcome.	Chapter 24
Organizational Culture Change	The process of changing the culture of an organization.	Chapter 20
Original Equipment Manufacturer	The rebranding of equipment and selling it under another name, or as a component of another product. OEM refers to the company that made the products (the "original" manufacturer), but with the growth of outsourcing, eventually became widely used to refer to the organization that buys the products and resells them. This term has two generally acceptable definitions which are actually opposites of each other and may vary by industry: 1) The OEM reseller is often the designer of the equipment (which is made to order). An example would be a computer manufacturer OEM which includes components built by other manufacturers, and 2) Companies that make products for others to repackage and sell, or to incorporate into a final assembly. An example would be an OEM manufacturing tires for use on automobiles.	Council of Supply Chain Management Professionals (CSCMP), 2020
Parcel Locker	Unattended delivery machines or systems of reception boxes, that enable to both receive and send parcels.	Chapter 25
Parcel Shipment	Parcels include small packages like those typically handled by providers such as UPS and FedEx.	Council of Supply Chain Management Professionals (CSCMP), 2020

Vocabulary	Definition	Definition Source
Passive Learning	In Passive Learning, learners are passive receivers of knowledge from a trainer. Passive learners were considered empty vessels or sponges that should be filled. They took notes and internalized the knowledge, while they did not receive feedback from the trainer. This style of learning is trainer-centered and contrasts to the learner-centered mode "Active Learning".	Chapter 21
Performance management	The continuous process of identifying, measuring, and developing the performance of individuals and teams.	Chapter 20
Physical Work	Work or tasks that require dexterity and strength.	Chapter 5
Platform as a Service	Pre-configured computing services that software developers can use to host applications without the need to configure the underlying server infrastructure.	Chapter 9
PLC	See "programmable Logic Controller"	
Predictive Analytics	A category of data analytics aimed at making predictions about future outcomes based on historical data and analytics techniques such as statistical modeling and machine learning.	Edwards, 2019
Prescriptive Analytics	Analyzed data (of the past) are used to detect indications that signal impending events in the future. For example if the system shows a certain amount of Things in the inbound to a warehouse and in parallel the internal data indicate that several workers are ill today then this will most likely lead to a delay or congestion in the goods receiving process (if no actions are taken).	Chapter 3
Private Blockchain	A private blockchain is a permissioned blockchain. Private blockchains work based on access controls which restrict the people who can participate in the network. There are one or more entities which control the network and this leads to reliance on third-parties to transact.	Sharma, 2020
process improvement	Designs or activities, which improve quality or reduce costs, often through the elimination of waste or non-value-added tasks.	Council of Supply Chain Management Professionals (CSCMP), 2020
Process Mining	An analytical discipline for discovering, monitoring, and improving real processes (i.e., not assumed processes) by extracting knowledge from event logs.	Celonis, 2020

Vocabulary	Definition	Definition Source
Procurement	Procurement, also known as Purchasing, can refer to a functional group (i.e., a formal entity on the organizational chart) as well as a functional activity (i.e., buying goods and services). The purchasing group performs activities to ensure it delivers maximum value to the organization, including supplier identification and selection, buying, negotiation and contracting, supply market research, supplier measurement and improvement, and purchasing systems development. Purchasing has been referred to as doing "the five rights": getting the right quality, in the right quantity, at the right time, for the right price, from the right source.	Monczka *et al.*, 2008
Programmable Logic Controller	An industrially hardened computer-based unit that performs discrete or continuous control functions upon equipment, processes, motion, and batch as to replace wires and limit human errors	Chapter 8
Protocol	A set of rules governing the exchange or transmission of data between devices.	Chapter 6
Protocol Stack	A complete set of network protocol layers that work together to provide complex networking and application capabilities.	Chapter 6
Protocol Suite	Created when two or more protocols are working efficiently together.	Chapter 6
Public Blockchain	A public blockchain is permissionless. Anyone can join the network and read, write, or participate within the blockchain. A public blockchain is decentralized and does not have a single entity which controls the network.	Sharma, 2020
QR Code	A two-dimensional barcode that is readable by smartphones. It allows to encode over 4000 characters in a two dimensional barcode. QR Codes may be used to display text to the user, to open a URL, save a contact to the address book or to compose text messages.	Braun, 2020
Radio Frequency Identification	The use of radio frequency technology including RFID tags and tag readers to identify objects. Objects may include virtually anything physical, such as equipment, pallets of stock, or even individual units of product. RFID tags can be active or passive. Active tags contain a power source and emit a signal constantly. Passive tags receive power from the radio waves sent by the scanner / reader. The inherent advantages of RFID over bar code technology are: 1) the ability to be read over longer distances, 2) the elimination of requirement for "line of sight" readability, 3) added capacity to contain information, and 4) RFID tag data can be updated / changed.	Council of Supply Chain Management Professionals (CSCMP), 2020

Vocabulary	Definition	Definition Source
Ransomware	A form of malware that encrypts a victim's files. The attacker then demands a ransom from the victim to restore access to the data upon payment.	Fruhlinger, 2020
Rate Management System	Software for storing and maintaining contracted rates for particular lanes with particular carriers.	Chapter 23
Reefer	A term used for refrigerated vehicles.	Council of Supply Chain Management Professionals (CSCMP), 2020
Request for Proposal	A document, which provides information concerning needs and requirements for a manufacturer. This document is created in order to solicit proposals from potential suppliers. For, example, a computer manufacturer may use a RFP to solicit proposals from suppliers of third party logistics services.	Council of Supply Chain Management Professionals (CSCMP), 2020
Request for Quote	A formal document requesting vendor responses with pricing and availability of products. RFQs are typically solicited from a broad group of suppliers from which a narrower group will be selected and asked to provide a more detailed Request for Proposal.	Council of Supply Chain Management Professionals (CSCMP), 2020
Return on Investment	The profit or loss resulting from an investment transaction, usually expressed as an annual percentage return. ROI is a popular metric for use in showing the value of an investment in new facilities, equipment or software vs. the cost of same.	Council of Supply Chain Management Professionals (CSCMP), 2020
RFID	See "Radio Frequency Identification"	
RFP	See "Request for Proposal"	
RFQ	See "Request for Quote"	
Robotic Process Automation	The application of technology that allows employees in a company to configure computer software or a "robot" to capture and interpret existing applications for processing a transaction, manipulating data, triggering responses and communicating with other digital systems.	Chapter 5
Robotics Labeling System	Robotics Labeling Systems, or Automatic Label Applicators use robots to place labels.	Chapter 25
ROI	See "Return on Investment"	
Rote and Repetitive Tasks	Tasks which are done the same way every time.	Chapter 5
RPA	See "Robotic Process Automation"	
SaaS	See "Software as a Service"	
Sales Enablement	The process of providing the sales organization with the information and tools that help salespeople sell more effectively.	Chapter 23

Vocabulary	Definition	Definition Source
SAP	A popular enterprise resource planning software created by the German company SAP AG.	SAP, 2021
Seasonality	A factor used in forecasting to reflect the seasonal variability in demand for certain products. Seasonality explains the fluctuation in demand for various recreational products which are used during different seasons.	Council of Supply Chain Management Professionals (CSCMP), 2020
Service Level Agreement	A part of a service contract where a service is formally defined. In practice, the term SLA is sometimes used to refer to the contracted delivery time (of the service or performance).	Descartes Systems Group, 2020
Service Oriented Architecture	A collection of loosely coupled services on a cloud platform sold by a software provider using a Software-as-a-Service (SaaS) model.	Chapter 14
Shipper	The party that tenders goods for transportation.	Council of Supply Chain Management Professionals (CSCMP), 2020
Six Sigma	Six-Sigma is a term coined to stress the continuous reduction in process variation to achieve near-flawless quality. When a Six Sigma rate of improvement has been achieved, defects are limited to 3.4 per million opportunities.	Council of Supply Chain Management Professionals (CSCMP), 2020
SLA	See "Service Level Agreement"	
Small and Medium Sized Enterprises	Businesses that maintain revenues, assets or a number of employees below a certain threshold. Each country has its own definition of what constitutes a small and medium-sized enterprise (SME). Certain size criteria must be met and occasionally the industry in which the company operates in is taken into account as well. For example, the European Union (EU), characterizes a small-sized enterprise as a company with fewer than 50 employees and a medium-sized enterprise as one with less than 250 employees.	Liberto, 2020
Smart Devices	Devices ('Things') everywhere collecting, transmitting, receiving data.	Chapter 3
Smart Port	A smart port is a port that uses automation and innovative technologies including Artificial Intelligence (AI), big data, Internet of Things (IoT) and blockchain to improve its performance.	Port Technology International Team, 2019
SME	See "Small and Medium Sized Enterprises"	
SOA	See "Service Oriented Architecture"	

Vocabulary	Definition	Definition Source
Soft Skills	Subjective skills that are hard to quantify, such as leadership skills, communication skills or motivation. They are also known as "people skills" or "interpersonal skills."	Doyle, 2020
Software as a Service	Software as a Service (SaaS) is a term which describes the use of computer systems provided by a remote third party, similar to what has traditionally been called a "Service Bureau" or "Application Service Provider (ASP)". In this setting the service provider maintains all of the computer hardware and software at their location, while the user accesses the systems via an internet connection and is charged a rate based on access time. It is also sometimes also referred to as "On Demand" services.	Council of Supply Chain Management Professionals (CSCMP), 2020
Spot Rates	Rates that are not based on predetermined prices created in a contract.	Chapter 16
Stakeholders	An individual or group who will be impacted in some way by a change. They have in interest (positive or negative) in how a project, initiative, or transformation will resolve itself.	Council of Supply Chain Management Professionals (CSCMP), 2020
Static routing	Static routing follows user defined routing and routing table is not changed until network administrator changes it.	Parahar, 2020
Stereolithography	An older method of additive manufacturing where light exposure hardens a certain CAD designed pattern.	Chapter 4
Strategic Alliance	A business relationship in which two or more independent organizations cooperate and willingly modify their business objectives and practices to help achieve long-term goals and objectives.	Council of Supply Chain Management Professionals (CSCMP), 2020
supply chain	1) Starting with unprocessed raw materials and ending with the final customer using the finished goods, the supply chain links many companies together. 2) the material and informational interchanges in the logistical process stretching from acquisition of raw materials to delivery of finished products to the end user. All vendors, service providers and customers are links in the supply chain.	Council of Supply Chain Management Professionals (CSCMP), 2020
Surcharge	An add-on charge to the applicable charges; motor carriers have a fuel surcharge, and railroads can apply a surcharge to any joint rate that does not yield 110% of variable cost.	Council of Supply Chain Management Professionals (CSCMP), 2020
Telematics	A general term that refers to any device which merges telecommunications and informatics. Telematics includes anything from GPS systems to navigation systems.	Teletrac Navman, 2020

Vocabulary	Definition	Definition Source
Terminal Operating System	The primary instrument of record-keeping, planning, control, and monitoring for the modern marine terminal.	Ward, 2013
TEU	See "Twenty foot equivalent unit"	
The Amazon Effect	The ongoing evolution and disruption of the retail market, both online and in physical outlets, resulting from increased e-commerce. The name is an acknowledgement of Amazon's early and continuing domination in online sales, which has driven much of the disruption.	Rouse, 2020
Third Party Logistics	The use of external companies to perform logistics functions that have traditionally been performed within an organization. The functions performed by the third party can encompass the entire logistics process or selected activities within that process.	Lieb, Millen and van Wassenhove, 1993
TMS	See "Transportation Management System"	
TOS	See "Terminal Operating System"	
Traceability	The ability to trace the history, application or location of that which is under consideration.	Descartes Systems Group, 2020
Trade Financing	Trade finance represents the financial instruments and products that are used by companies to facilitate international trade and commerce. Trade finance makes it possible and easier for importers and exporters to transact business through trade.	Murphy, 2020b
Transit Time	The total time that elapses between a shipment's pickup and delivery.	Council of Supply Chain Management Professionals (CSCMP), 2020
Transparency	The ability to gain access to information without regard to the systems landscape or architecture. An example would be where an online customer could access a vendor's web site to place an order and receive availability information supplied by a third party outsourced manufacturer or shipment information from a third party logistics provider.	Council of Supply Chain Management Professionals (CSCMP), 2020
Transportation Management System	A computer system designed to provide optimized transportation management in various modes along with associated activities, including managing shipping units, labor planning and building, shipment scheduling through inbound, outbound, intra-company shipments, documentation management (especially when international shipping is involved), and third party logistics management.	Council of Supply Chain Management Professionals (CSCMP), 2020
transportation method	A linear programming technique that determines the least-cost allocation of shipping goods from plants to warehouses of from warehouses to customers.	Council of Supply Chain Management Professionals (CSCMP), 2020

Vocabulary	Definition	Definition Source
Transportation Mode	The method of transportation: land, sea, or air shipment.	Council of Supply Chain Management Professionals (CSCMP), 2020
Transshipment	The shipment of cargo from one mode to another.	Chapter 15
Turing Complete	Describes programming languages that are computationally comprehensive and can thus simulate other real-world computations using data as inputs.	Chapter 6
Twenty foot equivalent unit	Standard unit for counting containers of various capacities and for describing the capacities of container ships or terminals. One 20 Foot ISO container equals 1 TEU. One 40 Foot ISO container equals two TEU. A 20-foot container is typically 8.5 feet tall and 8 feet wide outside and has an internal capacity of 1170 square feet.	Council of Supply Chain Management Professionals (CSCMP), 2020
UAV	See "Unmanned Aerial Vehicle"	
UI	User Interface	Council of Supply Chain Management Professionals (CSCMP), 2020
ULD	See "Unit Load Device"	
Unit Load Device	Refers to airfreight containers and pallets.	Council of Supply Chain Management Professionals (CSCMP), 2020
Unmanned Aerial Vehicle	An aircraft with no pilot on board. UAVs can be remote controlled aircraft (e.g. flown by a pilot at a ground control station) or can fly autonomously based on pre-programmed flight plans or more complex dynamic automation systems.	Omega, 2020
Upstream	Refers to the supply side of the supply chain. Upstream partners are the suppliers who provide goods and services to the organization needed to satisfy demands which originate at point of demand or use, as well as other flows such as return product movements, payments for purchases, etc.	Council of Supply Chain Management Professionals (CSCMP), 2020
Value Added	Increased or improved value, worth, functionality or usefulness.	Council of Supply Chain Management Professionals (CSCMP), 2020
Value Added Tax (VAT)	1) A general tax that applies, in principle, to all commercial activities involving the production and distribution of goods and the provision of services. 2) a consumption tax because it is borne ultimately by the final consumer. It is not a charge on businesses. 3) charged as a percentage of price, which means that the actual tax burden is visible at each stage in the production and distribution chain."	Descartes Systems Group, 2020

Vocabulary	Definition	Definition Source
Value chain	A series of activities, which combined, define a business process; the series of activities from manufacturers to the retail stores that define the industry supply chain.	Council of Supply Chain Management Professionals (CSCMP), 2020
Value Proposition	The value a company promises to deliver to customers should they choose to buy their product.	Twin, 2020
Vested Outsourcing	Vested outsourcing is rooted in a base of transparency and relationship scaling. The concept is materialized by a simple approach of first classifying the sourcing relationship and developing a plan to scale into what could be a true partnership of transparent rewards for all partners if those goals are received.	Chapter 24
Virtual Machine	In computing, a virtual machine (VM) is an emulation of a computer system. Virtual machines are based on computer architectures and provide functionality of a physical computer. Their implementations may involve specialized hardware, software, or a combination."	Descartes Systems Group, 2020
Virtualization	Migrating services that traditionally had a strong local component, be it through physical document flows, relationship sales, preferred communication means or business custom, to being conducted in the virtual world.	Chapter 11
Visibility	Supply chain visibility refers to the identity, location and status of objects moving through the supply chain, captured in timely messages about events, such as "waiting to be picked", along with the planned and actual dates/times for these events.	Vernon, 2008
Warehouse Management System	A key part of the supply chain that primarily aims to control the movement and storage of materials within a warehouse and process the associated transactions, including shipping, receiving, put away and picking. The systems also direct and optimize stock put away based on real-time information about the status of bin utilization. A WMS monitors the progress of products through the warehouse. It involves the physical warehouse infrastructure, tracking systems, and communication between product stations.	Descartes Systems Group, 2020
Wearable Device	Products controlled by electronic components and software that can be incorporated into clothing or worn on the body like accessories.	*Wearable Device - an overview*, 2020
Workflow Automation	The process of automating a workflow.	Chapter 9
XML	See "eXtensible Markup Language	

References

Andra, J. (2016). *What is Localization and Why Should You Care?* https://www.globaltrademag.com/what-is-localization-and-why-should-you-care/ (accessed 19 December 2020).

Arbitrage. (2020). *Dictionary.com.* https://www.dictionary.com/browse/arbitrage (accessed 19 December 2020).

Arena Solutions. (2021). *What is a Bill of Materials (BOM) and How Do You Create One?* https://www.arenasolutions.com/resources/category/bom-management/creating-a-bill-of-materials/ (accessed 2 January 2021).

Australian Border Force (2021). *HOW TO IMPORT.* https://www.abf.gov.au/imports/Pages/How-to-import/Import-declarations.aspx (accessed 10 February 2021).

Barrett, R. (2015). *Werteorientierte Unternehmensführung: Cultural Transformation Tools für Performance und Profit.* Springer Berlin Heidelberg. https://books.google.com/books?id=WiI3CwAAQBAJ.

Braun, M. (2020). *What is a QR Code?* https://www.the-qrcode-generator.com/whats-a-qr-code (accessed 19 December 2020).

Brodie, S. (2015). Internet+ turning into a buzzword in China, South China Morning Post. Available at: https://www.scmp.com/property/article/1889756/internet-turning-buzzword-china (Accessed: 11 February 2021).

Celonis SE. (2020). *What is Process Mining?, Corporate Website.* https://www.celonis.com/process-mining/what-is-process-mining/ (accessed 19 December 2020).

Chopra, S. (2019). *Supply Chain Management: Strategy, Planning, and Operation.* Pearson Education Limited. https://books.google.com/books?id=g2WvvAEACAAJ.

Council of Supply Chain Management Professionals (CSCMP). (2020). *CSCMP Supply Chain Management Definitions and Glossary.* https://cscmp.org/CSCMP/Educate/SCM_Definitions_and_Glossary_of_Terms.aspx (accessed 2 January 2021).

Descartes Systems Group (2020). *Glossary.* https://www.descartes.com/resources/glossary#/ (accessed 2 January 2021).

Digital Currency Group (2020). *Bitcoin.* https://www.coindesk.com/price/bitcoin (accessed 19 December 2020).

Dittrich, J. and Braun, M. (2004). *Business Process Outsourcing: Entscheidungs-Leitfaden für das Out-und Insourcing von Geschäftsprozessen.* Schäffer-Poeschel.

Dollarhide, M.E. (2020). *Mass Customization, Investopedia.* https://www.investopedia.com/terms/m/masscustomization.asp (accessed 19 December 2020).

Doppler, K. and Lauterburg, C. (2019). *Change Management: Den Unternehmenswandel gestalten.* 14th edn. Campus Verlag.

Downey, L. (2020). *Efficient Market Hypothesis (EMH).* https://www.investopedia.com/terms/e/efficientmarkethypothesis.asp (accessed 19 December 2020).

Doyle, A. (2020). *Hard Skills vs. Soft Skills: What's the Difference?* https://www.thebalancecareers.com/hard-skills-vs-soft-skills-2063780 (accessed 9 December 2020).

Edwards, J. (2019). *What is predictive analytics? Transforming data into future insights, Cio.* https://www.cio.com/article/3273114/what-is-predictive-analytics-transforming-data-into-future-insights.html (accessed 3 December 2020).

Faggella, D. (2020). *What is Machine Learning?* https://emerj.com/ai-glossary-terms/what-is-machine-learning/%0A (accessed 19 December 2020).

Feedvisor (2020). *Fulfillment By Amazon (FBA)*. https://feedvisor.com/university/fulfillment-by-amazon/ (accessed 2 December 2020).

Flash Global (2021). *Forward Stocking Locations*. https://flashglobal.com/global-supply-chain-solutions/forward-stocking-locations/ (accessed 2 January 2021).

Flexport (2020). *Container Yard (CY)*. https://www.flexport.com/glossary/container-yard/ (accessed 8 November 2020).

Vernon, F. (2008). Supply chain visibility: lost in translation? *Supply Chain Management: An International Journal*. 13(3): 180–184. doi: 10.1108/13598540810871226.

Fraunhofer Institute for Production Technology (2021). *Flexible production systems for "Batch Size1"*. https://www.ipt.fraunhofer.de/en/trends/industrie40/batchsizeone.html (accessed 2 January 2021).

Fruhlinger, J. (2020). *Ransomware explained: How it works and how to remove it, CSO*. https://www.csoonline.com/article/3236183/what-is-ransomware-how-it-works-and-how-to-remove-it.html (accessed 19 December 2020).

Gnoni, M.G., Andriulo, S., Maggio, G., and Nardone, P. (2013). "'Lean occupational" safety: an application for a near-miss management system design. *Safety Science* 53: 96–104.

Grupa, T. (2020). *Conditional Statements and Their Converse*. https://tutors.com/math-tutors/geometry-help/conditional-converse-statements (accessed 4 December 2020).

Hou, Z. (2020). *What Is a Center of Excellence and Why Do You Need One?* https://www.convinceandconvert.com/social-media-strategy/what-is-a-center-of-excellence/ (accessed 13 December 2020).

HRZone (2020). *What are Intangible Rewards?* https://www.hrzone.com/hr-glossary/what-are-intangible-rewards (accessed 12 December 2020).

Investopedia (2020). *A Beginner's Guide to Hedging*. https://www.investopedia.com/trading/hedging-beginners-guide/ (accessed 19 December 2020).

Krause, J. (2016). Multilateralismus in einer multipolaren Welt. *Politikum*, 4: 4–13.

Liberto, D. (2020). *Small and Mid-size Enterprise (SME)*. https://www.investopedia.com/terms/s/smallandmidsizeenterprises.asp (accessed 7 December 2020).

Lieb, R.C., Millen, R.A., and van Wassenhove, L.N. (1993). Third party logistics services: a comparison of experienced American and European manufacturers. *International Journal of Physical Distribution & Logistics Management*, 23(6): 35–44.

Lodamaster Group (2020). *Automated Guided Vehicles*. https://www.lodamaster.com/en/solutions/automated-guided-vehicles (accessed 5 November 2020).

Lokhande, S.A. (2017). *China's One Belt One Road Initiative and the Gulf Pearl chain*. https://www.chinadaily.com.cn/opinion/2017beltandroad/2017-06/05/content_29618549.htm (accessed 19 March 2020).

Maverick, J.B. (2020). *Capital Expenditures vs. Operating Expenses: What's the Difference? Investopedia*. https://www.investopedia.com/ask/answers/020915/what-difference-between-capex-and-opex.asp (accessed 3 December 2020).

Micro Shipping Cultural Communication (2020). *Ocean Network Express (ONE)*. https://www.micro-shipping.com/articles/detail/ocean-network-express-one.html (accessed 19 December 2020).

Monczka, R.M., Handfield, R.B., Giunipero, L.C., and Patterson, J.L. (2008). *Purchasing and Supply Chain Management*. 4th edn. South-Western Cengage Learning.

Murphy, C.B. (2020a). *Earnings Before Interest and Taxes - EBIT, Investopedia*. https://www.investopedia.com/terms/e/ebit.asp (accessed 2 January 2021).

Murphy, C.B. (2020b). *Trade Finance, Investopedia.* https://www.investopedia.com/terms/t/tradefinance.asp (accessed 2 January 2021).

Omega (2020). *The UAV.* https://www.theuav.com (accessed 2 January 2021).

OmniSci (2020). *Graphical User Interface Definition.* https://www.omnisci.com/technical-glossary/graphical-user-interface (accessed 2 January 2021).

Oracle (2020). *Company Website.* https://www.oracle.com/index.html (accessed 1 January 2021).

Parahar, M. (2020). *Difference between Static Routing and Dynamic Routing.* https://www.tutorialspoint.com/difference-between-static-routing-and-dynamic-routing (accessed 2 January 2021).

Port Technology International Team (2019). *What is a Smart Port?* https://www.porttechnology.org/news/what-is-a-smart-port/ (accessed 24 January 2020).

Raab, M. and Griffin-Cryan, B. (2011). *Creating Value-When Digital Meets Physical Digital Transformation of Supply Chains.* https://www.capgemini.com/wp-content/uploads/2017/07/Digital_Transformation_of_Supply_Chains.pdf (Accessed: 12 September 2018).

Rouse, M. (2020). *Amazon effect, Whatis.com.* https://whatis.techtarget.com/definition/Amazon-effect (accessed 2 January 2021).

SAP. (2021). *Corporate Website.* https://www.sap.com/index.html (accessed 2 January 2021).

Sharma, T.K. (2020). *Public Vs. Private Blockchain*: *A Comprehensive Comparison, Blockchain Council.* https://www.blockchain-council.org/blockchain/public-vs-private-blockchain-a-comprehensive-comparison/ (accessed 1 January 2021).

Shaw, K. (2019). *What is edge computing and why it matters, Network World.* https://www.networkworld.com/article/3224893/what-is-edge-computing-and-how-it-s-changing-the-network.html (accessed 16 November 2020).

Talend Inc. (2021). *What is Data Integration?* https://www.talend.com/resources/what-is-data-integration/ (accessed 2 January 2021).

Teletrac Navman (2020). *Telematics Overview.* https://www.telematics.com/telematics-overview/ (accessed 9 December 2020).

Twin, A. (2020). *Value Proposition, Investopedia.* https://www.investopedia.com/terms/v/valueproposition.asp (accessed 1 January 2021).

Van Der Aalst, W.M.P., Ter Hofstede, A.H.M., and Weske, M. (2003). Business process management: A survey, in *International Conference on Business Process Management*, 1–12.

Ward, T. (2013). *Terminal operating system selection, Port Technology.* https://www.porttechnology.org/technical-papers/terminal_operating_system_selection/%0A (accessed 6 February 2020).

Wearable Device - an overview (2020). *ScienceDirect Topics.* https://www.sciencedirect.com/topics/engineering/wearable-device (accessed 19 December 2020).

Win, A. (2008). The value a 4PL provider can contribute to an organisation. *International Journal of Physical Distribution & Logistics Management* 38(9), pp. 674–684.

Index

The Digital Transformation of Logistics: Demystifying Impacts of the Fourth Industrial Revolution,
First Edition. Edited by Mac Sullivan and Johannes Kern.
© 2021 by The Institute of Electrical and Electronics Engineers, Inc.
Published 2021 by John Wiley & Sons, Inc.